T0276076

DISTRIBUTED LEARNING

CHANDOS
INFORMATION PROFESSIONAL SERIES
Series Editor: Ruth Rikowski
(email: Rikowskigr@aol.com)

Chandos' new series of books is aimed at the busy information professional. They have been specially commissioned to provide the reader with an authoritative view of current thinking. They are designed to provide easy-to-read and (most importantly) practical coverage of topics that are of interest to librarians and other information professionals. If you would like a full listing of current and forthcoming titles, please visit www.chandospublishing.com.

New authors: we are always pleased to receive ideas for new titles; if you would like to write a book for Chandos, please contact Dr Glyn Jones on g.jones.2@elsevier.com or telephone +44 (0) 1865 843000.

DISTRIBUTED LEARNING

Pedagogy and Technology in Online Information Literacy Instruction

Edited by

Tasha Maddison

Maha Kumaran

AMSTERDAM • BOSTON • HEIDELBERG • LONDON
NEW YORK • OXFORD • PARIS • SAN DIEGO
SAN FRANCISCO • SINGAPORE • SYDNEY • TOKYO

Chandos Publishing is an imprint of Elsevier

Chandos Publishing is an imprint of Elsevier
50 Hampshire Street, 5th Floor, Cambridge, MA 02139, United States
The Boulevard, Langford Lane, Kidlington, OX5 1GB, United Kingdom

Copyright © 2017 T. Maddison and M. Kumaran. Published by Elsevier Ltd. All rights reserved.

No part of this publication may be reproduced or transmitted in any form or by any means, electronic or mechanical, including photocopying, recording, or any information storage and retrieval system, without permission in writing from the publisher. Details on how to seek permission, further information about the Publisher's permissions policies and our arrangements with organizations such as the Copyright Clearance Center and the Copyright Licensing Agency, can be found at our website: www.elsevier.com/permissions.

This book and the individual contributions contained in it are protected under copyright by the Publisher (other than as may be noted herein).

Notices
Knowledge and best practice in this field are constantly changing. As new research and experience broaden our understanding, changes in research methods, professional practices, or medical treatment may become necessary.

Practitioners and researchers must always rely on their own experience and knowledge in evaluating and using any information, methods, compounds, or experiments described herein. In using such information or methods they should be mindful of their own safety and the safety of others, including parties for whom they have a professional responsibility.

To the fullest extent of the law, neither the Publisher nor the authors, contributors, or editors, assume any liability for any injury and/or damage to persons or property as a matter of products liability, negligence or otherwise, or from any use or operation of any methods, products, instructions, or ideas contained in the material herein.

British Library Cataloguing-in-Publication Data
A catalogue record for this book is available from the British Library

Library of Congress Cataloging-in-Publication Data
A catalog record for this book is available from the Library of Congress

ISBN: 978-0-08-100598-9 (print)
ISBN: 978-0-08-100609-2 (online)

For information on all Chandos publications
visit our website at https://www.elsevier.com

Working together
to grow libraries in
developing countries

www.elsevier.com • www.bookaid.org

Publisher: Glyn Jones
Acquisition Editor: Glyn Jones
Editorial Project Manager: Harriet Clayton
Production Project Manager: Debasish Ghosh
Cover Designer: Mark Rogers

Typeset by MPS Limited, Chennai, India

CONTENTS

SECTION II PEDAGOGY

4. Designing Online Asynchronous Information Literacy Instruction Using the ADDIE Model   69

M. Allen

13. From Technical Troubleshooting to Critical Inquiry: Fostering Inquiry-Based Learning Across Disciplines Through a Tutorial for Online Instructors

M. Courtney

14. Embedding the Library in the LMS: Is It a Good Investment for Your Organization's Information Literacy Program?

E. Kline, N. Wallace, L. Sult and M. Hagedon

15. A Decade of Distributed Library Learning: The NOSM Health Sciences Library Experience

J. Dumond

SECTION IV CASE STUDIES

SECTION V INNOVATIONS

LIST OF CONTRIBUTORS

M. Allen
University of South Florida's Tampa Library, Tampa, FL, United States

E. Bellard
Adelphi University, Garden City, NY, United States

E. Berg
Austin Peay State University, Clarksville, TN, United States

D. Canevari de Paredes
University of Saskatchewan, Saskatoon, SK, Canada

C. Chester-Fangman
Austin Peay State University, Clarksville, TN, United States

M. Courtney
Indiana University, Bloomington, IN, United States

J. DeJonghe
Metropolitan State University, St Paul, MN, United States

M.R. Desilets
Metropolitan State University, St Paul, MN, United States

K. Detterbeck
Purchase College, Purchase, NY, United States

C. Doi
University of Saskatchewan, Saskatoon, SK, Canada

J. Dumond
Northern Ontario School of Medicine, Sudbury, ON, Canada

S. Eichenholtz
Adelphi University, Garden City, NY, United States

M.M. Filkins
Metropolitan State University, St Paul, MN, United States

M. Flaccavento
University of Toronto, Toronto, ON, Canada

D. Francis
University of Saskatchewan, Saskatoon, SK, Canada

G. Garber
Austin Peay State University, Clarksville, TN, United States

E. Getts
Johns Hopkins University, Baltimore, MD, United States

L. Glisson
City University of New York, New York, NY, United States

M. Hagedon
University of Arizona Libraries, Tucson, AZ, United States

R. Hall
Friends School of Baltimore, Baltimore, MD, United States

A.N. Hess
Oakland University Libraries, Rochester, MI, United States

S. Iverson
St. Michael's Hospital, Toronto, ON, Canada

C. Judd
Eastern Kentucky University Libraries, Richmond, KY, United States

M. Kelt
Glasgow Caledonian University, Glasgow, United Kingdom

E. Kline
University of Arizona Libraries, Tucson, AZ, United States

M. Kumaran
University of Saskatchewan, Saskatoon, SK, Canada

C.M. Larson
Metropolitan State University, St Paul, MN, United States

D. Lightfoot
St. Michael's Hospital, Toronto, ON, Canada

J. Logan
University of Toronto, Toronto, ON, Canada

J. Long
Miami University, Oxford, OH, United States

S. Lucky
University of Saskatchewan, Saskatoon, SK, Canada

T. Maddison
Saskatchewan Polytechnic, Saskatoon, SK, Canada

B. Marcum
Eastern Kentucky University Libraries, Richmond, KY, United States

B. Morant
St. Michael's Hospital, Toronto, ON, Canada

P. Ridlen
Fontbonne University, St. Louis, MO, United States

A. Roth
University of California—San Diego, La Jolla, CA, United States

M. Sciangula
Purchase College, Purchase, NY, United States

K.O. Secovnie
City University of New York, New York, NY, United States

L. Sult
University of Arizona Libraries, Tucson, AZ, United States

J.M. Theissen
Fontbonne University, St. Louis, MO, United States

D. Turnbow
University of California—San Diego, La Jolla, CA, United States

N. Wallace
University of Arizona Libraries, Tucson, AZ, United States

J. Webb
University of Toronto, Toronto, ON, Canada

C. Ziegler
St. Michael's Hospital, Toronto, ON, Canada

BIOGRAPHY

Maryellen Allen currently serves as the Assistant Director of Instructional Services at the University of South Florida's Tampa Library. Previously, she has worked at the University of Alabama as the Distance Learning Specialist and a Program Manager for online and distance degree programs. Her research interests include effective universal design for online information literacy instruction and the efficacy of online learning versus face-to-face instruction. She has Master's degrees in both Library and Information Sciences and Instructional Systems Design.

Eloise Bellard is an Associate Professor at Adelphi University, Garden City, New York, where she is a Coordinator of Instructional Services and Reference Librarian at the Swirbul Library Garden City Campus. Her extensive experience as a reference librarian and licensed teacher enables her to play an active role as an academic teaching librarian. As coordinator of instructional services, she is responsible for the coordination and development of the information literacy programs for the library. She earned a Master's in Library Science from St. Johns University in Jamaica, New York, and received a Master's in Education from Adelphi University in Garden City, New York. Professor Bellard earned her BA in Communication Arts from Marymount College in New York City. Her research interests include addressing graduate students' information literacy needs online and in the classroom and supporting ways to successfully integrate legal resources and research methodology into interdisciplinary programs at the graduate and undergraduate levels. She has presented at national and regional conferences and published in the field of librarianship and information literacy.

Elaine Berg is an Associate Professor of Library Administration and Coordinator of Access Services at Austin Peay State University's Felix G. Woodward Library in Clarksville, Tennessee. She serves as the Library Liaison to the Department of Languages and Literature and the Women's and Gender Studies Program. She enjoys working with her liaison faculty to find new ways to ensure that students have a valuable experience during library instruction. In addition, Professor Berg provides library instruction to the honors sections of the first-year experience course. She is a strong proponent of shared governance and is active in the APSU faculty senate and on the faculty handbook committee. She earned her Master's of Librarianship from the University of South Carolina.

Donna Canevari de Paredes is a senior Humanities Librarian with the University Library at the University of Saskatchewan. Her professional practice has focused on academic librarianship in the humanities, with emphasis on information literacy, collection development, and research support. A particular area of interest within information literacy is the knowledge and utilization of resources for international studies, particularly Spanish and Latin American studies. In the latter context she has spoken at meetings of the Seminar on the Acquisition of Latin American Library Materials (SALALM), and at the Canadian Association of Hispanists, and has also provided workshops for students of library and information studies outside of North America. Donna holds a BA in History from Skidmore College and an MS in Library Science from Columbia University, as well as Library Media and Social Studies Teaching Certification from the State University of New York.

Christina Chester-Fangman is an Associate Professor of Library Administration and the Coordinator of Research and Instruction at Austin Peay State University's Felix G. Woodward Library in Clarksville, Tennessee. She also serves as the designated First-Year Experience Librarian and has a great deal of interest in information literacy for students in transition. She has been a member of the Curriculum Committee for APSU 1000—Transition to the University, APSU's first-year experience course. Professor Chester-Fangman enjoys collaborating with her colleagues to produce informative and engaging video tutorials about library services, the research process, and issues of academic honesty for Austin Peay's unique student population. A native of Clarksville, she is an alumna of Austin Peay, where she majored in History with a minor in Women's Studies. She then went on to earn a Master's of Library Science from the University of Kentucky. Her recent publications include a "recipe" on evaluating websites in the Association of College & Research Libraries' *The Library Instruction Cookbook* and a coauthored article in *Reference Services Review* regarding "The Librarian's Role in Combating Plagiarism."

Michael Courtney is the Outreach and Engagement Librarian for the Indiana University Libraries, where he also serves as the Online Learning Librarian. Michael is an adjunct Associate Professor in the Department of Information and Library Science, School of Informatics and Computing (IUB), where he has taught the reference course. Prior to coming to IUB, he worked in many facets of librarianship in both public and technical service positions within public and academic libraries over the past 20 years. Current areas of scholarship and publishing include future trends in academic reference service, library history, instructional design in online learning, and the connections between service learning and libraries. He credits his formative years at Barningham CEVC Primary School (Suffolk) as the foundation for a lifetime of inquiry.

Jennifer DeJonghe is a Librarian and Professor at Metropolitan State University. She teaches both online and in-person courses in research and information studies, as well as courses within the graduate liberal studies program. Jennifer also leads the library web team and is involved in user experience testing and digital initiatives. She holds a Master of Library and Information Science from Dominican University, and a Master of Arts in Liberal Studies from Metropolitan State University.

Michelle R. Desilets is a Reference and Instruction Librarian and Assistant Professor at Metropolitan State University in Brooklyn Park, Minnesota. She teaches credit-bearing information literacy courses in person and online, as well as bibliographic instruction sessions for a variety of disciplines. She has a particular interest in online teaching and learning and instructional design. Michelle holds a Master's in Library and Information Science from the University of Texas at Austin and is currently pursuing a Master of Science in Education Technology from Minnesota State University, Mankato.

Kimberly Detterbeck is the Art Librarian at Purchase College, State University of New York (SUNY). In her work, she oversees collection development, reference, and library instruction for art history, art and design, new media, and arts management, as well as participates in librarywide projects such as digitization, assessment, outreach and communications, and e-learning across multiple disciplines. Kim is an active member of the Metropolitan New York Library Council (METRO) as a member of the social media, bibliographic instruction, and collective access special interest groups and a member of the myMETRO Researchers Pilot Project. She is also active in the Art Libraries Society of North America (ARLIS/NA) as former chair of the membership committee and secretary of the New York chapter. Kim has been published in *Art Documentation* and *The Global Librarian* and regularly presents at local and national conferences including SUNY's Conference on Instruction and Technology (CIT), SUNY Librarians Association annual conferences, METRO's annual conferences, ARLIS/NA, and Association of College and Research Libraries. She received an MLS from the University of Maryland, an MA in Art History from Syracuse University, and a BA in Art History from Rutgers University.

Carolyn Doi is an Assistant Librarian at the University Library at the University of Saskatchewan, where she specializes in music, art and art history, and education librarianship. She has experience teaching in online environments, including designing and delivering an undergraduate music research methods class using a blended classroom model with the flipped classroom teaching methodology. Her research and professional practice looks at the management of special music collections, particularly in online environments for study, preservation, and access to community histories. She received an AVCM diploma in piano performance and pedagogy from the Victoria Conservatory of Music in 2004, and BMus degree in Music History from the University of Victoria (2007). She went on to receive an MLIS from McGill University in 2010, where she completed a research project designing a mobile-based audio-streaming solution for course-based listening requirements.

Jennifer Dumond is the Education Services Librarian at the Northern Ontario School of Medicine (Lakehead University campus). In this capacity, she coordinates information literacy instruction for faculty, staff, and learners in all NOSM programs. She also oversees the librarian liaison program, eReserves, and subject guides. Her research interests include embedded librarianship, evidence-based medicine instruction, and instructional design.

Susan Eichenholtz is an Associate Professor at Adelphi University, Garden City, New York, where she teaches graduate students in the educational leadership and technology program. This program satisfies the certification requirements for aspiring administrators at the building and district level. She earned her Doctoral degree in Educational Administration and Leadership from Hofstra University, Hempstead, New York. She has also received, an MS in School Counseling from Wright State University in Dayton, Ohio, and a BA in Psychology and Education from the State University of New York at Stony Brook. She has always been in the field of education and during her career served as an elementary level teacher, elementary, middle and high school counselor, and central and building administrator in K−12 school districts. Her research interests include supporting school leaders in their use of instructional technology, integrating technology with K−12 school curriculum and developing e-portfolios. She has presented at national and regional conferences. In addition, she has publications in the area of educational administration and the application of educational technology.

Michelle M. Filkins is a Reference and Instruction Librarian and Professor at Metropolitan State University. She teaches credit-bearing information literacy courses in person and online, as well as course-integrated and single-session instruction in several disciplines. Her interests include books and publishing, poetics, and information literacy instruction. In addition, Michelle is an editor for Spout Press, a nonprofit literary press based in Minneapolis. She holds a Master's of Library and Information Science from Dominican University and a Master's in English from the University of St. Thomas.

Monique Flaccavento is the Acting Director at the Ontario Institute for Studies in Education (OISE) Library at the University of Toronto Libraries. She is U of T's Education Liaison Librarian and Adjunct Lecturer at the U of T's faculty of information, where she has taught design and evaluation of information literacy programs. She is also a former elementary school teacher. Monique's research interests include student engagement in online learning environments and improving instructional practice through peer observation and mentorship.

David Francis has enjoyed a long career in instructional design and educational leadership in Canada and the United States. His contributions include technological innovations in teaching and learning, faculty development, and the use of educational research and applied research to advance institutional goals. David holds Undergraduate degrees in

Mathematics (University of New Brunswick) and Education (Mount Allison University), a Master's of Education degree (Memorial University of Newfoundland), and a Doctorate in Educational Administration (University of Saskatchewan). He is a board member of the Canadian Network for Innovation in Education (CNIE) and the local chapter of the American Society for Quality (ASQ) in Saskatoon, Saskatchewan.

Gina Garber is an Associate Professor of Library Administration and the Coordinator of Digital Services at Austin Peay State University's Felix G. Woodward Library. She teaches the first-year experience course and an enhanced remedial course for at-risk students. She has served as the library liaison, fellow, and founding member of the planning and development committee for the first-year experience program. Professor Garber enjoys collaborating and teaching in her library liaison areas for the Department of History and Philosophy and the Department of Art and Design. She earned her Master's of Information Science from the University of Tennessee. Publications include an article in *Career Strategies for Librarians*, "Making the Leap from Paraprofessional to Professional in an Academic Library," and contributed chapters "Academic Honesty" and "Library Research and Journal Assignment" to the *APSU 1000- Liberal Arts in University Life: Providing the Foundations for University Success* textbook.

Erica Getts began working as a Distance Education Librarian at Johns Hopkins University in 2014. She received her Master's in Library Science from the University of Maryland and Bachelor's degree in Education from the Kutztown University of Pennsylvania. Previously, Erica has worked as a high school librarian in Pennsylvania and as an access services staff member at the University of Maryland.

Lane Glisson, MA, MLS, is the E-learning and Instruction Librarian at the A. Philip Randolph Memorial Library, Borough of Manhattan Community College, City University of New York, where she is a Subject Liaison to the English, Modern Languages, and Music and Art departments. Her teaching interests include the integration of research into students' writing, scaffolded instruction techniques, and online course design methods. Her scholarly research has focused on the literature of migration and marginality, particularly as it relates to memory and war, including a study of W. G. Sebald's *The Emigrants*. Currently, her research examines the audacious use of candor to express dissent in women's writing from a world literature perspective.

Michael Hagedon is the Web Development Work Team Leader at the University of Arizona libraries. Previously, he was a Senior Software Engineer in the same library, focusing on the development of instructionally focused web applications, including the award-winning "Guide on the Side" and Library Tools tab. He has presented and lectured on various topics including "Guide on the Side", agile software development, and object-oriented programming. He is currently focusing on innovative ways of helping the library's web designers and developers make sites and tools that are useful—through the encouragement of user-experience best practices—and sustainable—through the use of agile software-development techniques. He completed a BS in Computer Science at Oklahoma Baptist University and an MA in Classics at the University of Arizona.

Renee Hall holds a Master's degree in Library Science from Simmons College and a Bachelor of Arts degree from Colorado State University. She spent the previous 10 years as a Distance Education Librarian in the entrepreneurial library program at Johns Hopkins University where she gained valuable experience working with a small team of librarians to maintain and host a required online information literacy course. Renee is currently an upper school librarian at the Friends School of Baltimore, working in partnership with other faculty members and 9th through 12th graders to hone students' information literacy skills in preparation for college and lifelong learning.

Amanda Nichols Hess is the E-learning Instructional Technology and Education Librarian at Oakland University libraries. In this role, she works with her colleagues to develop the libraries' diverse and user-focused online learning offerings; she is also responsible for delivering professional learning offerings aimed at building librarians' capacity to integrate instructional design and technology into information literacy instruction. Amanda is the Liaison Librarian to OU's School of Education and Human Services, where she maintains an active teaching presence. Her research focuses on information literacy instruction, instructional design and technology, and the intersection of these practices into faculty development. She has shared her research on these ideas at conferences

such as ALA and ACRL, and in journals, including *portal, College &
Research Libraries*, and the *Journal of Academic Librarianship*.

Sandy Iverson is the Manager of the Health Information and Knowledge
Mobilization Program at St. Michael's Hospital in Toronto, where she has
been employed since 2010. She holds graduate degrees in Library Science
and Adult Education. Her career in library and information services has
included positions as diverse as managing learning centers for English as a
second language students, selling library information and technology services to libraries, writing e-learning curriculum for public librarians,
managing academic and health sciences libraries, and providing information and communication consulting services to nonprofit organizations.
Her research interests include health and information literacy, measurement and evaluation, and bibliotherapy. Sandy is also a registered
Psychotherapist in the province of Ontario.

Cindy Judd is an Associate University Librarian for Eastern Kentucky
University libraries in Richmond, Kentucky. She serves as the team leader
for the learning resources center, overseeing the area of the library that
provides a variety of services and diverse collections for the College of
Education. Cindy has worked in the field of education for more than 20
years and has conducted workshops at the Association of College and
Research Libraries conference, as well as state library and teaching conferences in Kentucky. She earned her Bachelor's degree in English
Education from the University of Central Florida and a Master's Degree
in Library and Information Science from the University of Kentucky.

Marion Kelt has been working as a librarian since 1982. Her career spans
several sectors, starting in the public library service, then moving on to
secondary school, further education college, and university-level information provision. She has worked on abstracting, indexing, and database
management. She also has experience of technical marketing consultancy
work in the private sector. Her work in higher education includes subject
librarianship and information literacy. She is committed to improving services to the library user, especially by using blended learning, effective
website design, and appropriate web technologies. At present she is working in the area of repository provision and research support. In her spare
time she enjoys yoga, pilates, gardening, and conversing with her cat.

Elizabeth Kline is an Associate Librarian on the research and learning department at the University of Arizona libraries. Her past work and research interests focused on reference services, online pedagogy, and organization and management of online learning objects. She will undertake a sabbatical next year to explore the needs of graduate students and align them with current or new services. She earned her Master's in Information Sciences from the University of Tennessee.

Maha Kumaran is currently the Liaison Librarian for the College of Nursing and the Division of Nutrition at the University of Saskatchewan, Saskatoon, Canada. She graduated from School of Library, Archival and Information Studies (SLAIS) at the University of British Columbia, British Columbia, Canada. She started her profession at the Saskatoon Public Library where she served as the Adult and Young Adult Librarian and later as the Virtual Reference Services Librarian. In 2010, she moved to the University of Saskatchewan with her first position as the coordinator of the Saskatchewan Health Information Resources Program (SHIRP), a virtual library of health-care resources. In this position, she traveled all across the province to instruct health-care professionals on using the SHIRP Library resources that included health databases and online book and journal collections. Currently as the liaison librarian she collaborates with faculty to teach undergraduate and graduate students. Students from the College of Nursing are located all over the province, the country and beyond. Depending on the locality of the students, she either teaches in-person or via Web-Ex, Skype, or Blackboard Collaborate or by using the remote desktop connection. She has recorded videos using Camtasia and has made it available for students through her nursing research guide. Some of her instructional materials are recorded through WebEx and relayed later. She would like to explore how best to teach nursing students in a distributed learning environment. Her research interests are in the areas of diversity, visible minorities, and multiculturalism.

Christine M. Larson is Assistant Professor and Reference and Instruction Librarian at Metropolitan State University in St. Paul, Minnesota. She teaches online and in-person information literacy courses, as well as single-session and course-integrated research instruction in a variety of disciplines. Her interests include teaching critical thinking skills, marginalized knowledges, and religious studies bibliography. She holds a Master's of Science in Library and Information Science from the University of Illinois at Urbana—Champaign and a Master's of Arts in Counseling and Psychological Services from St. Mary's University of Minnesota.

David Lightfoot is an Information Specialist at St. Michael's Hospital in Toronto, Ontario, where in addition to working with clinicians, staff, and students affiliated with global health, trauma, mobility, and perioperative services, he teaches monthly workshops on database searching. He participates actively in the Health Sciences Information Consortium of Toronto (HSICT), where he has chaired a subcommittee devoted to continuing education for professional practice in health sciences. In addition to online instruction and flipped classrooms, his research interests include measuring the impact of hospital libraries on health care. Prior to becoming a librarian, David taught languages and language arts to children, as well as at a handful of universities. In 2016 he will serve on the board of CHLA/ABSC as the coordinator of continuing education. He holds a BA in English and Education, as well as an MISt and a PhD from the University of Toronto.

Judith Logan is a User Services Librarian at the University of Toronto libraries. In addition to her work in the reference and research services unit of the largest library on campus, John P. Robarts Library, she also works with the library's information technology services unit on web content development and management. Her research interests include online learning, preprofessional work experiences, web writing, web-content management, assessment, and service design.

Jessica Long is the Public Services Librarian at the Gardner-Harvey Library on the Middletown regional campus of Miami University in Ohio. She received an MLIS from Kent State University and a BA in Anthropology from the University of New Mexico. Jessica teaches a course related to using library resources effectively with a focus on completing research online, and she has transitioned the course from face-to-face to hybrid and online offerings. She has presented talks on online learning courses and tools at the Distance Library Services Conference, Ubiquitous Learning Conference, ACRL, LOEX, ECIL, and various regional and state conferences. Jessica has also served as co-chair of the Distance Learning Interest Group (DLIG) of the Academic Library Association of Ohio (ALAO) and is interested in how we offer library services to students who spend more time online than in the library or a physical classroom.

Shannon Lucky is an Information Technology Librarian at the University of Saskatchewan. She has a passion for open-source technology, new media, and the maker movement. Her curiosity about how technology can empower users to build their information literacy and do better research led her to a career in library systems and information technology. She has been involved in the upgrade and customization of LibGuides for library resource discovery at the University of Saskatchewan, and her research involves digital archives and interface design based on user experience principles. With a background in the digital humanities, Shannon is passionate about designing systems for real users. She completed a BA in History at the University of Saskatchewan, a BFA in Photography at Concordia University in Montreal, and the joint MLIS and Digital Humanities MA program at the University of Alberta, where she specialized in digital technologies for knowledge sharing and new media design.

Tasha Maddison is a librarian with Saskatchewan Polytechnic. She has also worked as a Science Liaison Librarian at the University of Saskatchewan. As a librarian, she has been active on many committees involving online library instruction, including a team that looked at integrating library learning objects into Desire2Learn, and a task force studying the first-year experience and how best to meet student's needs. She spent several years researching and implementing flipped classrooms into undergraduate engineering courses, and she has also explored teaching effectiveness of online tutorials when compared to in-person instruction. She completed a BA in English and Drama at the University of Saskatchewan, Saskatoon, Saskatchewan, Canada. She then completed the online MLIS program at Wayne State University in Detroit in 2012, where she specialized in academic libraries, reference and instruction. Previously she worked for the training and development team at ProQuest in the K12 market, where she worked for nine years. She traveled across North America offering training on ProQuest's K12 database solutions. She also spent considerable time creating online tutorials and offering webinars for clients. She has extensive knowledge of online training, including the use of both synchronous and asynchronous learning options. Her current research interests include resource discovery, distributed learning environments, online teaching, and utilizing active learning techniques.

Brad Marcum is an Associate University Librarian and the Distance and Online Education Program Officer at Eastern Kentucky University libraries in Richmond, Kentucky. Brad has worked in libraries since 1998. Brad splits his time between working with EKU's regional campuses and online students and faculty, and he is an active presenter at regional and national conferences and author on distance-librarianship issues. He received his MLS from the University of Kentucky in 2003 and has worked in distance education and distance librarianship since that time.

Bridget Morant is an Information Specialist for Education and Consumer Health at St. Michael's Hospital in Toronto. She facilitates the library's instruction and marketing initiatives and oversees the hospital's patient and family learning center. In 2013, Bridget was involved in the implementation of flipped classrooms for health sciences at St. Michael's Hospital along with her colleagues. She is passionate about new ways of engaging with library users through instruction and outreach and has presented at various conferences on library marketing. Bridget holds a BA and Honors Specialization in Media and the Public Interest and an MLIS, both from Western University in London, Ontario.

Peggy Ridlen is Reference Librarian and Instructional Liaison at the Jack C. Taylor Library of Fontbonne University in St. Louis, Missouri. She oversees the instructional services of the library and is also Professor of information literacy for higher education and the first-year seminar "Culture and Common Good," both general education courses at Fontbonne. In addition to information literacy, her work includes curriculum development focusing on global poverty, the U.N. sustainable development goals, and catholic social teaching. She has served on various library committees on the local, state, and national levels.

Amanda Roth is an Instructional Technologies Librarian at the University of California—San Diego. Amanda uses more than 5 years of experience with website design, information architecture, and knowledge of user experience best practices in the corporate world to create and deliver information literacy instruction through the use of e-learning objects. Amanda received her MLIS from San Jose State University in 2013 and has since worked in academic libraries providing reference and instruction services to undergraduate students.

Marie Sciangula is the Assistant Director of the Teaching, Learning, and Technology Center (TLTC) at Purchase College, SUNY. In this capacity she manages and administers Moodle, the campus's chosen learning management system, and promotes innovative uses of academic technologies and pedagogical approaches across campus. She offers a variety of workshops for faculty and staff geared toward effectively integrating technology into teaching and learning and encourages the use of open-source applications. Marie is an active member of the Metropolitan New York Library Council (METRO) as a member of the social media, bibliographic instruction, and collective access special interest groups and a member of the myMETRO researchers pilot project. She regularly presents at local and national conferences, including SUNY's Conference on Instruction and Technology (CIT), SUNY Librarians Association conference, and METRO's annual conferences. She has also presented at ARLIS/NA. Marie has been published in *Art Documentation* and *The Global Librarian*. She received a Bachelor's degree in Women's Studies from Purchase College, SUNY and her MLIS from Long Island University's Palmer School of Library and Information Science.

Kelly O. Secovnie, PhD, is an Assistant Professor of English at the Borough of Manhattan Community College, City University of New York, where she has taught writing and literature courses since 2009. Her research interests include West African literature, particularly Anglophone drama, and the roles that African American characters and culture play in African literature. Her next research project will explore the depictions (in works of both fiction and nonfiction) of Malcolm X and other black radical writers in West African literature and culture. She has also collaborated on a project to integrate library research into online composition and literature courses with the goal of helping students successfully enhance their information literacy skills within the context of English courses.

Leslie Sult is a Research and Learning Librarian at the University of Arizona library. She earned her MLS from the School of Information and Library Science at the University of North Carolina at Chapel Hill. Over the last 13 years, she has worked with numerous campus colleges and programs, including the university's English composition program and the College of Education to develop and improve scalable teaching models that enable the library to reach and support many more students than was possible earlier through traditional one-shot instructional sessions. She also developed the libraries' first fully online credit-granting course and helped overhaul the manner in which course and subject guides are created and delivered to the campus community. Leslie spent several years collaborating with University of Arizona programmer Mike Hagedon to develop and expand the "Guide on the Side" tutorial creation software, which was released by the University of Arizona libraries as an open-source tool in July 2012. The "Guide on the Side" software was named a 2013 Cutting Edge Technology Service by the American Library Association's Office for Information Technology Policy and has received the 2013 Association of College and Research Libraries' Instruction Section Innovation Award.

Jane M. Theissen holds a Master's degree in Library and Information Science (MLIS) from the University of Missouri—Columbia and has been a Reference Librarian at Fontbonne University since 2004. In addition to reference duties, she oversees the Information Commons, collaboratively teaches Fontbonne's information literacy course, manages the library's print periodicals, and supervises four part-time reference librarians. Jane has served on numerous committees both at Fontbonne and in library organizations at the local, state, and national levels.

Dominique Turnbow, Instructional Design Librarian, University of California—San Diego, combines her expertise in instructional design with more than a decade of experience working in academic libraries to deliver information literacy instruction effectively in online environments. Dominique received her MLIS from the University of California—Los Angeles in 2002 where she began her career at the biomedical library. After several years in the profession, she pursued her Instructional Design degree at San Diego State University while providing instruction, reference, and outreach services to undergraduate biology students at the University of California—San Diego. After completing her MEd degree in 2013, she began coordinating the development of online library tutorials across all disciplines.

Niamh Wallace is a librarian in the Research and Learning Department and Liaison to the School of Social and Behavioral Sciences at the University of Arizona libraries. She has experience creating information literacy instruction tutorials and is currently working on a project to reduce the costs of required course materials for students. She received her MLS from the University of Arizona in 2010.

Jenaya Webb is a Public Services Librarian at the Ontario Institute for Studies in Education (OISE) library at the University of Toronto libraries. She provides information literacy instruction and research support for students, faculty, and staff at OISE and the U of T's Factor-Inwentash Faculty of Social Work. Her research interests include wayfinding in library spaces and librarian−user interactions in online environments.

Carolyn Ziegler is an Information Specialist in the Health Sciences Library at St. Michael's Hospital, Toronto. She supports the research, educational, and clinical information needs of the Departments of Emergency Medicine, Mental Health, Pediatrics, Family Medicine, and the Centre for Research on Inner City Health. She has contributed to research on a wide range of medical and social determinants of health topics. She holds an MA and MISt from the University of Toronto.

FOREWORD

The world of information has become exceedingly complex, and the landscape of educational opportunity is broader than it has ever been. In the midst of a rapidly changing educational environment, from traditional to hybrid to online to massive open online courses and whatever is coming after that, most students continue to lack the basic knowledge and skills to handle information well. Increasingly they are being asked to move beyond the role of knowledge consumer to that of knowledge creator, yet they generally lack the ability to do either effectively.

Decades after the term *information literacy* came into vogue, librarians, its main proponents, struggle to put the development of students as researchers and skilled information users into the curriculum. Students mistakenly see themselves as skilled information handlers, and faculty members and administrators have thus far not given information literacy enough of a priority to do the work needed to develop information-literate students.

Even if information literacy were a priority in academia, higher education now has a maze of delivery options such that the student in the live classroom is only one scenario among several. Today's student is just as likely to be participating in an evening forum online or doing asynchronous assignments through Blackboard or Canvas as sitting through an hour-long lecture in a real classroom. Education has become "distributed"—that is, disseminated through a wide variety of models, platforms, and delivery options. Nontraditional students, often far from their home campus, are becoming the majority.

While it adds a level of challenge to information literacy instruction (ILI), distributed education also provides many opportunities, often using the same technology that made distributed education possible. In an era in which information literacy has not yet achieved the prominence it needs in higher education, vastly expanded occasions to support ILI through technology are welcome.

For the average information literacy instructor, just navigating through the options, let alone optimizing them, is a daunting task. Most of us in the information literacy field have embraced at least some technological tools while avoiding others for which the learning curve seems too steep.

Yet we are nagged by the thought that better opportunities may well be available.

This volume—*Distributed Learning: Pedagogy and Technology in Online Information Literacy Instruction*—is no mere tech manual showing instructors how to use relevant tools. Rather, it is an introduction to a multitude of technological solutions through the lens of sound pedagogy while maintaining a human touch. Further, it provides options and cases to enable all of us to expand our horizons and broaden our use of many more opportunities than we have believed were possible.

Distributed Learning presents an impressive amount of educational and instructional design theory applied to ILI through the use of technology. It also gives us a wide variety of options for ILI within a distributed educational environment. A clear advantage is the book's accessibility to readers who want to enhance their own instruction. In a time in which information literacy is attempting to push its way further into the varied forms of higher education, this volume provides an opportunity to make significant advances. We live in exciting times.

William Badke
Trinity Western University, Langley, BC, Canada

CHAPTER 1

Introduction

Libraries must meet the information and scholarship needs of their users. The norm now seems to be that many users are away from the physical library or institutional location but remain entitled to receive equitable access and services just like those who are in the physical campus. Distributed library services are offered to the distance learning community that cannot for various reasons be present at a physical location. Fortunately, electronic resources and technology allow libraries to provide equitable - services to all users. Early literature on the topic distinguishes between *distributed* and *distance learning* (Fleming & Hiple, 2004; Hawkins, 1999), with *distributed learning* being defined as a delivery mode in a distance learning environment (Shimoni, Barrington, Wilde, & Henwood, 2013); in most of the references found throughout this book, however, there is some overlap and situations in which the two terms are used interchangeably. Distance learning focuses on reaching remote and probably employed populations who still wish to continue to attend school. Distributed learning and instruction happens independently of time and location. It is about learning from anywhere and anytime using synchronous or asynchronous models. Immediacy and intimacy are two major concepts that could deter a student or an instructor in an online learning environment. *Intimacy* refers to nonverbal factors, and *immediacy* refers to the psychological distance between the instructor and the student (Sung & Mayer, 2012). In a well-designed distributed learning model, intimacy and immediacy can be embedded with the faculty being available to students through various modes, including email, a content management system (CMS), a learning management system (LMS), WebEx, Skype, and other means.

Whether synchronous or asynchronous instruction is employed, the methodology of distributed learning expects a certain level of responsibility and accountability from the student. Students who participate in these environments need to be self-motivated learners; they must demonstrate a willingness to adapt to and learn new technology, as well as engage with the instructor and other learners online. Without commitment and

Distributed Learning. © 2017 T. Maddison and M. Kumaran.
Published by Elsevier Ltd. All rights reserved.

engagement from the both the student and the instructor, distributed learning environments will not thrive.

Some of the terms in this book are used synonymously such as CMS and LMS and distributed and online instruction. These terms tend to be used interchangeably in much of the library literature on the topic. When one term is favored over the other, it is generally dependent on an institution or a geographical location. All of our contributors are engaged in some form of distributed or distance learning, agree that the use of technology is here to stay, and understand that its ever-evolving nature makes it challenging. The two major components of distributed learning are pedagogy and technology; for that reason, we have chosen to divide the book into sections that address these components as well as case studies that discuss a specific implementation for a class or population at a specific institution. Distributed learning as we know it today would not be possible without new technology and easy access to it—either because the technology is free or is available through the institution. Several of our chapters address the latest information literacy framework and how programs have been adapted accordingly. Most of our chapters address the challenge of assessing instructional initiatives, and other chapters help provide a solution by including their rubric or by discussing their assessment tools and providing evidence of student engagement, learning, and retention. The following paragraphs summarize our chapters.

Chapter 2, Literature Review of Online Learning in Academic Libraries, by Maddison et al. sets the foundation for the rest of the book in its discussion of the current state of distributed and online learning. The chapter focuses on 55 case studies as presented in peer-reviewed scholarly journals between 2010 and 2015 with a concentration on online information literacy instruction delivered to undergraduate student populations within academic institutions. The value of this chapter lies in the meticulous listing of each category and the linking of the reviewed articles to the types of technology used (to create online learning objects and to deliver the material), the challenges faced, the learning environment (synchronous or asynchronous), assessment tools, best practices, and more are discussed. Librarians will be able to make excellent use of the tables when seeking articles to conduct their own study. The thorough investigation of online learning will be valuable to librarians and researchers who are interested in conducting further research on the topic.

Chapter 3, Using Theory and Practice to Build an Instructional Technology Tool Kit, by Amanda Hess presents a "theoretical foundation on which librarians can create an instructional technology toolkit" (p.47).

This chapter successfully integrates pedagogical practice, Bloom's taxonomy, and the principles of the Association of College and Research Libraries (ACRL) for using information literacy in higher education. This chapter is critical for librarians as they develop and enrich their online teaching with active learning techniques and the integration of technology. The focus on Bloom's taxonomy, ACRL standards (2000), and the ACRL framework (2015) will be tremendously valuable for librarians who are investigating distributed learning or are exploring the use of the framework in their current online teaching practices. Readers will benefit from Hess's immense knowledge of technology tools and how they can be implemented in the teaching environment to ensure that effective learning has taken place. Hess's visuals help to illustrate these concepts by setting the learning environment, noting the activity or strategy, and then suggesting a potential tool that could be implemented for of each learning outcome as guided by the framework.

Chapter 4, Designing Online Asynchronous Information Literacy Instruction Using the ADDIE Model, by Maryellen Allen is a perfect example of how resource-challenged librarians are adapting and finding alternate means of providing information literacy instruction in the online environment. Her library evaluated various instructional design models before successfully implementing the analysis, design, development, implementation, and evaluation model that she explains in the chapter. Every model has its strengths and weaknesses, and each library and librarian must adapt the best model for their users. Creating one's own suite of resources does not come without challenges, and as Allen demonstrates, it is absolutely acceptable to borrow videos from the web as long as they meet the instructional objectives of you and your clients.

Chapter 5, Enhancing Kuhlthau's Guided Inquiry Model Using Moodle and LibGuides to Strengthen Graduate Students' Research Skills, by Susan Eichenholtz and Eloise Bellard focuses on graduate students participating in an education law class. Their research began with the observation that students were not aware that specialized library resources were available and instead used popular search engines for research. The authors discuss how the development of a successful collaboration with faculty, the use of Kuhlthau's Guided Inquiry Model, and a redesigned library instruction module assisted students in successfully completing a research assignment based on educational law. The authors contribute to the library literature by noting their incorporation of Kuhlthau's learning theory in class and in online tools (Moodle and LibGuides). A significant finding of this chapter is that the librarians' availability in the distributed learning environment encouraged the students to study independently and seek assistance when required. The authors include details about the mandatory research assignment along with the library course handout and corresponding research guide. This is an important chapter for librarians who are looking to broaden their use of learning theories or to incorporate these tools.

Chapter 6, A Model for Teaching Information Literacy in a Required Credit-Bearing Online Course, by Renee Hall and Erica Getts shares the experiences of delivering one-credit information literacy modules that are self-paced and embedded within Blackboard. The authors discuss how they integrated technical elements into the course and enhanced the content through the use of audio, video, and visual elements. Readers will benefit from the unique perspective offered by Hall and Getts because they work in an institution that only offers programs online. The authors address both synchronous and asynchronous learning options that were used to teach students basic skills such as evaluating resources and the ethical uses of information. This chapter will be useful to librarians who are beginning to embed their instruction online because the authors discuss the importance of mapping learning objectives to an assessment that accurately reflects what the students have learned. Authors offer advice on formulating and delivering assessments and course evaluations as well as setting up a sustainable schedule for updates and tutorial maintenance.

Dominique Turnbow and Amanda Roth weave several different learning theories throughout Chapter 7, Engaging Learners Online: Using Instructional Design Practices to Create Interactive Tutorials. The authors describe the creation and implementation of an interactive tutorial created for undergraduate students at a large academic university. The chapter offers an example of how to use instructional design theories and approaches to deliver effective online instruction. Readers will benefit from detailed illustrations about the process and workflow, as well as the descriptions about how to incorporate engaging activities in online tutorials.

In Chapter 8, Developing Best Practices for Creating an Authentic Learning Experience in an Online Learning Environment: Lessons Learned, Cindy Judd and Brad Marcum introduce readers to the concept of the librarian as a "meddler in the middle," an "approach that brings the librarian into student interaction as an active and equal participant" (p. 152). This educational concept moves beyond both the "sage on the stage" and the "guide on the side" philosophies of teaching by proposing active learning activities for students, with the instructor acting as a facilitator of knowledge. The authors provide details on befriending gatekeepers and teaching in the trenches, along with practical advice for maintaining authority and creating assessment that demonstrates the relevance and value of instruction. Readers will benefit from the authors' vast knowledge of technology and how it can be integrated into classroom activities such as using Padlet, an online whiteboard, to "encourag[e] student experimentation and interaction" (p. 138).

In Chapter 9, Delivering Synchronous Online Library Instruction at a Large-Scale Academic Institution: Practical Tips and Lessons Learned, University of Toronto librarians Jenaya Webb, Judith Logan, and Monique Flaccavento explore web-conferencing software as an alternative to traditional face-to-face instruction for engaging new undergraduate students. This chapter touches on several questions that challenge teaching librarians today, including "How can a small number of librarians support a large population of undergraduate students?" and "Is there a perfect time to offer library orientations?" Readers will benefit from the discussion of challenges encountered when implementing and delivering synchronous online instruction. The authors provide a checklist for teaching librarians who are preparing to use synchronous instruction. They also provide a thoughtful discussion on lessons learned, including technological barriers experienced by the instructors and participants, the timing of the sessions, and whether this was an appropriate delivery method for the target population.

In Chapter 10, Making Library Research Real in the Digital Classroom: A Professor—Librarian Partnership, Lane Glisson and Kelly Secovnie share their success story using scaffolded information literacy instruction with research and writing assignments in online courses. Their approach provides a way to ensure students really understand and synthesize what they learn and show their acquired skills in their assignments. Some of the conclusions from their chapter include making sure library resources are well represented in the course-management site, with librarians and faculty working closely and collaborating every step of the way to integrate information literacy into writing assignments. The authors also recommend that librarians should have a presence on the course-management site to answer questions, and partner with faculty members to design appropriate tasks and useful assessment strategies; that student abilities, knowledge, skills, or background inform course content and delivery; and finally that the assessment of content and collaboration before every term leads to changes that improve the course.

In Chapter 11, Forging Connections in Digital Spaces: Teaching Information Literacy Skills through Engaging Online Activities by Michelle Desilets, Christine Larson, Michelle Filkins, and Jennifer DeJonghe use multiple literacies to develop and stretch their students experience with technology to support meaningful and engaging instruction. This chapter details their work integrating information literacy instruction into assignments and other learning opportunities throughout the coursework at Metropolitan State University. The strength of this chapter lies in the examples that are included throughout the text that illustrate an activity and then tie it directly to the appropriate ACRL frame. The authors also note feedback from the professors and students that they collaborated with, which adds an authority to the paragraphs, and their anecdotal evidence that is both compelling and interesting. The chapter will be of use to librarians who are curious about levels of student engagement, the use of social media in the classroom and the adoption of learning activities in order to forge strong connections online and enhance student learning. Many of the examples discussed in the chapter could be easily implemented at other institutions.

In Chapter 12, Innovation Through Collaboration: Using an Open Source Learning Management System to Enhance Library Instruction and Student Learning, Kimberly Detterbeck and Marie Sciangula move beyond a purely instructional treatment of the LMS to outline many unique and innovative possibilities with Moodle. They recommend including librarians as course collaborators, providing reference assistance through the 'service desk staff customized role,' reviewing syllabi as a collection-development strategy, using quiz results to tailor instruction specifically to student needs, and transforming and digitizing the culminating student project process. A highlight of this chapter details the collaboration between the library and the Teaching, Learning, and Technology Center at Purchase College, SUNY, which has enhanced both the teaching and learning experience. Librarians reading this will benefit from the assignment outline and rubric that the authors' have shared within the chapter. The rubric adds great value and is licensed under Creative Commons; we are sure that librarians will make excellent use of this resource.

In Chapter 13, From Technical Troubleshooting to Critical Inquiry: Fostering Inquiry-Based Learning Across Disciplines Through a Tutorial for Online Instructors, Michael Courtney examines moving beyond providing information access and collaborating with faculty through LMS and maximizing librarian and information literacy outreach, challenges that will resonate with most librarians. It is true that some faculty members, including librarians, may be rightfully hesitant to adopt new ways of delivering content. The fact that they are not trained in using technology, that they cannot gauge how well it is received and retained at the user end, that it is time consuming particularly as technology evolves and interfaces change, and the difficulty in converting instruction sessions to viewable and archival formats for future use are all legitimate reasons why they hesitate to adopt new ways of delivering content. Courtney encourages librarians to be harbingers of technology evolution, to assuage faculty hesitations by first training themselves on online course creations and implementation, and to share and use this knowledge by collaborating through LMS with other faculty members.

If research guides are to be effective, they should be used as course guides and embedded in the LMS. This is what Elizabeth Kline et al. found in Chapter 14, Embedding the Library in the LMS: Is it a Good Investment for Your Organization's Information Literacy Program? These librarians started with a widget that finally evolved into a Library Tools Tab, and through this process they also realized the importance of a programmatic approach to instruction. Through trial and error and by working with various units such as campus administrators, the information technology unit, the office of instruction and assessment, and subject faculty members, they were successful in embedding a "superwidget" in a highly accessible location within the LMS. This tool made it easier to collaborate with faculty and customize library information and allowed them to reach students on a wider scale. Their analysis shows that the tool is well used and that students are learning about library resources available to them. Like many others in this book, the authors of this chapter also reiterate the importance of collaboration and a willingness to change and adapt as needed.

Most librarians engaged in distributed learning will relate to Chapter 15, A Decade of Distributed Library Learning: The NOSM Health Sciences Library Experience, and Jennifer Dumond's description of the Northern Ontario School of Medicine's health sciences library experience. As institutions embrace the distance or distributed education model and students demand synchronous and asynchronous instruction sessions, librarians are devoting time to finding the right technology to enable such sessions. Finding and training in new technology, overcoming challenges, producing and distributing instruction materials with it, and encouraging student uptake are all challenges that librarians everywhere now face. Because technology is ever changing and nebulous only adds to the challenge. Librarians have to choose technology that is cost-effective, easy to use, malleable enough to meet the needs of varied purposes, and almost everlasting.

In Chapter 16, Parallel Lines: A Look at Some Common Issues in the Development, Repurposing and Use of Online Information Literacy Training Resources at Glasgow Caledonian University, Marion Kelt writes about designing and producing an online information literacy training resource. She highlights eight themes from the literature that include pedagogy and technology. Glasgow Caledonian University has adapted to its national guidelines and ever-changing technology by creating different training packages from Study Methods and Information Literacy Exemplars (SMILE) to Small Mobile InfoLit Realworld Knowledge (SMIRK). As Kelt notes, planning and having a clear sense of the audience will help create usable, relevant, and accessible training resources. The importance of user feedback to revise content is another crucial takeaway from this chapter.

In Chapter 17, Concept to Reality: Integrating Online Library Instruction Into a University English Curriculum, Donna Canevari de Paredes and David Francis take us through their journey of developing an online literacy course from their initial idea to building a compulsory online program. They stress the importance of collaboration among all working parties: librarians, instructional designers, faculty, and information technology staff. The importance of collaboration is evident from this chapter; without it, librarians cannot successfully proceed in the university environment.

The collaboration between Chester-Fangman, Garber, and Berg produced two case studies that are included in this book. In Chapter 18, A Successful Reboot: Reimagining an Online Information Literacy Tutorial for a First-Year Experience Program, they reinvent the library information literacy tutorial in a first-year experience program that is a required for-credit interdisciplinary course. The study compares student achievement using two different online modules: video tutorials and the traditional web-based option. The authors pay careful consideration to the best practices originally outlined in Dewald's work on design principles of web tutorials. The working group also discusses the implementation of a flipped classroom that included active learning opportunities for students to further engage with library instruction. Readers will enjoy the clever use of subheadings throughout the chapter, as well as the use of evaluation tools such as focus groups, one-minute papers, quizzes, and a faculty survey.

In Chapter 19, Rethinking Plagiarism in Information Literacy Instruction: A Case Study on Cross-Campus Collaboration in the Creation of an Online Academic Honesty Video Tutorial, the same authors discuss the creation of a plagiarism tutorial. Again, their work was guided by an outdated tutorial that desperately needed a significant overhaul. The team faced the challenges of creating a video that is fun and engaging for students on a topic that is often viewed negatively. The video also needed to supply critical information on responsible behavior without focusing solely on the punitive. Readers will appreciate the opportunity to discover connections between the authors' two case studies—the workflow of the projects and team development—as well as their creative process. Their use of visuals is excellent and can help readers envision how these librarians chose to illustrate their content, including the process of storyboarding the script and the final creation of the video and accompanying quizzes.

In Chapter 20, Adapting to the Evolving Information Landscape: A Case Study, Peggy Ridlen and Jane Theissen, librarians at Fontbonne University in St. Louis, highlight the evolution of their online information literacy course from before their arrival in 2004 through their most recent revision. They take us through the ongoing process of regularly updating the course, including adopting the new ACRL frameworks, adding modules on social media, finding a textbook for the course, and developing an assignment using visual thinking strategies. Finally, they explain the benefits of using assessment data to revise course content.

In Chapter 21, Gaming Library Instruction: Using Interactive Play to Promote Research as a Process, Jessica Long highlights gaming technology as a means to engage students in learning the research process. Along with other librarians at the Gardner-Harvey Library at Miami University regional campus in Ohio, she uses game-based learning for their Choose Your Own Adventure tutorial. In this chapter, she takes readers through a journey beginning with Google Sites and how these librarians have currently settled on *Adventr*, an online platform that allows for creating interactive videos by simply dragging and dropping video clips into a platform. The Choose Your Own Adventure tutorial is a blend of text, videos, and images in which students follow the story line to move from one level to the next. Technology is a challenge, and every software program has its pros and cons; features change, and website interfaces evolve; additional help may be needed from colleagues, software support, or students who are willing to help. Some challenges are foreseeable, but others are not. The success or failure of building any resource is based on the creator's ability to adapt by trying new things and abandoning what does not work. Student needs and technology are constantly diversifying, and libraries must transform with them.

Health professionals have demanding schedules, and they often do not have the luxury of investing significant work time for professional development in information literacy instruction. To remedy this issue, Sandy Iverson et al., librarians at St. Michael's Hospital in Toronto, began a pilot project using flipped classroom methodology. Most literature on flipped classrooms in libraries has focused on academic environments to this point, so Chapter 22, Implementing Flipped Classroom Model Utilizing Online Learning Guides in an Academic Hospital Library Setting, fills a gap in the research by focusing on academic hospital environments. Readers will benefit from the discussion of the pilot project, the lessons learned from the instructional committee, feedback from participants and instructors, and recommendations for practice.

Distributed learning in today's world of library instruction is at the intersection of information technology, information, and instructional design. The most prevalent themes that have arisen from our contributors, as well as from our own research include the need to address librarian faculty workload in environments with large student populations, the desire to provide students with timely and effective training, the need to develop collaborations in which all parties are engaged with curriculum-driven instruction or assignments, the financial advantage that online instruction offers, and finally the willingness of all involved—individuals and institutions—to continually learn and evolve. Librarians are trained to be lifelong learners and quickly adapt to these changes by learning new technology, collaborating with faculty, learning from student assessments, and adapting future content, as well as by networking with other librarians and other experts in the field. Librarians are creative in their instructional practice, bringing gaming and flipped classrooms to their learners. Librarians involved with instruction are learning about instructional design and both using and presenting with technology.

We hope our book will be a useful resource to all librarians who currently provide or are considering providing online instruction, particularly in a distributed environment. Our contributing authors have shared many ideas that worked or did not work for them, as well as the lessons learned, their best practices, designs and concepts developed, and frameworks and models used. We would like to express our sincere gratitude to all of our authors for their patience in working through the proposal, peer review, and draft process. We would also like to thank our peer reviewers for the significant contribution that they made to the individual chapters as well as to the book as a whole.

REFERENCES

Fleming, S., & Hiple, D. (2004). Distance education to distributed learning: Multiple formats and technologies in language instruction. *CALICO Journal, 22*(1), 63.

Hawkins, B. L. (1999). Distributed learning and institutional restructuring. *Educom Review, 34*(4), 42−44.

Shimoni, R., Barrington, G., Wilde, R., & Henwood, S. (2013). Addressing the needs of diverse distributed students. *International Review of Research in Open and Distance Learning, 14*(3), 134−157.

Sung, E., & Mayer, R. E. (2012). Five facets of social presence in online distance education. *Computers in Human Behaviour, 28*(5), 1728−1747.

Foundations of Distributed Learning

CHAPTER 2

Literature Review of Online Learning in Academic Libraries

2.1 INTRODUCTION

Online learning grows out of the literature on distributed and distance learning services, which dates to 1963 when the Association of College and Research Libraries created a series of guidelines for extended campus services (Association of College and Research Libraries, 2015). Today online learning is used interchangeably with other terms such as "web-based learning," "e-learning," "computer-assisted instruction," and "Internet-based learning" (Ruiz, Mintzer, & Leipzig, 2006). Distributed learning evolved from distance learning (Dede, 1996). Now enabled by new technology, it has a much broader application than distance learning ever did. While distance learning referred to students who were physically away from a given campus location, the definition of distributed learning "refers to delivery anytime and anywhere, including formal settings such as classrooms and schools but also homes, workplaces, museums, libraries, and community centers" (Fletcher, Tobias, & Wisher, 2007, p. 96). It is also used to describe "everything beyond ... face-to-face encounters such as webinars, tutorials, LibGuides, and video conferencing" (Maddison, 2013, p. 266). In this chapter, the term *online learning* refers to the delivery of and access to instructional materials from anywhere, at any time, and in any of a variety of different formats that can include both distance education and distributed learning.

While online learning is no longer a new concept, the way in which instruction is delivered in this format is still very much in development. Universities have been working for some time to develop both free and paid online learning experiences to students at a distance as sparked by the "Our History" the Massachusetts Institute of Technology's OpenCourseWare initiative in 2001 (Massachusetts Institute of Technology). Since then, universities have focused on offering equitable learning experiences in the online environment for learners both on and off campus. Meanwhile, students are looking for opportunities to learn where they live and at their own time, pace, and convenience. Current technologies allow for such opportunities, and academic libraries are in a position to take advantage of these technologies to improve the content and delivery of information literacy instruction (ILI) online.

2.2 BACKGROUND

In 2013, the university library at the University of Saskatchewan set out to investigate one of four strategies in the library strategic plan known as the Learner and Teacher Strategy. This strategy focused on "transform[ing] our services, collections, and facilities to contribute to the success of our learners and teachers and the library as a learning organization" (University Library, 2014). The goals of this strategy were "to foster individual and collaborative teaching that employs optimal methods of delivery and models best practices to support the University's goal of information literate students" and "develop as teachers through sharing our teaching experiences and resources with each other and through other methods of professional development" (University Library, 2014). To achieve these goals, three action items were determined, one of which was to "explore and develop distributed learning methods." A working group with the authors of this chapter formed in Oct. 2014 to explore distributed learning methods used in the wider academic library community by conducting a literature review on this topic and by consulting with teaching librarians on campus. The results of the literature review are presented in this chapter.

2.3 METHODOLOGY

The literature review focused on academic libraries, including articles that were published from 2010 through 2015. Search parameters were limited to peer-reviewed scholarly publications from North America, the United Kingdom, and Australia. A search string was used that contained the following keywords combined with the Boolean operator "OR": online learning, electronic learning, distributed learning, distance learning, synchronous, asynchronous, educational technology, blended, embedded, and integrated. Where appropriate, keywords were truncated with an asterisk to garner a broader list of results. Search results were limited to undergraduate students (college or university) and ILI, ensuring the results focused on the delivery of library instruction and not online learning as applied in other disciplines. Articles were sought in the following databases: Library and Information Science Abstracts (LISA), Library, Information Science, and Technology Abstracts (LISTA), Education Resources Information Center (ERIC), ProQuest Education Journals, Medline (OVID), Cumulative Index to Nursing and Allied Health Literature (CINAHL), Scopus, Web of Science, and Journal Storage (JSTOR). In addition to that strategy, the team carefully reviewed the reference list of each article to discover further relevant articles that met the search criteria as previously listed. In a few cases, older articles were included in the review because of their significant contribution to the research literature on the topic. Scopus and Web of Science were used for citation analysis.

Abstracts were scanned for initial impressions and relevance to the project and then added to a master list. The list was deduped and a section was assigned to each member of the group to read and analyze. A total of 89 articles were reviewed. The working group members analyzed each article using the following criteria:

- the type of article (e.g., case study, literature review, pilot project, environmental scan);
- whether the study addresses ILI for a specific subject area or a general audience;
- how study authors made online tools accessible to learners (e.g., a learning management system (LMS) such as Blackboard, video hosting such as YouTube, webpages, a library guide);
- the technology used to create the online learning object;
- the type of learning environment employed within the article (e.g., synchronous, asynchronous, face-to-face (f2f), online, blended);
- assessment tools mentioned in the article (e.g., pre- and post-tests, citation analysis, or embedded quizzes); and
- outcomes, recommendations, and key takeaways.

From the literature review, 55 articles were case studies. The remaining articles were either literature reviews or research articles that did not provide enough information on current online ILI practices; these were excluded for the purposes of this chapter. The small sample size of 55 articles limits the viability of any statistically significant claims about the use of technology for online learning, but clear trends do emerge. The following sections are an analysis of the 55 case studies that identify trends and outliers relating to the disciplines using online technology, creation, access, and learning environment.

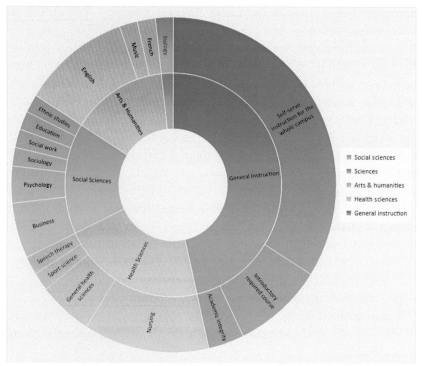

Figure 2.1 Distribution of subject concentrations with the case study literature on online learning and information literacy instruction.

2.4 SUBJECT DISTRIBUTION

The literature on online learning in ILI in academic libraries represents a variety of disciplinary concentrations. Twenty-six of the reviewed articles (47.3%) focused on instruction for a general audience, while the remaining 29 (52.7%) made reference to a specific discipline (see Fig. 2.1). Of these, 19 (34.6%) referenced instruction intended for the entire campus community, whether through a widely accessible content management system tutorial, module, or set of self-serve tools. Five articles (9.1%) spoke to cases in which a required course was offered to all students at a first-year level that covered basic library concepts. Two articles (3.6%) were case studies that offered content appropriate for all students on campus on the topic of academic integrity. In the literature that was situated within a specific discipline, health sciences represented the highest percentage of articles reviewed: 12 (21.8%). Of these, the majority (seven articles, 58.3%) discussed information literacy strategies for nursing students. The second highest disciplinary area was in the social sciences, where eight articles (14.6%) came from various subjects. Meanwhile, case studies from disciplines in arts and humanities were discussed in seven of the reviewed articles (12.7%). Of these, the majority of cases were situated in English, with remaining articles covering single instances from music and French. By comparison, online ILI within the sciences was relatively low and underrepresented in the literature (1.8% of the literature reviewed) (see Table 2.1).

2.5 TECHNOLOGY

The 55 case studies were used to examine what kinds of technology were used to create online learning content for ILI, where students encounter this instructional material, and what kinds of learning environments were being used. The group identified popular creation and dissemination technologies and predominant learning environments in online instruction programs by counting each aspect in these articles. However, there were complicating factors inherent in identifying these three aspects of each case study. While many articles referred to specific technologies and tools by name—such as Blackboard, YouTube, or PowerPoint—others generally referenced unnamed video technologies, LMSs, websites, or other types of technology. In some cases, the authors did not identify what kinds of technologies were used to create the course content or did not describe how students engaged with it, whether it was embedded in a library website, was part of a school's LMS, or was found somewhere else entirely.

In some cases, where an article described or compared multiple projects or, where categories crossed over, multiple entries for an individual article were created. For example, a project might use Camtasia and Prezi to create content for a Blackboard course that was also posted on YouTube and embedded in the library website. This information might be used for an asynchronous online course as well as in the f2f portion of a flipped classroom's instruction session. All of these points of information would be recorded in reference to that article.

In some cases, the line between a tool for content creation and content access was unclear. For example, is subject guide software such as LibGuides a creation or an access platform? To clarify these inconsistencies, the group described the article content as specifically as possible, using product names whenever that information was available, as well as describing the type of technology in generic media and platform terms (e.g., website, video, web conferencing). For technologies such as LibGuides where

Table 2.1 Disciplines within the literature on online instruction presented in order of frequency

Number of articles	Subject	References
19	Instruction for the whole campus or general instruction modules, programs, or tools	Adebonojo (2011), Alleyne and Rodrigues (2011), Bolorizadeh, Brannen, Gibbs, and Mack (2012), Chisholm and Lamond (2012), Dewan and Steeleworthy (2013), Easter, Baily, and Klages (2014), Ergood, Padron, and Rebar (2012), Gustavson, Whitehurst, and Hisle (2011), Hahn and Bussell (2012), Hemmig and Montet (2010), Hess (2013, 2014b), Lockerby and Stillwell (2010), McLean and Dew (2006), Mee (2013), Meredith and Mussell (2014), Mestre (2012), Scales, Nicol, and Johnson (2014), Vander Meer, Perez-Stable, and Sachs (2012)
7	Nursing	Baker (2014), Brettle and Raynor (2013), Dee and Reynolds (2013), Schroeder (2010), Turnbull, Royal, and Purnell (2011), Walters et al. (2015), Xiao (2010)
5	English	Archambault (2011), Bishop, Yonekura, and Moskal (2013), Domínguez-Flores and Wang (2011), Mery, Newby, and Peng (2012), Striwinter (2013)
5	General introductory required first-year course for all students	Anderson and May (2010), Clapp, Johnson, Schwieder, and Craig (2013), Ganster and Walsh (2008), Gray and Montgomery (2014), Long, Burke, and Tumbleson (2012)
3	General health sciences	Appelt and Pendell (2010), Blevins, Deberg, and Childs (2014), Ritchie (2011)
3	Business	Arguello (2013), Gunn, Hearne, and Sibthorpe (2011), Tooman and Sibthorpe (2012)
2	Academic integrity	Germek (2012), Greer et al. (2012)
2	Psychology	Henrich and Attebury (2012), Boden, Neilson, and Seaton (2013)
1	Biology	Craig and Friehs (2013)
1	Ethnic studies	Lo and Dale (2009)
1	French	Usova (2011)
1	Music	Kern (2013)
1	Physical education	Chaiyama (2015)
1	Speech pathology and audiology	Trail and Hadley (2010)
1	Social work	Gall (2014)
1	Sociology	Hess (2014a)
1	Sports science	Bruce-Low et al. (2013)

textual content might be generated in that platform, the group looked at where the digital content would be initially created compared to where that content might be assembled and viewed by the user. As an example, a LibGuide might include an instructional video, a typed transcript, and screenshots. The video could have been created using some kind of screen-capture or video-editing software, the transcript created in a text editor, and the screenshots made in an image-editing program. These parts would then be assembled in a LibGuide for viewing by the student. This differentiates the creation tools from the access platform.

2.6 CREATION TECHNOLOGY

Within the sample, screen-capture software is the platform most frequently used to create online ILI, with close to half of all articles (43.6%) using this type of technology (see Table 2.2). Screen-capture programs are frequently used to demonstrate how to use websites and online resource interfaces. The commercial software package Camtasia is the dominant program in this category with 17 of the 24 mentions (70.8%) of screen-capture projects using this widely adopted tool. Jing, the second most popular screen-casting tool, is free to use but has limitations on video length and editing features; it was mentioned in only four screen-capture projects. Non—screen-capture video such as recordings of lectures or tours of library spaces are used in 9.1% of all of the articles surveyed but without showing an obvious bias for a particular software.

The strong preference for video tools to create online learning content is illustrated by the tremendous gap in representation between projects using video and screen-capture technology and those using all other creation tools. Presentation software for sharing multimedia slides and Adobe Captivate, an e-learning content creation tool for software demonstrations, simulations, and quizzes, were the next most commonly used creation tools, but both categories were only referenced in 8 of the 55 articles. The remainder of this category was populated by a range of technologies with three or fewer examples in our sample set. Survey tools, audio editors, quiz software, image-editing software, Articulate Storyline, and Dreamweaver for website creation complete all of the tools and technologies cataloged in this section of the analysis. Table 2.2 details the tools and programs used and the articles in which they are described. In addition to these numbers, the group noted 19 articles in this sample

Table 2.2 Technology used to create online learning content

Number of articles	Technology type and name	References
24	Screen capture (all tools)	
	Camtasia (17 articles)	Baker (2014), Blevins et al. (2014), Boden et al. (2013), Bolorizadeh et al. (2012), Clapp et al. (2013), Craig and Friehs (2013), Ergood et al. (2012), Gray and Montgomery (2014), Gustavson et al. (2011), Henrich and Attebury (2012), Hess (2014a, 2014b), Kern (2013), Mestre (2012), Scales et al. (2014), Schroeder (2010), Trail and Hadley (2010)
	Jing (4 articles)	Kern (2013), Hess (2013, 2014b), Vander Meer et al. (2012)
	Other (3 articles) SnapzPro, Screencast-O-Matic, Adobe Presenter, unnamed	Bolorizadeh et al. (2012), Ganster and Walsh (2008), Hess (2014b)
5	Video (nonscreen capture): Final Cut Pro, iMovie, Movie Maker, Media Site, unnamed	Bolorizadeh et al. (2012), Clapp et al. (2013), Greer et al. (2012), Kern (2013), Turnbull, Royal, and Purnell (2011)
8	Presentation software PowerPoint, Prezi	Bolorizadeh et al. (2012), Clapp et al. (2013), Gustavson et al. (2011), Hahn and Bussell (2012), Hess 2014b), Kern (2013), Lockerby and Stillwell (2010), Stiwinter (2013)
8	Adobe Captivate	Germek (2012), Hemmig and Montet (2010), Hess 2014b), Mestre (2012), Stiwinter (2013), Usova (2011), Vander Meer et al. (2012), Xiao (2010)
3	Survey software	Brettle and Raynor (2013), Gustavson et al. (2011),

(Continued)

Table 2.2 (Continued)

Number of articles	Technology type and name	References
	ConstantContact, Google Forms, Survey Monkey	Henrich and Attebury (2012)
2	Audio-editing software: Audacity, Garage Band	Bolorizadeh et al. (2012), Gustavson et al. (2011)
2	Quiz software CourseBuilder, unnamed	Tooman and Sibthorpe (2012), Walters et al. (2015)
2	Image-editing software Adobe Creative Suite (Illustrator, Photoshop)	Bolorizadeh et al. (2012), Kern (2013)
2	Articulate Storyline (online course creation software)	Kern (2013), Walters et al. (2015)
1	Dreamweaver (website creation software)	Brettle and Raynor (2013)

set of 55 that did not refer specifically to the creation of online instructional content or platforms; these are not represented in this table.

2.7 ACCESS TECHNOLOGY

The online environments where students accessed and interacted with instructional content also showed a strong preference for one technical platform: the LMS. LMS's were utilized in 56.3% of all articles, with Blackboard being the single most commonly used platform. Although Blackboard had the largest representation in this category and was mentioned in 35.5% of LMS, there is more diversity in software representation here than in the creation category discussed previously. Moodle, another commercial product and a relatively close second, was used in 22.6% of all LMS projects; the remaining 42.0% of this subcategory comprises custom LMS and other proprietary systems (see Table 2.3).

Like the findings about the creation tools, there is a significant gap between the popularity of LMS and the next most frequently used platform, video hosting systems, which was represented in 31.9% of all articles. YouTube was the most frequently named video-hosting platform, but there was a diversity of named and unnamed options represented in our sample. Other moderately popular platforms for presenting ILI were library guide software (particularly Springshare's LibGuides), websites, and web-conferencing software (Adobe Connect, Skype, etc.).

Table 2.3 Technology used to access online learning content

Number of articles	Technology type and name	References
31	Learning Management System (LMS): All Tools	
	Blackboard (11 articles)	Arguello (2013), Baker (2014), Bolorizadeh et al. (2012), Brettle and Raynor (2013), Easter et al. (2014), Ganster and Walsh (2008), Henrich and Attebury (2012), Kern (2013), Mee (2013), Turnbull, Royal, and Purnell (2011), Xiao (2010)
	Moodle (seven articles)	Chisholm and Lamond (2012), Easter et al. (2014), Gray and Montgomery (2014), Greer et al. (2012), Kern (2013), McLean and Dew (2006), Meredith and Mussell (2014)
	Other (13 articles): ANGEL, Canvas, Desire2Learn, ELearning, Sakai, HealthStream, Custom tools, unnamed	Adebonojo (2011), Bishop et al. (2013), Chaiyama (2015), Clapp et al. (2013), Dee and Reynolds (2013), Hahn and Bussell (2012), Kern (2013), Long et al. (2012), Mee (2013), Ritchie (2011), Schroeder (2010), Tooman and Sibthorpe (2012), Vander Meer et al. (2012)
17	Video Hosting: All Tools	
	YouTube (seven articles)	Blevins et al. (2014), Bolorizadeh et al. (2012), Ergood et al. (2012), Hemmig and Montet (2010), Hess (2013, 2014b), Kern (2013)
	Other (10 articles)JW Player, QuickTime Streaming Server, Windows Movie Player, unnamed	Adebonojo (2011), Arguello (2013), Bolorizadeh et al. (2012), Bruce-Low et al. (2013), Clapp et al. (2013), Gall (2014), Gustavson et al. (2011), Scales et al. (2014), Trail and Hadley (2010), Vander Meer et al. (2012)

	Library Guide Creation Software: All Tools	
9	LibGuides (seven articles)	Archambault (2011), Bolorizadeh et al. (2012), Easter et al. (2014), Gustavson et al. (2011), Hemmig and Montet (2010), Kern (2013), Mee (2013)
	Other (two articles): "Guide on the Side", SubjectPlus, custom course pages	Arguello (2013), Hess (2013)
9	Websites	Bolorizadeh et al. (2012), Ergood et al. (2012), Gall (2014), Hahn and Bussell (2012), Scales et al. (2014), Schroeder (2010), Stiwinter (2013), Trail and Hadley (2010), Vander Meer et al. (2012)
	Web Conferencing: All Tools	
5	Adobe Connect (three articles)	Germek (2012), Lockerby and Stillwell (2010), Mee (2013)
	Other (two articles)	Alleyne and Rodrigues (2011), Mee (2013)
2	Global Collaboration Grids, Skype, unnamed	
	Repository of online learning resourcesi Tunes U, MERLOT, Trace IR	Adebonojo (2011), Bolorizadeh et al. (2012)
2	Clicker Technology	Bolorizadeh et al. (2012), Vander Meer et al. (2012)
	i > Clicker, Turning Technologies	
2	Hardware and devices	Bruce-Low et al. (2013), Hahn and Bussell (2012)
	iPad, Samsung Netbook	
1	Apps	Hahn and Bussell (2012)
1	Blog posts	Schroeder (2010)
1	Discussion boards and forums	Vander Meer et al. (2012)
1	Email	Schroeder (2010)
1	Poll Everywhere (SMS polling software)	Bolorizadeh et al. (2012)
1	Online learning community (one article)	Domínguez-Flores and Wang (2011)
1	Presentation software	Gustavson et al. (2011)
	Photo Story 3, SlideShare	

The scope of available platforms to present online information and frequent references to multiple technologies in many of the articles resulted in a long list of examples that had three or fewer articles mentioning each type. Repositories of online learning resources, clicker technology, and specific hardware such as iPads each appear in two articles. Several technology types and brands were mentioned only once: apps (unspecified), blog posts, discussion boards, email, online learning communities, Poll Everywhere polling software, and presentation software. Table 2.3 provides a complete accounting of the types of access technology represented in our sample.

2.8 LEARNING ENVIRONMENTS

In the sample of 55 articles, online learning instruction predominantly occurred in an asynchronous environment (30 articles, 54.5%). Asynchronous instruction was identified by programs that utilized learning materials that students accessed outside of class or without the instructor present, either in person or online in real time. Examples of these kinds of asynchronous ILI are posted videos (YouTube, etc.), webpages, and instructional apps. Perhaps this is not surprising because of the growing need to provide just-in-time instruction to globally dispersed learners. Having materials that can be accessed by learners when they require it on their own schedules is becoming a critical function of online learning systems.

Blended learning environments that incorporate real-time engagement with instructors and f2f instruction in person or via web-conferencing software were both represented in eight articles (14.5%). Synchronous instruction, occurring either in the classroom using online tools or via live web conferencing was utilized in seven articles (12.7%). Ten articles (18.2%) did not clearly indicate the learning environment of the information literacy curriculum they described.

2.9 CHALLENGES WITH ONLINE INSTRUCTION

In the reviewed literature, 28 articles (50.9%) mentioned challenges and limitations that arose in the course of the case study. Within the literature on online ILI, articles addressed challenges related to the creation of online instruction tools and content, to how users engage with online learning content, and how faculty members perceive and integrate online learning content from the library into their classrooms (see Table 2.4).

Table 2.4 Summary of challenges and limitations identified in case study literature

Identified challenges and limitations	References
Assessment was not incorporated into the case study, so it was difficult to determine the impact of the work	Alleyne and Rodrigues (2011), Blevins et al. (2014), Chisholm and Lamond (2012), Ergood et al. (2012), Hess (2014b), Lo and Dale (2009), Long et al. (2012), Schroeder (2010)
Engaging with faculty and ensuring faculty buy in	Appelt and Pendell (2010), Arguello (2013), Chishold and Lamond (2012), Easter et al. (2014), Hemmig and Montet (2010), Henrich and Attebury (2012), Hess (2014a), Vander Meer et al. (2012)
Content creators experienced a steep learning curve with the technology.	Adebonojo (2011), Button et al. (2014), Ergood et al. (2012), Gustavson et al. (2011), Hess (2014b), Kern (2013)
Design, presentation, and accessibility	Gray and Montgomery (2014), Greer et al. (2012), Henrich and Attebury (2012), Long et al. (2012), Meredith and Mussell (2014), Schroeder (2010), Usova (2011)
Users experienced difficulties with technological aspects of the instruction	Adebonojo (2011), Baker (2014), Greer et al. (2012), Hahn and Bussell (2012), Mestre (2012), Domínguez-Flores and Wang (2011)
Content creators experienced time constraints when developing or updating instructional content	Adebonojo (2011), Arguello (2013), Button et al. (2014), Ergood et al. (2012), Hemmig and Montet (2010), Henrich and Attebury (2012), Kern (2013)
Content creators noticed difficulty keeping users engaged in the learning content	Alleyne and Rodrigues (2011), Easter et al. (2014), Hemmig and Montet (2010), Meredith and Mussell (2014)
Creating new online learning materials required additional and unanticipated funding	Ergood et al. (2012), Mee (2013)
Organizational structures need to be updated	Dewan and Steeleworthy (2013), Kern (2013)

For libraries and library staff, several mentioned challenges to the process of adopting, managing, and upgrading online learning technologies. Several papers spoke to the steep learning curve that comes with mastering new creation tools, especially for those with little experience with online learning technologies (Adebonojo, 2011; Button, Harrington, & Belan, 2014). They also said these inconsistent technology skills could indicate that the quality of the end learning product might be inconsistent, depending on which staff member had produced it (Hess, 2014b). Among the areas of training required for someone to begin actively developing new online learning tools, Button et al. (2014) mentioned staff development should focus on "online course development, assessment and monitoring the quality of online courses in addition to improving their own ICT [information and communications technology] skill base" (p. 11). Another challenge with managing learning technologies, specifically video and audio production, is finding appropriate spaces within the workspace to record video tutorials (Ergood et al., 2012; Kern, 2013). In Kern's case study (2013), the library ended up funding the development of a video-creation space that had soundproofed walls and specialty software available. In some cases, there were unexpected costs related to developing online learning content in the library, one in relation to additional online resources in the library collection that needed to be purchased to meet the demands of students (Mee, 2013) and another in relation to software that needed to be purchased or upgraded (Ergood et al., 2012).

Libraries also found that the development and maintenance of online learning resources required significant time from librarians and library staff (Adebonojo, 2011; Arguello, 2013; Button et al., 2014; Ergood et al., 2012; Hemmig & Montet, 2010; Henrich & Attebury, 2012; Kern, 2013). In some cases, new workflows needed to be developed or new communications methods implemented to ensure work was not duplicated within the library system (Kern, 2013). Henrich and Attebury (2012) remarked that "the videos themselves were time consuming to draft, create, and edit, while production required librarians to learn new, dedicated media software" (p. 175), and they recommended dividing video tutorials into smaller sections to allow for easier revisions.

Seven of the reviewed articles mentioned challenges related to the design and accessibility of online learning materials (Gray & Montgomery, 2014; Greer et al., 2012; Henrich & Attebury, 2012; Long et al., 2012; Meredith & Mussell, 2014; Schroeder, 2010; Usova, 2011).

In relation to accessibility, both Schroeder (2010) and Meredith and Mussell (2014) commented on the importance of developing online learning content that uses accessible design standards so as not to exclude any learner from retrieving the content. Some cases made specific reference to challenges with presenting content that was traditionally taught in an f2f environment; they sought to have it redesigned for a blended or online learning environment (Long et al., 2012; Usova, 2011). In particular, attention needs to be paid to the order in which the information is presented and when elements of online tutorials are released to the user (Greer et al., 2012; Henrich & Attebury, 2012).

Some of the literature mentioned challenges from the perspective of user engagement, technology challenges, and effectiveness of the learning materials. Four articles mentioned the difficulty of keeping users engaged with the learning material and ensuring uptake of the content. In particular, some mentioned that there were challenges getting students to participate in online discussions and there was lower uptake with online learning materials that were not required to fulfill assigned elements (Alleyne & Rodrigues, 2011; Easter et al., 2014; Hemmig & Montet, 2010; Meredith & Mussell, 2014). Other technological challenges for students included the hurdle of getting the content to load or play back (Adebonojo, 2011; Domínguez-Flores & Wang, 2011; Hahn & Bussell, 2012). The video tutorials presented certain challenges for participants who did not feel they could "easily go back and find the spot that would help them with a step, even with chapter markers, and that it was difficult to do the steps along with the tutorial" (Mestre, 2012, p. 266). In some cases, students did not want to spend the required time viewing the video tutorial, preferring to "quickly scan the webpage and get to the task at hand" (p. 266). Design of video tutorials may have a significant impact on whether the content is found useful by students. Baker (2014) commented on the usability of video tutorials in her paper "Students' Preferences Regarding Four Characteristics of Information Literacy Screencasts," pointing out that "length, pace, number of callouts, and frequency of zooming all contribute to the viewability of screencasts. ... Screen casts that are too long, too fast, or slow or with too many special effects such as circling, highlighting, or zooming may distract viewers from the message of the video and may potentially even cause students to stop watching" (p. 76). This is particularly important to note because videos and screen casts are the most popular content type identified in this literature review. While required elements may force students to spend time completing online learning content, content that is poorly designed and not part of a course requirement is less likely to be used.

Some of the literature mentioned that once the online learning content was created it was a challenge to get faculty members on campus to integrate it into their classrooms (Appelt & Pendell, 2010; Arguello, 2013; Chisholm & Lamond, 2012; Easter et al., 2014; Hemmig & Montet, 2010; Henrich & Attebury, 2012; Hess, 2014a,b; Vander Meer et al., 2012). To address this problem, designing content in collaboration with faculty will ensure that it is specific enough to meet the needs of the curriculum and course content (Appelt & Pendell, 2010). Once online learning content is created, having it added correctly to an LMS may mean that the librarian will have to work with the instructor to have the content uploaded properly (Hemmig & Montet, 2010; Henrich & Attebury, 2012).

Finally, many of the articles in the reviewed cases mentioned that they had not developed any assessment tools or that the assessment that was done was not robust enough to get a sense of the impact of the online learning materials on student learning outcomes (Alleyne & Rodrigues, 2011; Blevins et al., 2014; Chisholm & Lamond, 2012; Ergood et al., 2012; Hess, 2014b; Lo & Dale, 2009; Long et al., 2012; Schroeder, 2010). Many articles recommended that further work be done to add an assessment piece to the project, a crucial step in evaluating the effectiveness and usefulness of the time, effort, and resources devoted to creating and implementing online learning materials for library instruction.

2.10 BEST PRACTICES

Several reviewed articles included recommendations of best practices for online ILI. These recommendations help to further the research on distributed learning by providing a starting point for researchers and as a guide to future projects.

Dewald's (1999) seminal article addresses the needs of online students. She provides recommendations on the effective delivery of ILI for librarians venturing into online learning environments such as ensuring that the instruction is directly tied to the curriculum or to an assignment because "students will have more incentive to study the tutorial carefully when they know they will be required to produce something as a result of what they have learned" (p. 30). Further best practices as noted in Dewald's article include the use of active learning techniques and collaborative learning as two methods for engaging learners with the materials. Graphics and other forms of media should be used to enhance understanding of the materials because "they will make the learning experience more pleasant for the student by breaking up the text and by stimulating different types of thought processes" (p. 30). Objectives should be clearly stated with students identifying a "conceptual understanding of the skill being learned. When students understand why something is done, they are more likely to remember it and to be able to apply the concept in a context" (p. 30). Dewald concludes her paper with a timeless reminder that "the offer of help by a librarian, whether that help is provided online, by phone, or in person, is a vital part of any instruction and should be included in web tutorials" (p. 31).

Maddison's (2013) paper outlines the delivery of online instruction in discipline-specific teaching discovered through interviews with professors at the University of Saskatchewan. Recommendations from this paper include building relationships with faculty members to ensure adequate placement of the library session within the curriculum and connection to an assignment. The use of active learning techniques assist with students' engagement with the learning materials while improving "their understanding of the materials and retention of critical information" (p. 275). The technology used to deliver the online instruction should be familiar to the student whenever possible, "ensuring that the technology assists in the effective facilitation of the learning process and does not distract from it. There has to be an educational purpose for the application that is chosen and an understanding of how it will help to engage the learning" (p. 275). Finally, assessment is a key component of teaching, "ongoing assessment and evaluation ensure that learning has occurred, as well as informing improvements in course delivery and instruction" (p. 275).

Since Dewald published her article in 1999, many researchers who focus on and discuss distributed learning initiatives include lists of best practices in their works. These recommendations were typically discovered and noted while the researchers were conducting their case studies. For example, the team of Crawford-Ferre and Wiest (2012) outline several advantages to the implementation of online instruction, including the ability to offer instruction regardless of the student's physical location and time zone. Although time is generally noted as an advantage (e.g., attendance at a time convenient for the student), it can also be a significant disadvantage for the instructor because students may expect an immediate response from an always-connected instructor. It is therefore critical that instructors set boundaries (their availability, virtual office hours, etc.).

Funding can be an advantage, as well as a disadvantage. Funds that are generated from tuition for online courses can be higher because more students can attend an online class at one time when compared to an f2f class that is limited by the number of desks in the classroom. But the tutorials themselves can require a huge investment of capital as additional software or hardware or both are often necessary. Librarians should also consider the time and the investment required to create, implement, evaluate, and possibly update materials on an annual basis. The work of Crawford-Ferre and Wiest is supported by Dewan and Steeleworthy (2013), who note that "budget cuts have forced institutions to re-evaluate strategies and find alternatives to the exclusive face-to-face learning models" (p. 281). Both case studies comment on the benefits of online learning for international learners, as "online instruction allows for self-pacing, which is helpful for struggling learners and students whose mother tongue is not English" (p. 282). Yet Crawford-Ferre and Wiest are quick to point out that the lack of visual language clues may lead to difficulties in comprehension for some students. The role of instructors also does not truly change in an online learning environment because they are responsible for managing the classroom and creating a positive and safe environment so that all participants are encouraged to engage with the materials and their peers because "students with greater teacher involvement collaborated more with online peers" (Crawford-Ferre & Wiest, 2012, p. 13).

Dewan and Steeleworthy (2013) postulate that the learning "objectives must support the values of its parent institution, be rooted in best practices, and gain wide acceptance by all instructional librarians" (p. 279). Their article provides an extensive list of best practices that can be summarized as follows: develop strong learning outcomes and then implement an evaluation tool such as a rubric to assess the level of student learning and also student satisfaction; create activities that are centered on active learning principles so that students learn through doing instead of passively viewing the materials; adopt the constructivist learning theory, which promotes student-centered learning; use the principles of universal design to accommodate all learners wherever possible; create peer-learning activities such as discussion boards and group work; scaffold learning within programs; and consider the logistics of providing video content such as posting the length and providing a variety of play options (rewind, pause, etc.) so that students can review the content at their own pace. The implementation of a novel use of technology can be relatively easy in a one-shot library instruction session because the single course provides the librarian and the instructor with an opportunity to experiment without having to redesign the entire course.

Mestre (2012) provides a comprehensive list of best practices for online tutorials based on a well-designed usability study on the impact of student preferences and learning. This list can be summarized as follows:

- "create a clear outline and concise navigation";
- "provide clear and detailed images";
- "use appropriate multimedia and remove unnecessary graphics, text, and audio";
- "highlight salient points";
- "keep text to a minimum";
- ensure that the tutorial flows from one activity to another in a meaningful way;
- "make the experience personal and relevant";
- "present information in chunks and in multiple formats";
- "provide options for novices and advanced learners"; and
- "provide ongoing and relevant feedback" (pp. 271–273).

Hess (2013) created MAGIC guidelines in order to ensure that Oakland University Library's online tutorials were focused on users' needs. Tutorials should be easy for students to discover because they will not seek homework they were not assigned or cannot easily locate; a search box is vital "because it allowed users to more quickly find online objects than a scroll-through list" (p. 341). The tutorial needs to identify who created the video, an estimated running time, and the date of the last update, and these elements should be reviewed often. As for the content, "shorter, more focused online learning objects demonstrate consideration of both the learner and the learning objectives" (p. 338). Chunking strategies are useful to students because they "break up information reduc[ing] user's cognitive load" (p. 338), as well as ensure that "updating content [is] more manageable and the process maintainable" (p. 338). Finally, she identified that the use of naming conventions and the generation of keywords can be an important part of the process because they also help "users by providing a common vocabulary across web tutorials" (p. 343).

The literature on best practices include the following recommendations: collaborate with library or external colleagues; provide clear learning objectives and outcomes that are measurable; consider the length, pace, and ease of use of the tutorial; consider the type of technology used; incorporate active learning techniques; and, finally, determine the scalability of the project.

2.11 ASSESSMENT

Some librarians tend to cram one-shot sessions with everything students may need to know throughout their academic careers, relying on "just in case" instead of "just in time" instruction. Yet students can quickly become overwhelmed when the information provided is not targeted to an assignment or tied directly to the curriculum. So the questions become, how much information is required to meet the learning outcomes of the class? and how can librarians assess how much knowledge the student has retained during and following the library session?

A wide variety of information on assessment was included in the 55 scholarly articles consulted by the group. Most notably, 26 articles (47.3%) made use of pre- or post-tests as well as quizzes to assess knowledge and how well information literacy skills were retained. Twenty articles (36.4%) noted the use of a pre- and post-test as part of their research project. When a quiz was used, it may have followed individual modules, was a self-assessment, or was used as a form of capstone assignment for the students. The use of quizzes was seen in six articles (10.9%).

Feedback for session participants was also sought as a form of assessment in 12 articles (21.8%), including reviews that used a rubric and studies where general feedback was solicited through polls to generate targeted feedback, optional feedback, "one thing that you learned today" information, faculty feedback, or self-assessment surveys. A further study noted the use of anecdotal evidence, and one article described the use of a formal written evaluation.

Eight articles (14.5%) mentioned the use of assignments as an assessment tool. Examples of assignments included the use of annotated bibliographies, worksheets, user surveys, and general library assignments.

The final eight categories determined by the team's analysis of the use of assessment methodology include five articles (9.1%) that used focus groups or participant interviews, four articles (7.3%) that used statistics generated from video usage, and three articles (5.5%) that indicated their online project involved for-credit course work either as a sole library instruction course or as part of a discipline specific course (e.g., credit for completion of a library module within a course). Two articles (3.6%)

focused on a user-needs assessment, and two articles (3.6%) were from teams of researchers assessing a usability study. Finally, a team of researchers used a validated pre- and post-session checklist that evaluated students' use of keywords, Boolean operators, truncation options, and appropriate use of synonyms. One article outlined the use of talk-aloud observations. Table 2.5 provides further information on the specific articles and the types of assessment that were noted.

Table 2.5 Summary of the types of assessment identified in the case study literature

Number of articles	Type of assessment used	References
20	Pretest, post-test, or both	Anderson and May (2010), Bishop et al. (2013), Boden et al. (2013), Bruce-Low et al. (2013), Chaiyama (2015), Dee and Reynolds (2013), Domínguez-Flores and Wang (2011), Gall (2014), Gray and Montgomery (2014), Henrich and Attebury (2012), Hess (2014a); Lo and Dale (2009), Long et al. (2012), Mery et al. (2012), Mestre (2012), Scales et al. (2014), Schroeder (2010), Stiwinter (2013), Trail and Hadley (2010), Xiao (2010)
6	Quizzes	Arguello (2013), Ganster and Walsh (2008), Germek (2012), Gustavson et al. (2011), Tooman and Sibthorpe (2012), Usova (2011)
12	Feedback	Adebonojo (2011), Arguello (2013), Craig and Friehs (2013), Easter et al. (2014), Ganster and Walsh (2008), Gray and Montgomery (2014), Hahn and Bussell (2012), Meredith and Mussell (2014), Stiwinter (2013), Tooman and Sibthorpe (2012), Turnbull, Royal, and Purnell (2011), Usova (2011)
8	Assignments	Archambault (2011); Clapp et al. (2013), Craig and Friehs (2013), Gray and Montgomery (2014), Greer et al. (2012), Gunn et al. (2011), Hess (2013), Trail and Hadley (2010)

(*Continued*)

Table 2.5 (Continued)

Number of articles	Type of assessment used	References
5	Focus groups, participant interviews, or both	Appelt and Pendell (2010), Boden et al. (2013), Bruce-Low et al. (2013), Domínguez-Flores and Wang (2011), Hahn and Bussell (2012)
4	Usage statistics	Germek (2012), Kern (2013), Mee (2013), Stiwinter (2013)
3	For-credit instruction	Arguello (2013), Greer et al. (2012), Gunn et al. (2011)
2	Usability study	Boden et al. (2013), Mestre (2012)
2	User-needs assessment	McLean and Dew (2006)
1	Environmental scan	Walters et al. (2015)
1	Anecdotal evidence	Adebonojo (2011)
1	Validated checklist	Brettle and Raynor (2013)
1	Talk-aloud observations	Mestre (2012)

Several important themes on assessment arose from the research literature, including method of delivery, collaboration, student satisfaction and success, and the connection of assessment tools to the use of learning outcomes or ILI standards or both.

2.12 METHODS OF DELIVERY

A great deal of the literature on distributed learning in academic libraries compares delivery methods of online instruction. Authors of these case studies most typically compare the effectiveness of the teaching and learning activity in an f2f, a blended and fully online classroom. The majority of these comparisons look at engagement, pre- and post-test scores, and whether students retained knowledge following the class.

Anderson and May (2010) investigated whether the method of delivery influenced retention rates through the use of pre- and post-tests, as well as performance on in-class assignments. They concluded that the method did not affect the rate of retention yet noted the importance of collaborating with faculty as discussion "results have led to a re-evaluation of the IL curriculum used at this institution and in this specific course to include media and visual literacy" (p. 499). They concluded that instruction should be tied directly to the curriculum and assignments, therefore acknowledging the crucial role that the librarian plays in successfully delivering ILI. Mery et al. (2012) also evaluated different delivery methods, inferring that "information literacy is best taught through a well-designed online course where students have multiple opportunities to engage with information literacy concepts" (p. 375). Mery, Kline, Sult, and DeFrain's (2014) case study analyzed three groups of students and their use of tutorials created using either "Guide on the Side" or another general screencasting tool. Their results indicate that the type of instruction may not matter because "instruction can be successfully taught online in a number of ways from static tutorials to highly interactive ones" (p. 78). Mestre (2012) contributes to the research on the delivery of ILI instruction by comparing different types of tutorials, concluding "that regardless of learning style, students prefer static tutorials with screenshots over screencast tutorials. They want the ability to control their pace and progress and see options at a glance, especially through visuals" (p. 273).

When comparing f2f instruction with online delivery, a variety of perspectives can be found. For instance, Clark and Chinburg (2010) noted in their study that "there were virtually no differences between the performances of the online students versus the face-to-face students" (p. 538) based on the results of the citation analysis. Boden et al. (2013) state that "participants in both groups revealed their preference for in-person instruction as the interview progressed. However, they saw the screen-cast tutorials as a good alternative and likely a pragmatic option for a busy health professional if in-person instruction was not available" (p. 323).

Gall (2014) concludes his article by stating that the f2f session "seemed to make students more comfortable with the library but did not necessarily provide better retention of information" (p. 286) and that "online orientation seemed to increase students' self-efficacy" (p. 286). In contrast, Domínguez-Flores and Wang (2011) compared four groups of undergraduate students: a control group that received one f2f instruction session, an online community group (OLC), an online tutorial (OT) group, and a group that combined online tutorials and participated through online community. She found significant differences in the test scores between OT and OLC groups, with groups that participated in online community securing high scores.

A variety of delivery methods are presented in the literature with contradictory results. Most educators believe that no two classes are the same, so instructors should consider the needs of the students and the viability of using an online environment to deliver instruction.

2.13 COLLABORATION WITH FACULTY AND OTHER STAKEHOLDERS

Projects involving collaborations between internal library staff and external colleagues often have the greatest chance of success. When a librarian is able to work with a faculty member on the delivery of library instruction that is tied directly to the curriculum or an assignment, the rates of student engagement and success are typically amplified.

Appelt and Pendell (2010) ascertain that involving faculty in the assessment of online library tutorials is a beneficial way to improve and promote the service because "students are more likely to use and complete library online tutorials when they are required in their classes" (p. 246). Content development can be complicated, and "librarians who design online library tutorials with little input from faculty (whose recommendation is vitally important for the success of library tutorials) could risk creating faculty ambivalence or even disregard for tutorial content" (p. 251).

In academic libraries, collaboration is thought of as librarians working with faculty members within their institution or perhaps teams of librarians working together, but in distance education collaboration can occur between institutions and in some cases between different types of libraries. Alleyne and Rodrigues (2011) provided details of a case study in which distance students depended on their local libraries for research assistance and not necessarily the academic institution through which they were attending classes.

In Turnbull, Royal and Purnell's (2011) study, liaison librarians, unit coordinators, and multimedia staff members worked together to create six online modules for nursing students that showed enhanced positive relationships between faculty and library staff and in which "librarians are seen as core members of the academic team" (p. 128), providing students with a learning tool that builds skills and confidence. This collaborative process between librarians and faculty brings together various professional and complementary skills that "illustrate[s] the potential of interdisciplinary partnerships" (p. 128).

Gonzales (2014) notes the difficulty of measuring the success of an ILI session due to the nature of offering a one-shot session, often in the middle of a course. How can librarians determine that student learning was directly related to what was presented in the session when they will likely never see the students again? Assessment is an issue for most ILI delivery because the librarian cannot really test students and determine if the learning outcome was met, particularly when the assignment might be due well after the librarian was in the class. Statements such as these speak to the importance of partnering with faculty members on assignments that require a research component. This idea is supported by the work of Xiao (2010), who notes that communication can be facilitated by an LMS such as Blackboard that can fully integrate library tools and services. To further support the faculty and librarian collaboration of course content, Gonzales (2014) proposed that a librarian can be responsible for analyzing citations, evaluating the quality of the resources used, and determining their relevance to the topic. She concludes that "the use of online tutorials can provide for more consistency in the information being conveyed to students than face-to-face instruction sessions taught by multiple librarians with varied experience and teaching abilities" (p. 53). Even sessions in which one instructor delivers all of the content will vary slightly because the students are different, the questions they will ask are different, and how they choose to interact with the materials will be different.

Collaborative projects not only benefit from increased student engagement and rates of success but also help to improve and strengthen the relationship between the library and the course instructor. When a partnership is formed, library instruction becomes a vital component in the overall course delivery rather than a once-a-semester activity.

2.14 LEARNING OUTCOMES

Student satisfaction and rates of success are tied to projects that have clearly defined learning objectives and outcomes and where library instruction is tied to an assignment. When students see immediate value for what they are learning, they are likely to use what they have learned. As Craig and Friehs (2013) point out in the material that follows, most researchers are quick to point out that students may enjoy a tutorial but do not find it useful, so careful considerations should be given to how the librarians measure teaching effectiveness.

Lo and Dale (2009) confirm that the "new learning paradigm ... create[s] environments and experiences that bring students to discover and construct knowledge for themselves" (p. 148−149). The impetus for their project was the desire to make a previous f2f class available online while guaranteeing faculty buy in and tailoring of the instruction to a class assignment. They endeavored to create a step-by-step guide that was divided into seven modules, with each module requiring less than 15 minutes to complete. They felt that it was also important to use evaluation tools to measure whether or not the information would be useful to students by "assess(ing) the users' expectations and level of satisfaction in order to make improvement to the tutorials" (p. 153). They go on to note that "creating modules that are flexible, embeddable, and scalable is the most efficient use of resources" (p. 155).

Craig and Friehs (2013) note that many previous studies indicate that "just because students like a tutorial does not mean they actually learn from it" (p. 294). As such, survey instruments should go beyond questioning student satisfaction and instead concentrate on achievement of learning outcomes. This study compared student's performance on a quiz following f2f instruction, f2f instruction combined with a video, and f2f instruction combined with a text-based HTML tutorial. Students in the video test group scored higher on the quiz than the other two groups. Craig and Friehs (2013) conclude that "the video tutorial appeared to be more effective at teaching information about a database than the HTML tutorial and the live instruction" (p. 299).

Gonzales (2014) provides important information on student preference and levels of satisfaction, concluding "that given the choice, students are more likely to choose online instruction if only for the convenience, the self-paced nature, and student comfort with the technology" (p. 53). Gonzales goes on to state that when "measuring student satisfaction ... students are generally more satisfied overall with classroom instruction or with a hybrid of online and classroom instruction than with online instruction alone" (p. 53). In other words, according to this study, students are more likely to choose the second-rate option merely because it is more convenient for them, yet she cautions practitioners to be thoughtful when considering moving away from f2f or hybrid instruction entirely because students maintain greater satisfaction with this delivery method. This study is supported by the work of Boden et al. (2013) as noted previously in "Methods of Delivery."

Directly related to the topic of student satisfaction is the discussion of student success, of which assessments tools are a key component in recognizing the rates of effectiveness of distributed learning. The raw data generated in Trail and Hadley's (2010) study from pre- and post-test measures revealed no significant statistical difference, yet when an item analysis was completed, gains were demonstrated in some of the skill areas. Based on the results of the study, students were able to apply information literacy skills in the development of a review of literature after participating in tutorial instruction using screen casting. The results support combining the use of authentic measures of assessment such as rubrics to evaluate course projects in addition to static assessments such as tests of knowledge and skills. Ergood et al. (2012) also list assessment as a key takeaway from their project: "Assessment must be considered when developing tutorials in order to measure effectiveness and usability" (p. 106).

The final theme that arose from the literature on assessment is the use of outcomes or ILI standards to inform practice. Germek (2012) states the importance of starting any instructional output by creating strong learning objectives that can be successfully matched with measureable outcomes. Germek suggests partnering with a faculty member to create outcomes that address not only the ACRL ILI framework but also the needs of that department or discipline.

Long et al. (2012) support a programmatic approach to ILI by noting the differences in ability between first-year and capstone students, as well as their use of varied types of materials and the types of results they retrieve. Faculty should be encouraged to build assignments around core literacy competencies that necessitate the use of increasingly sophisticated databases along with search techniques as students progress through their education, thus "developing competence in information literacy requires intention and instruction" (p. 388). The researchers recognized a need to develop online course materials in response to the nontraditional student body by first creating a hybrid course and then adapting course content and delivery methods in the development of an entirely online course. To support 21st-century information literacy skills the researchers developed a hybrid for-credit course embedded in an LMS, with components taught in person that focused on the research process. Students were responsible for readings, viewing lectures, and being active on the discussion boards before class; the in-class session then focused on lectures and offered time to ask and answer questions. Assessment was completed through an annotated bibliography, a presentation, and a research reflection, as well as online quizzes.

2.15 LIMITATIONS

While conducting this literature review, the working group encountered limitations within the body of research on ILI. The literature reveals an inconsistency in the way librarians write and disseminate their findings in online learning case studies and in library and information science generally, making it difficult to compare individual case studies. The literature also tends to lack focus on rigorous research methods for assessing online teaching and learning techniques, as demonstrated in the many case studies that mention methods for future improvements but negate teaching theories, formal assessment of teaching effectiveness, or the learner's retention of knowledge.

Research literature on emerging practices has not been discussed in this review, and informal publishing venues such as blogs, forums, LISTSERVs, and conference proceedings were not reviewed for this project, although the information may be both relevant and useful in documenting emerging trends and current practices. An area for further research is the use and impact of implementing the ACRL Framework for Information Literacy for Higher Education in online learning environments.

2.16 CONCLUSION

This literature review included 55 scholarly case studies on online learning initiatives in the delivery of ILI in academic libraries. The majority of the research literature focused on first-year or undergraduate students or other general audiences. In case studies mentioning specific disciplines, nursing programs and English department were most typically the focus of the case study. Science disciplines are underrepresented in the library literature. Screen-capture software is the technology most frequently used to create online instructional materials, and students engage with this content predominantly through LMSs or video-hosting platforms. Most online instruction occurred asynchronously within our sample.

Identified challenges for libraries, librarians, and other staff members include both time considerations and the need for skill development in video creation and the use of specific technologies. When designing online learning materials, librarians should take into account good design practices and aim to collaborate with faculty to ensure that content is relevant to students. Instruction should be timely and focus on a specific curriculum-based assignment. The literature also shows that students may run into technological difficulties when trying to access online learning content. Making the content easy to access, compatible with a variety of devices, and clear and concise will ensure uptake by users.

An assessment or evaluation tool was not incorporated into all of the case studies. Where it was missing, it was not possible to show that the online learning materials had an impact on users' learning despite the fact that many students were anecdotally satisfied or felt positive about the instructional intervention. Studies that did include a form of assessment such as pre- and post-tests, feedback surveys, focus groups, interviews, and the analysis of usage statistics were able to verify that their use of technology was to a certain extent effective and useful and that students retained their knowledge and were able to apply their skills in projects or assignments.

The literature shows that the practice of incorporating online instruction into information literacy is growing in popularity and in constant development and exploration. Many institutions are trying to reach large numbers of students and improve the quality of the learning resources. While the online learning environment presents new challenges for ILI, librarians still face some of the same concerns and challenges: making content accessible, useful, and engaging. That being said, new teaching technologies provide opportunities to reflect on the current state of library instruction in higher education, challenging libraries to improve and develop content to meet the unique learning needs of today's students.

REFERENCES

Adebonojo, L. G. (2011). A way to reach all of your students: The course management system. *Journal of Library & Information Services in Distance Learning*, *5*(3), 105–113. Available from http://dx.doi.org/10.1080/1533290X.2011.605936.

Alleyne, J. M., & Rodrigues, D. (2011). Delivering information literacy instruction for a joint international program: An innovative collaboration between two libraries. *College & Undergraduate Libraries*, *18*(2–3), 261–271.

Anderson, K., & May, F. A. (2010). Does the method of instruction matter? An experimental examination of information literacy instruction in the online, blended, and face-to-face classrooms. *The Journal of Academic Librarianship*, *36*(6), 495–500.

Anderson, S. A., & Mitchell, E. R. (2012). Life after TILT: Building an interactive information literacy tutorial. *Journal of Library & Information Services in Distance Learning*, *6*(3–4), 147–158.

Appelt, K. M., & Pendell, K. (2010). Assess and invest: Faculty feedback on library tutorials. *College & Research Libraries*, *71*(3), 245–253.

Archambault, S. G. (2011). Library instruction for freshman English: A multi-year assessment of student learning. *Evidence Based Library and Information Practice*, *6*(4), 88–106.

Arguello, N. (2013). Secondary marketing research certificate: Library collaboration with the college of business and marketing faculty. *Journal of Business & Finance Librarianship*, *18*(4), 309–329.

Armstrong, D. (2010). *A qualitative study of undergraduate students' approaches, perceptions, and use of online tools*. Retrieved from ProQuest Digital Dissertations. (ERIC ED517913).

Armstrong, D. A. (2011). Students' perceptions of online learning and instructional tools: A qualitative study of undergraduate students use of online tools. *Turkish Online Journal of Educational Technology-TOJET, 10*(3), 222–226.

Association of College and Research Libraries. (2015). *Standards for distance learning library services*. Retrieved from < http://www.ala.org/acrl/standards/guidelinesdistancele arning#provenance >.

Baker, A. (2014). Students' preferences regarding four characteristics of information literacy screencasts. *Journal of Library & Information Services in Distance Learning, 8*(1–2), 67–80.

Bell, B. S., & Federman, J. E. (2013). E-learning in postsecondary education. *The Future of Children, 23*(1), 165–185.

Bishop, C., Yonekura, F., & Moskal, P. (2013). Pilot study examining student learning gains using online information literacy modules. *Proceedings from Association of College & Research Libraries 2013: Imagine, Innovate, Inspire, 466*, 471.

Blevins, A. E., Deberg, J., & Childs, C. (2014). Developing a best practices plan for tutorials in a multi-library system. *Medical Reference Services Quarterly, 33*(3), 253–263.

Boden, C., Neilson, C. J., & Seaton, J. X. (2013). Efficacy of screen-capture tutorials in literature search training: A pilot study of a research method. *Medical Reference Services Quarterly, 32*(3), 314–327.

Bolorizadeh, A., Brannen, M., Gibbs, R., & Mack, T. (2012). Making instruction mobile. *The Reference Librarian, 53*(4), 373–383.

Bottorff, T., & Todd, A. (2011). Making online instruction count: Statistical reporting of web-based library instruction activities. *College & Research Libraries, 73*(1), 33–46.

Brettle, A., & Raynor, M. (2013). Developing information literacy skills in pre-registration nurses: An experimental study of teaching methods. *Nurse Education Today, 33*(2), 103–109.

Bruce-Low, S. S., Burnet, S., Arber, K., Price, D., Webster, L., & Stopforth, M. (2013). Interactive mobile learning: A pilot study of a new approach for sport science and medical undergraduate students. *Advances in Physiology Education, 37*(4), 292–297.

Bulger, M. E., Mayer, R. E., & Metzger, M. J. (2014). Knowledge and processes that predict proficiency in digital literacy. *Reading and Writing, 27*(9), 1567–1583.

Button, D., Harrington, A., & Belan, I. (2014). E-learning & information communication technology (ICT) in nursing education: A review of the literature. *Nurse Education Today, 34*(10), 1311–1323.

Chaiyama, N. (2015). The development of blended learning management model in developing information literacy skills (BL-ILS Model) for undergraduate students. *International Journal of Information and Education Technology, 5*(7), 483–489.

Chisholm, E., & Lamond, H. M. (2012). Information literacy development at a distance: Embedded or reality? *Journal of Library & Information Services in Distance Learning, 6* (3–4), 224–234.

Clapp, M. J., Johnson, M., Schwieder, D., & Craig, C. L. (2013). Innovation in the academy: Creating an online information literacy course. *Journal of Library & Information Services in Distance Learning, 7*(3), 247–263.

Clark, S., & Chinburg, S. (2010). Research performance in undergraduates receiving face to face versus online library instruction: A citation analysis. *Journal of Library Administration, 50*(5–6), 530–542.

Craig, C. L., & Friehs, C. G. (2013). Video and HTML: Testing online tutorial formats with biology students. *Journal of Web Librarianship, 7*(3), 292–304.

Crawford-Ferre, H. G., & Wiest, L. R. (2012). Effective online instruction in higher education. *The Quarterly Review of Distance Education, 13*(1), 11–14.

Dede, C. (1996). The evolution of distance education: Emerging technologies and distributed learning. *American Journal of Distance Education, 10*(2), 4–36.

Dee, C. R., & Reynolds, P. (2013). Lifelong learning for nurses—Building a strong future. *Medical Reference Services Quarterly, 32*(4), 451−458.

Dewald, N. H. (1999). Transporting good library instruction practices into the web environment: An analysis of online tutorials. *The Journal of Academic Librarianship, 25*(1), 26−31.

Dewan, P., & Steeleworthy, M. (2013). Incorporating online instruction in academic libraries: Getting ahead of the curve. *Journal of Library & Information Services in Distance Learning, 7*(3), 278−296.

Domínguez-Flores, N., & Wang, L. (2011). Online learning communities: Enhancing undergraduate students' acquisition of information skills. *Journal of Academic Librarianship, 37*(6), 495−503.

Dow, M. J., Algarni, M., Blackburn, H., Diller, K., Hallett, K., Musa, A., & Valenti, S. (2012). Infoliteracy@ a distance: Creating opportunities to reach (instruct) distance students. *Journal of Library & Information Services in Distance Learning, 6*(3−4), 265−283.

Easter, J., Bailey, S., & Klages, G. (2014). Faculty and librarians unite! How two librarians and one faculty member developed an information literacy strategy for distance education students. *Journal of Library & Information Services in Distance Learning, 8*(3−4), 242−262.

Ergood, A., Padron, K., & Rebar, L. (2012). Making library screencast tutorials: Factors and processes. *Internet Reference Services Quarterly, 17*(2), 95−107.

Eva, N. C. (2012). Improving library services to satellite campuses: The case of the University of Lethbridge. *Journal of Library & Information Services in Distance Learning, 6*(2), 53−66.

Fletcher, J. D., Tobias, S., & Wisher, R. A. (2007). Learning anytime, anywhere: Advanced distributed learning and the changing face of education. *Educational Researcher, 36*(2), 96−102.

Gall, D. (2014). Facing off: Comparing an in-person library orientation lecture with an asynchronous online library orientation. *Journal of Library & Information Services in Distance Learning, 8*(3−4), 275−287.

Ganster, L. A., & Walsh, T. R. (2008). Enhancing library instruction to undergraduates: Incorporating online tutorials into the curriculum. *College & Undergraduate Libraries, 15*(3), 314−333.

Gawalt, E. S., & Adams, B. (2011). A chemical information literacy program for first-year students. *Journal of Chemical Education, 88*(4), 402−407.

Germek, G. (2012). Empowered library eLearning: Capturing assessment and reporting with ease, efficiency, and effectiveness. *Reference Services Review, 40*(1), 90−102.

Gonzales, B. M. (2014). Online tutorials and effective information literacy instruction for distance learners. *Journal of Library & Information Services in Distance Learning, 8*(1−2), 45−55.

Gray, C. J., & Montgomery, M. (2014). Teaching an online information literacy course: Is it equivalent to face-to-face instruction?. *Journal of Library & Information Services in Distance Learning, 8*(3−4), 301−309.

Greer, K., Swanberg, S., Hristova, M., Switzer, A. T., Daniel, D., & Perdue, S. W. (2012). Beyond the web tutorial: Development and implementation of an online, self-directed academic integrity course at Oakland University. *The Journal of Academic Librarianship, 38*(5), 251−258.

Gunn, C., Hearne, S., & Sibthorpe, J. (2011). Right from the start: A rationale for embedding academic literacy skills in university courses. *Journal of University Teaching & Learning Practice, 8*(1), 1−10.

Gustavson, A., Whitehurst, A., & Hisle, D. (2011). Laying the information literacy foundation: A multiple-media solution. *Library Hi Tech, 29*(4), 725−740.

Hahn, J., & Bussell, H. (2012). Curricular use of the iPad 2 by a first-year undergraduate learning community. *Library Technology Reports*, *48*(8), 42−47.

Hemmig, W., & Montet, M. (2010). The "just for me" virtual library: Enhancing an embedded eBrarian program. *Journal of Library Administration*, *50*(5−6), 657−669.

Henrich, K. J., & Attebury, R. I. (2012). Using blackboard to assess course-specific asynchronous library instruction. *Internet Reference Services Quarterly*, *17*(3−4), 167−179.

Hess, A. N. (2013). The MAGIC of web tutorials: How one library (re)focused its delivery of online learning objects on users. *Journal of Library & Information Services in Distance Learning*, *7*(4), 331−348.

Hess, A. N. (2014a). Online and face-to-face library instruction: Assessing the impact on upper-level sociology undergraduates. *Behavioral & Social Sciences Librarian*, *33*(3), 132−147.

Hess, A. N. (2014b). Web tutorials workflows: How scholarship, institutional experiences, and peer institutions' practices shaped one academic library's online learning offerings. *New Library World*, *115*(3/4), 87−101.

Julien, H., Tan, M., & Merillat, S. (2013). Instruction for information literacy in Canadian academic libraries: A longitudinal analysis of aims, methods, and success/ L'enseignement visant les compétences informationnelles dans les bibliothèques universitaires Canadiennes: Une analyse longitudinale des objectifs, des méthodes et du succès obtenu. *Canadian Journal of Information and Library Science*, *37*(2), 81−102.

Kern, V. (2013). Actions speaking louder than words: Building a successful tutorials program at the University of Washington libraries. *Fontes Artis Musicae*, *60*(3), 155−162.

Leonard, E., & Morasch, M. J. (2012). If you can make it there, you can make it anywhere: Providing reference and instructional library services in the virtual environment. *Journal of Electronic Resources Librarianship*, *24*(4), 257−267.

Li, J. (2014). Greeting you online: Selecting web-based conferencing tools for instruction in e-learning mode. *Journal of Library & Information Services in Distance Learning*, *8*(1−2), 56−66.

Lindorff, M., & McKeown, T. (2013). An aid to transition? The perceived utility of online resources for on-campus first year management students. *Education + Training*, *55*(4/5), 414−428.

Lo, L. S., & Dale, J. M. (2009). Information literacy "learning" via online tutorials: A collaboration between subject specialist and instructional design librarian. *Journal of Library & Information Services in Distance Learning*, *3*(3−4), 148−158.

Lockerby, R., & Stillwell, B. (2010). Retooling library services for online students in tough economic times. *Journal of Library Administration*, *50*(7−8), 779−788.

Long, J., Burke, J. J., & Tumbleson, B. (2012). Research to go: Taking an information literacy credit course online. *Journal of Library & Information Services in Distance Learning*, *6*(3−4), 387−397.

Lyons, T., & Warlick, S. (2013). Health sciences information literacy in CMS environments: Learning from our peers. *The Electronic Library*, *31*(6), 770−780.

Maddison, T. (2013). Learn where you live: Delivering information literacy instruction in a distributed learning environment. *Journal of Library & Information Services in Distance Learning*, *7*(3), 264−277.

Massachusetts Institute of Technology. "Our History." MIT OpenCourseWare. < http:// ocw.mit.edu/about/our-history/ > . Accessed 08.11.15.

McLean, E., & Dew, S. H. (2006). Providing library instruction to distance learning students in the 21st century: Meeting the current and changing needs of a diverse community. *Journal of Library Administration*, *45*(3−4), 315−337.

Mee, S. (2013). Outreach to international campuses: Removing barriers and building relationships. *Journal of Library & Information Services in Distance Learning*, 7(1−2), 1−17.

Meredith, W., & Mussell, J. (2014). Amazed, appreciative, or ambivalent? Student and faculty perceptions of librarians embedded in online courses. *Internet Reference Services Quarterly*, 19(2), 89−112.

Mery, Y., DeFrain, E., Kline, E., & Sult, L. (2014). Evaluating the effectiveness of tools for online database instruction. *Communications in Information Literacy*, 8(1), 70−81.

Mery, Y., Newby, J., & Peng, K. (2011). Why one-shot information literacy sessions are not the future of instruction: A case for online credit courses. *College & Research Libraries*, 74(3), 366−377.

Mery, Y., Newby, J., & Peng, K. (2012). Performance-based assessment in an online course: Comparing different types of information literacy instruction. *Portal: Libraries and the Academy*, 12(3), 283−298.

Mestre, L. S. (2010). Matching up learning styles with learning objects: What's effective? *Journal of Library Administration*, 50(7−8), 808−829.

Mestre, L. S. (2012). Student preference for tutorial design: A usability study. *Reference Services Review*, 40(2), 258−276.

Montgomery, S. E. (2010). Online webinars! Interactive learning where our users are: The future of embedded librarianship. *Public Services Quarterly*, 6(2−3), 306−311.

Nielsen, J. (2014). Going the distance in academic libraries: Identifying trends and innovation in distance learning resources and services. *Journal of Library & Information Services in Distance Learning*, 8(1−2), 5−16.

Ritchie, A. (2011). The library's role and challenges in implementing an elearning strategy: A case study from Northern Australia. *Health Information & Libraries Journal*, 28(1), 41−49.

Ruiz, J. G., Mintzer, M. J., & Leipzig, R. M. (2006). The impact of e-learning in medical education. *Academic Medicine*, 81(3), 207−212.

Scales, B. J., Nicol, E., & Johnson, C. M. (2014). Redesigning comprehensive library tutorials. *Reference & User Services Quarterly*, 53(3), 242−252.

Schroeder, H. (2010). Creating library tutorials for nursing students. *Medical Reference Services Quarterly*, 29(2), 109−120.

Stiwinter, K. (2013). Using an interactive online tutorial to expand library instruction. *Internet Reference Services Quarterly*, 18(1), 15−41.

Summey, T. P., & Valenti, S. (2013). But we don't have an instructional designer: Designing online library instruction using ISD techniques. *Journal of Library & Information Services in Distance Learning*, 7(1−2), 169−182.

Tewell, E. (2010). Video tutorials in academic art libraries: A content analysis and review. *Art Documentation*, 29(2), 53−61.

Tooman, C., & Sibthorpe, J. (2012). A sustainable approach to teaching information literacy: Reaching the masses online. *Journal of Business & Finance Librarianship*, 17(1), 77−94.

Trail, M. A., & Hadley, A. (2010). Assessing the integration of information literacy into a hybrid course using screencasting. *Journal of Online Teaching and Learning*, 6(3), 647−654.

Tuñón, J., & Ramirez, L. (2010). ABD or EdD? A model of library training for distance doctoral students. *Journal of Library Administration*, 50(7−8), 989−996.

Turnbull, B., Royal, B., & Purnell, M. (2011). Using an interdisciplinary partnership to develop nursing students' information literacy skills: An evaluation. *Contemporary Nurse*, 38(1−2), 122−129.

University Library. (2014). *University library strategic plan 2014—15*. Unpublished internal document, University of Saskatchewan.

Usova, T. (2011). Optimizing our teaching: Hybrid mode of instruction. *Partnership: The Canadian Journal of Library and Information Practice and Research, 6*(2), 1—12.

Vander Meer, P. F., Perez-Stable, M. A., & Sachs, D. E. (2012). Framing a strategy. *Reference & User Services Quarterly, 52*(2), 109—122.

Walters, K., Bolich, C., Duffy, D., Quinn, C., Walsh, K., & Connolly, S. (2015). Developing online tutorials to improve information literacy skills for second-year nursing students of University College Dublin. *New Review of Academic Librarianship, 21*(1), 7—29.

Williams, S. (2010). New tools for online information literacy instruction. *The Reference Librarian, 51*(2), 148—162.

Xiao, J. (2010). Integrating information literacy into blackboard: Librarian-faculty collaboration for successful student learning. *Library Management, 31*(8/9), 654—668.

Yang, S. Q., & Chou, M. (2014). Promoting and teaching information literacy on the internet: Surveying the websites of 264 academic libraries in North America. *Journal of Web Librarianship, 8*(1), 88—104.

CHAPTER 3

Using Theory and Practice to Build an Instructional Technology Tool Kit

3.1 INTRODUCTION

As librarians seek to integrate technology into information literacy instruction, whether face-to-face, blended, or online, they experience both challenges and opportunities. While technology can engage learners in new and innovative ways, it can also provide a stumbling block for learners and librarians alike. Identifying and integrating appropriate technology tools into library instruction can be challenging, especially when these interactions are confined to single instances with learners (i.e., one-shot instructional sessions). Adding the layer of online or distributed learning further complicates this situation. How can librarians use instructional technology tools to effectively convey content and engage learners in essential information literacy ideas?

This chapter presents an innovative way for librarians to consider instructional technology use in their teaching. Through both a theoretical foundation and practical technology integration suggestions, librarians can build an instructional technology tool kit grounded in pedagogy that is rich in its use of both active-learning strategies and technology tools. This two-pronged approach to teaching using technology tools can help librarians integrate these resources into their information literacy instruction in novel ways.

3.2 THEORY: A REVIEW OF THE LITERATURE

Three important areas of the literature are presented here to construct a theoretical foundation on which librarians can create an instructional technology tool kit for either distributed or traditional learning environments.

© 2017 T. Maddison and M. Kumaran.
Published by Elsevier Ltd. All rights reserved.

First we examine the scholarship of active learning, specifically in higher education. Next we consider the research on effective instruction with technology tools. Finally, these two components are outlined using learning objectives as a scaffold.

3.3 THE SCHOLARSHIP OF ACTIVE LEARNING

An important consideration in the literature on teaching and learning involves engaging students in the learning process through active learning. Chickering and Gamson's (1987) *Seven Principles for Good Practice in Undergraduate Education* represents a starting point to consider active learning. In this text they assert that "learning is not a spectator sport" (Chickering & Gamson, 1987, p. 5). Instead, they argue that students need to connect learning to their past and present experiences to engage with learning in the future. While Chickering and Gamson are widely cited in academic literature on active learning, Gogus (2012) argued that this notion of engaging students in their learning reaches back to the Socratic method, which tasks learners with creating their own knowledge through a question-and-answer dialogue with an instructor. Bonwell and Eison (1991) further expounded on how active learning can function in the higher education environment in *Active Learning: Creating Excitement in the Classroom*. They noted that instructors who are designing active-learning opportunities need to create learning activities that involve "students doing things and thinking about the things they are doing" (p. 1). Active learning, then, means connecting students with the learning content, their peers, and the instructor through participatory engagement.

Active-learning techniques can be employed in many different ways in many different environments. Much literature on active learning exists outside of library scholarship, which can provide a different perspective on information literacy instructional practices. For instance, active learning may involve including individual activities, pairs, small groups, group projects, and even all-class collaboration (Zayapragassarazan & Kumar, 2012). Some studies focus on how active learning impacts student performance. For instance, Powner and Allendoerfer (2008) examined whether active-learning activities affected students' retention of short-term course knowledge. They found that different active-learning techniques led to different classroom impacts: for instance, engagement in role-play activities improved performance on multiple-choice assessments, while active discussion boosted outcomes on short-answer questions. Campisi and Finn (2011) studied whether active learning impacted student attitudes toward conducting discipline-grounded research. They found that students' active participation in small group research and online group discussions led to dramatic improvement in those areas. The idea of an instructional strategy "working," though, can also be seen more broadly. For instance, Braxton, Jones, Hirschy, and Hartley III (2008) found that an institution's use of active learning "influences students' beliefs about how much their institution cares about their success" and can play "a significant role in the retention of first-year college students" (pp. 80–81). Active learning, then, can impact learners' academic and affective outcomes.

3.3.1 Active Learning in Library Instruction

While there is broader research on active learning in instruction, a specific segment of research on librarianship focuses on how this instructional strategy has been or can be used in information literacy instruction. In perhaps one of the earliest examples of this scholarship, Dabbour (1997) examined how a self-guided small-group exercise on using an academic library's online system could take the place of more traditional lecture and demonstration styles of instruction in a seminar for first-year students. From her examination of this practice across several sections of this introductory course, she found that students considered this instructional technique as valuable. Subsequent research on active learning in library instruction has often focused on the use of specific tools such as student response systems or polling resources to help students focus on content, participate in learning activities, engage in discussion, and retain knowledge (Hoffman & Goodwin, 2006; Hoppenfeld, 2012). Other literature has addressed how broader active-learning frameworks such as problem-based learning can be applied in one-shot library sessions to enrich learners' experiences and aid in knowledge creation (Kenney, 2008). It is important that much of this library-centric research on active learning emphasizes using learner feedback to shape these kinds of activities in library instruction. Jacklin and Robinson (2013) pointed out how critical this feedback can be in the ongoing development of active-learning strategies, especially in distributed learning. Using learners' experiences to inform practices can help librarians design and deliver active learning effectively and meaningfully.

3.3.2 Limitations in Using Active-Learning Strategies

Other research on active learning, however, illustrates how it can be most effectively implemented. Andrews, Leonard, Colgrove, and Kalinowski (2011) examined how active-learning activities functioned in a large undergraduate biology class; they found that just doing these kinds of activities did not lead to student learning gains. Similarly, Ross and Furno (2011) studied how integrating student response systems into face-to-face library instruction impacted students' performance on a final assessment. They, too, found that the use of "clickers" did not lead students to outperform their peers who had engaged in learning in more of a traditional, lecture-style classroom setting. These studies illuminate several important points: First, Andrews et al. (2011) noted that these activities were more effective when constructivist learning principles provided active-learning activities. And Ross and Furno (2011) found that both the traditional and active learning classrooms outperformed students' scores on previous assessments. Active-learning activities, then, should not be thought of as instructional tools to use in every instructional interaction. Instead, they should be employed when instructors want to engage students in "structured exercises, challenging discussions, team projects, and peer critiques" (Chickering & Gamson, 1987, p. 4). Librarians looking to engage learners should not try to fit active-learning strategies into their instruction but should instead consider what they want learners to be able to do and identify if any active-learning techniques can help them achieve these ends.

3.4 THE SCHOLARSHIP OF EFFECTIVE INSTRUCTIONAL TECHNOLOGY USE

When considering active learning in distributed learning environments or incorporating active learning using technology tools, it is also important to consider the scholarship on using instructional technology effectively. Research from the Pew Research Center (Perrin & Duggan, 2015) has illustrated that today's Millennial and Generation X students have the highest percentage of Internet use across generational groups, and this data suggest that learners expect to use technology for both social and educational processes. To reach these students, Brown (2002) argued that instruction and student learning need to shift to be more discovery based, which capitalizes on the affordances of the Internet that allow individuals and groups to construct shared social meaning. Moreover, he suggested that instructors should use the Internet and online resources for both informational and social purposes to engage students in learning. Distributed learning seeks to do just that.

Research on instructional technology addresses the different ways librarians can do what Brown recommended while connecting students with content, their peers, and the instructor. For instance, Williams and Chinn (2009) examined how tools such as social media, blogs, and online discussions engaged students while building their digital and information literacy skills. They found that using these tools increased student engagement while also building recognition of how such tools could be used in real-world environments. Obenland, Munson, and Hutchinson (2012) examined active learning through student response clickers in large general chemistry classrooms, and they found these activities engaged the silent students who did not normally speak up in class. Roehl, Reddy, and Shannon (2013) used the flipped-classroom model in which content is delivered outside of class to identify ways students could be engaged in more collaborative learning during traditional instructional time. While videos are commonly considered to be a core component of flipping a classroom, these authors noted that computer-aided instructional modules could also engage students in the content. Technology tools and active learning, then, can be used in tandem to design meaningful learning scenarios, and librarians can consider how these kinds of practices can shape distributed information literacy instruction in particular.

3.5 DESIRED LEARNING OUTCOMES: AN IMPORTANT CONSIDERATION

To meaningfully implement active-learning strategies and technology tools into distributed library instruction, librarians may need an overarching structure to guide their work. Bloom's Revised Taxonomy (Anderson, Krathwohl, & Bloom, 2001) provides a scaffold frequently employed by educators for integrating active learning into a variety of instructional scenarios. The six levels of learning described therein—remembering, understanding, applying, analyzing, evaluating, and creating—can help instructional librarians consider the different types of learning and understanding that can occur and use this understanding to fit technology tools to achieve these desired learning goals and enable active learning. While there are resources on the Internet that reference Bloom's Revised Taxonomy with technology tools (see, for instance, Schrock, 2015), applying this scaffold cross-layered with technology tools and active-learning principles can be done innovatively in distributed information literacy instruction. Also, using this particular structure as a way to address active learning can connect librarians' practices with other educators, both in K—12 and higher education, who employ Bloom's ideas to design meaningful learning interactions.

3.6 PRACTICE: MEANINGFULLY INTEGRATING TECHNOLOGY TOOLS IN INFORMATION LITERACY INSTRUCTION

Using Bloom's Revised Taxonomy (Anderson et al., 2001), librarians can intentionally develop learning activities—whether online, face-to-face, or in a blended format—that use technology tools to engage learners in making skills and concepts personally relevant. The six hierarchical levels of the taxonomy can be aligned to both the Association of College and Research Libraries' (ACRL) *Information Literacy Competency Standards for Higher Education* (2000) and *Framework for Information Literacy for Higher Education* (2015); both documents may be in use, or at least in discussion, in distributed information literacy instruction. Connecting Bloom's Revised Taxonomy with these guiding library instruction documents can ground instructional strategies, and the technology that can be used to support it in accepted information literacy principles.

3.7 REMEMBERING INFORMATION

3.7.1 Alignment With Information Literacy

In Bloom's Revised Taxonomy (Anderson et al., 2001), the task of remembering asks learners to "retrieve relevant knowledge from long-term memory" (p. 31). This idea aligns with the guidelines presented in the *Framework for Information Literacy for Higher Education* (ACRL, 2015) and *Information Literacy Competency Standards for Higher Education* (ACRL, 2000). The *Framework*'s six frames address learners' need to remember and recall information, stating that learners who recognize that research is a process of inquiry "formulate questions for research based on information gaps or on reexamination of existing, possibly conflicting, information" (ACRL, 2015, p. 7). Also, learners who view searching as strategic exploration "determine the initial scope of the task required to meet information needs" (ACRL, 2015, p. 9). Implied in these practices is that learners are recalling and employing information they already know to design their questions of inquiry and to determine their search scope. Similarly, the idea of remembering and recalling information is embedded throughout the *Standards* (ACRL, 2000): the standard that states that an "information literate student defines and articulates the need for information" indicates that learners who do this will recognize their "existing information can be combined with original thought, experimentation, and/or analysis to produce new information" (p. 8). Also, as learners engage in "evaluat[ing] information and its sources critically," they will engage in reading texts, restating "textual concepts," and identifying "verbatim material" (ACRL, 2000, p. 11) that can be quoted or used.

3.7.2 Tools to Use

While remembering information, especially prior knowledge, is a task that falls primarily on learners, librarians can use technology tools to aid in this information retrieval. In particular, note-taking and collection tools can help automate information recall and engage learners with the information. **Diigo**, **Zotero**, and **Evernote** are three free online tools that learners can use to cull, organize, and share information from a wide variety of Internet sources. Of these tools, **Zotero** is the most scholarly focused, allowing integration of scholarly PDFs from libraries and generation of scholarly citations. However, **Diigo** and **Evernote** have more advanced note-taking features, where additional important information about something can be written and referred to later. For instance, **Diigo** allows its users to annotate a webpage with highlighting and notes. And **Evernote** allows users to snap photos with a mobile device and save them to an account. For librarians looking to integrate these tools in an active-learning context, they can design opportunities for students to use them in the context of a classroom or on their own for remembering and engaging in content. For instance, they could add information or research to a shared class group or class folder, which can add to the collective memory of all learners (see Fig. 3.1).

3.8 UNDERSTANDING INFORMATION

3.8.1 Alignment With Information Literacy

In Bloom's Revised Taxonomy (Anderson et al., 2001), the task of understanding asks learners to "construct meaning from instructional messages" in a variety of formats (p. 31). In the ACRL (2015) *Framework*, understanding is a key component throughout several of the information literacy frames.

Remembering Information with Zotero

Setting

- In a one-shot or extended learning interaction with students
- Online, in a blended classroom, or face-to-face

Activity

- Create a shared Zotero group with a few sample resources
- Invite students to participate in the group and add resources based on the course content or requirements
- Demonstrate Zotero's collaborative potential and illustrate how citations can be generated

Figure 3.1 A sample information literacy instructional activity for *remembering* that can be used in online, face-to-face, or blended learning interactions.

For example, learners developing their recognition of searching as strategic exploration "understand how information systems . . . are organized to access relevant information" (ACRL, 2015 p. 9). Moreover, grasping the idea that scholarship is a conversation requires learners to "understand the responsibility that comes with entering [these kinds of] conversations through participatory channels" (ACRL, 2015, p. 8). Similarly, learners practicing the idea that information has value "understand how the commodification of their personal information and online interactions [affect] the information they receive and the information they produce or disseminate online" (ACRL, 2015, p. 6). The *Standards* (ACRL, 2000) addressed understanding information as well, specifically by "defin[ing] and articulat[ing] the need for information" (p. 8) in the first place. For learners to be able to do so, they must understand the information they are seeking and the information at hand.

3.8.2 Tools to Use

In information literacy instruction, building understanding of information often connects with assessing that understanding: measuring learners' comprehension of ideas, concepts, strategies, and skills can help librarians identify when and whether instructional strategies are working.

Librarians can use free student response systems to assess students' understanding both online and in person. While many institutions have student response systems on campus, there are free options that students and instructors can connect to via a mobile device or computer. Using these free tools ensures that everyone has access to the technology (especially if the instructional interaction is held in person in an instructional lab or other space where computers can be accessed). **Poll Everywhere** and **Socrative** are two free systems that librarians can use to ask students to respond to a single question or a series of questions. With both of these systems, learners can respond using their cell phones, computers, tablets, or other Internet-connected devices. Both tools allow librarians to precreate multiple choice and free response questions to which learners can respond. While both **Poll Everywhere** and **Socrative** are free, **Socrative** is more feature-rich and gives those with teacher accounts options for understanding and analyzing data for free (while these options must be paid for in Poll Everywhere). This resource allows its "teachers" to prebuild multiquestion quizzes and pull response data into an Excel spreadsheet, which can be useful for tracking as well as more formal assessment purposes (see Fig. 3.2).

Understanding Information with Socrative

Setting

- Before the interaction, the librarian creates a classroom and learning activity for learners to work through
  - Types of activities:
    - Quick question—single comprehension-focused task
    - Quiz—multiquestion assessment (informal or formal)
    - Exit ticket—a question students answer before leaving class

Activity

- Ask learners to respond to the predesigned activity individually or as a group
  - This can be done synchronously or asynchronously
- Use the data generated to identify:
  - What learners know
  - What learners want to know
  - Where learners are confused or need help

Figure 3.2 A sample information literacy instructional activity for *understanding* that can be used in online, face-to-face, or blended learning interactions.

3.9 APPLYING INFORMATION

3.9.1 Alignment With Information Literacy

In Bloom's Revised Taxonomy (Anderson et al., 2001), the task of applying information asks learners to "carry out or use a procedure in a given
situation" (p. 31). In the *Framework* (ACRL, 2015), application is a key
component: as learners grapple with the idea that authority is contextual
and constructed, they must apply this idea when they "use research tools
and indicators of authority to determine the credibility of sources" (p. 4).
Also, learners need to apply the concept that information has value to
"decide where and how their information is published" and to "give
credit to the original ideas of others through proper attribution and citation" (ACRL, 2015, p. 6). The frame that presents research as inquiry
also asks learners to apply this concept by "seek[ing] multiple perspectives
during information gathering and assessment" and "seek[ing] appropriate
help when needed" (ACRL, 2015, p. 7). In the *Standards* (2000), the
information-literate student engages in applying information by accessing
"needed information effectively and efficiently" (ACRL, 2000, p. 9)
through the search process. Here application involved using strategies,
tools, and resources to the information-seeking process. Moreover, the
Standards (2000) outlined that information literate students also use
"information effectively to accomplish ... specific purpose[s]" (p. 13)
while learning to apply "various search systems," "various classification
schemes," and other "specialized online or in person services" (p. 10) to
help them find necessary information.

3.9.2 Tools to Use

As different librarians work with different groups of learners, applying
information may take myriad forms. For example, a group of lower-
division undergraduates may be tasked with applying citation styles on
their own or with a citation generator; in contrast, a group of upper-
division undergraduates or graduate students may be asked to demonstrate their mastery of information literacy concepts in different or more
complex ways. There are, however, two broad categories of tools that
can help librarians consider how they may design distributed learning
opportunities where learners can apply information literacy ideas. First,
screen-casting tools can be used by instructors—and students—to record
short videos of something on their computer. Librarians can ask students
to use these tools to apply a search process, show their work in using

a database, or demonstrate how they select information sources that are right for their needs or for an assignment. Free screen-casting tools include **Screencast-O-Matic**, **Screenr**, and **Jing**. **Screencast-O-Matic** and **Screenr** utilize web browsers without the need to download software.

Another category of tools that can be used for learners to apply and demonstrate their knowledge involves the use of online wall or discussion board tools. **Padlet** is one example of an online wall tool, and participants can post video, image, and hyperlinked content to share with others. With these kinds of tools, librarians can pose questions or prompts that learners can respond to with explanations as to how they would apply the information in a given situation or scenario. This type of activity can be done during an online or face-to-face instruction session, but it could also be designed as a supplemental activity after a one-shot instructional interaction or an online learning module. Both of these groups of tools engage learners with content and, in the case of the online wall and discussion board tools, can also connect students with the instructor and each other (see Fig. 3.3).

Applying Information with Screencast-O-Matic

Setting

- In a one-shot or extended learning interaction with students
- Online, in a blended classroom, or face-to-face
  - If in a face-to-face environment: ensure computers have Java installed to run Screencast-O-Matic
  - If online: ensure students know ahead of time the tool to be used so they can ensure computers can run Screencast-O-Matic

Activity

- At designated points in the learning interaction, ask students to create short videos that capture applications of skills, processes, or tactics for finding information
- Invite students to share these videos with their classmates and talk about how they applied knowledge
  - If online, ask students to record brief narrations to describe what they're doing

Figure 3.3 A sample information literacy instructional activity for *applying* that can be used in online, face-to-face, or blended learning interactions.

3.10 ANALYZING INFORMATION

3.10.1 Alignment With Information Literacy Standards

In Bloom's Revised Taxonomy (Anderson et al., 2001), analyzing information involves learners "break[ing] material into constituent parts and determin[ing] how parts relate to one another and an overall structure or purpose" (p. 31). The *Framework* (ACRL, 2015) addresses these concepts in several ways. In the frame of scholarship as a conversation, learners need to "recognize they are … entering into an ongoing scholarly conversation and not a finished conversation" (ACRL, 2015, p. 8), which involves identifying constituent parts within an overall scholarly structure. Searching as strategic exploration also requires learners to break material into pieces and then fit these pieces into a flexible structure by "utiliz[ing] divergent … and convergent … thinking," "match[ing] information needs and search strategies to tools," and "design[ing] and refin[ing] needs and search strategies, based on search results" (ACRL, 2015, p. 9). In the *Standards* (ACRL, 2000), learners engaging with evaluation of information need to analyze "new knowledge [in light of] prior knowledge to determine the value added, contradictions, or other unique characteristics" (p. 11), thereby engaging in analytical processes. Moreover, they need to analyze "the structure and logic of supporting arguments or methods" (ACRL, 2000, p. 11) in this evaluative process. As they seek information, learners also need to assess "the quantity, quality, and relevance of the search results to determine whether alternative … methods should be used" (ACRL, 2000, p. 10), which also connects to the idea of analyzing information.

3.10.2 Tools to Use

As with applying information, analyzing information in information literacy instruction can take different forms, depending on the level of the learner and the concepts being addressed. For instance, if learners are using information sources to find data, librarians may ask them to use online chart and graph tools such as **Online Chart Tool** and **Creately** to visually represent and analyze the information they have found. By creating charts and graphs from collected data, librarians can see how learners are understanding, interpreting, and analyzing numerical information.

However, what may be more useful and applicable to a wider swath of librarians is to ask learners to analyze the information in the research they find. By reading through scholarly articles, trade publications, books, and other sources, students can analyze the information they have and identify

Analyzing Information with Text 2 Mind Map

Setting

- In a one-shot or extended learning interaction with students
  - Before, during, or after the learning interaction takes place
- Online, in a blended classroom, or face-to-face

Activity

- Demonstrate how to use Text 2 Mind Map
- Ask students to create a mind map about their research question or topic, or reflecting the "big ideas" found in their research so far
- Allow students to share their mind maps and highlight their thinking to others
  - Online: share links for review

Figure 3.4 A sample information literacy instructional activity for *analyzing* that can be used in online, face-to-face, or blended learning interactions.

the information they need to find. Online mind–mapping and concept–mapping tools can facilitate this analysis. Free tools include **Bubbl.us**, **Text 2 Mind Map**, and **MindMeister**. These resources allow students to create simple maps in which they can illustrate the connections they see, identify issues or themes, ask questions, and highlight gaps in their knowledge. Another tool that can serve a similar purpose is **Prezi**; the flow of a **Prezi** can duplicate the mind–mapping effect (see, e.g., Greer, 2014). And these mind-mapping tools are often collaboration tools, which can allow groups of learners to work together on analyzing information (see Fig. 3.4).

3.11 EVALUATING INFORMATION

3.11.1 Alignment With Information Literacy Standards

In Bloom's Revised Taxonomy (Anderson et al., 2001), evaluating information tasks learners with making "judgments based on criteria and standards" (p. 31). This practice is a key component of many of the *Framework*'s (ACRL, 2015) information literacy frames. For instance, learners who are developing their understanding that authority is contextual and constructed need to evaluate "authoritative content [that] may be packaged formally or informally and may include sources of all media types" (ACRL, 2015, p. 4). Developing an "awareness of the importance of assessing content with a skeptical stance" (ACRL, 2015, p. 4) is a key component to creating this knowledge. As students work with the concept of information creation as a process, they need to "assess the fit between an information product's creation process and a particular information need" (ACRL, 2015, p. 5). As they engage with the idea of research as inquiry, learners need to "monitor gathered information and assess for gaps or weaknesses" (ACRL, 2015, p. 7) to ensure their process matches the desired product. Finally, learners working through their understanding of scholarship as conversation need to "critically evaluate contributions made by others in participatory information environments" (ACRL, 2015, p. 8) to develop their understanding of this idea. Evaluation is a key component in the *Standards*, (ACRL, 2000), with one standard focused solely on the information literate student's ability to evaluate "information and its sources critically" so as to incorporate "selected information into his or her knowledge base and value system" (p. 11). From this overarching standard, learners need to then articulate and apply evaluation criteria to both information and information sources (ACRL, 2000).

3.11.2 Tools to Use

In focusing on evaluating information, many of the tools discussed so far in this chapter can be used. For instance, a poll can be designed to ask learners to assess a specific resource; students could also create a short screen capture demonstrating how they would evaluate a specific resource, or librarians can pose questions or scenarios to learners via an online discussion wall that asks them to evaluate a specific information source for a specific purpose. In addition, librarians can use resources such as **SurveyMonkey** and **Google Forms** to create free survey-type instruments that provide a structure students can use to take a position and evaluate an information source. These forms, which generate responses that the librarian can then review, allow learners to engage both with the library instructor and the information literacy content. Evaluation may also happen in more collaborative ways through group chat or back-channel tools. Back channels are popular on **Twitter** using specific hashtags or through the use of a tool such as **Chatzy**, which allows anyone to create a closed chat room where a back-channel discussion can happen. If librarians have access to students through a learning management system (e.g., Moodle, Blackboard, Canvas, or D2L), they may be able to set up similar chats there. In creating such closed discussions, librarians can enhance instruction in a distributed learning interaction (see Fig. 3.5).

Evaluating Information with Google Forms

Setting

- In a one-shot or extended learning interaction with students
  - Before, during, or after the learning interaction takes place
- Online, in a blended classroom, or face-to-face

Activity

- Create a form that asks students essential questions for evaluating an information resource
  - Currency, Relevance, Accuracy, Authority, Appropriateness, Purpose
- Ask students to work through the form individually or in groups
- Share aggregated responses, or ask students or groups to share their thinking processes in completing the form

Figure 3.5 A sample information literacy instructional activity for *evaluating* that can be used in online, face-to-face, or blended learning interactions.

3.12 CREATING INFORMATION

3.12.1 Alignment With Information Literacy Standards

In Bloom's Revised Taxonomy (Anderson et al., 2001), creation involves learners "put[ting] elements together to form a coherent or functional whole" or "reorganiz[ing] elements into a new pattern or structure" (p. 31). This highest level of Bloom's Revised Taxonomy exists throughout the *Framework*'s (ACRL, 2015) information literacy frames. For example, learners grappling with the idea of research as inquiry need to "synthesize ideas gathered from multiple sources" to "draw ... conclusions based on the analysis and interpretation of information" (ACRL, 2015, p. 7). Learners who are engaging with the idea of scholarship as conversation can "contribute to scholarly conversation at an appropriate level, such as local online community, guided discussion, undergraduate research journal, conference presentation [or] proposal session" (ACRL, 2015, p. 8). And learners who are working to understand information creation as a process "transfer [the] knowledge [they gained] of capabilities and constraints to new types of information products" (ACRL, 2015, p. 5) that they create. Similarly, the *Standards* (ACRL, 2000) noted that, in evaluating information, learners who are information literate need to synthesize "main ideas to construct new concepts" (p. 11). Moreover, the standard that addresses using information charged learners with "manipulat[ing] digital text, images, and data, as needed, transferring them from their original locations and formats to a new context" (ACRL, 2000, p. 13). Central to this development is the standard that considers how learners incorporate new knowledge into their knowledge bases and value systems indicated learners' synthesis of "main ideas to construct new concepts" (ACRL, 2000, p. 11). Creating new information, then, is a critical component of both the *Framework* and the *Standards*.

3.12.2 Tools to Use

As librarians consider how they can work with students on creating information, a diverse set of tools and resources can be used, depending on the intended purpose. Generally, though, these resources can be categorized as writing, visualization, or media creation tools. If librarians are working with students online or face-to-face to create written products, many free resources are available to encourage creation and collaboration in writing. These tools include wikis such as **Wikispaces**, in which librarians can set up a collaborative space where students can write and post media; blogs such as **Blogger**; and collaborative document tools such as **Google Docs**. Free online infographic tools such as **Piktochart**, **Learnist**, and **infogr.am** allow students to create visual representations of data or information that can be easily shared with others. Finally, media creation tools include website creation tools such as **Google Sites** and **Weebly**, where learners can create e-portfolios or resource guides; and online presentation tools such as **Prezi**, **Glogster**, a virtual poster creation resource, and **Animoto**, an easy video creator (see Fig. 3.6).

3.13 NEXT STEPS AND CONCLUSIONS

As librarians seek to build instructional technology tool kits for use in online, face-to-face, and hybrid learning environments, assessing the impact of these resources—and the active-learning strategies that accompany them—is important. Measuring the effectiveness of these strategies can occur as formative or summative evaluation. To formatively measure the effectiveness of specific active-learning strategies, librarians need to gauge how a tool is working in a given environment as it is being used. This kind of assessment may include observation of learners in process (especially in a face-to-face, one-shot instructional interaction), or it may ask for learners' feedback on what is and is not going well as they work. This information

Creating Information with Animoto

Setting

- During or at the conclusion of the learning interaction
- Create a sample Animoto video to illustrate output

Activity

- Ask students to create short (30-second) videos on an information literacy skill or concept they learned in library instruction
  - Target these videos to specific groups or audiences to ground in real-world needs or uses
  - Share videos with the class and post somewhere public (with students' permission)

Figure 3.6 A sample information literacy instructional activity for *creating* that can be used in online, face-to-face, or blended learning interactions.

can be collected from casual conversations, but it can also be pulled together more formally using form or survey tools. Summative evaluation of these strategies and tools, which occurs after the completion of a project or instructional task, can be a bit more challenging for librarians, especially those engaged in one-shot learning interactions. However, collaborating with the course instructor to review—or even grade—learners' work can provide librarians with an understanding of whether active learning through technology tools impacted students' understanding of information literacy concepts.

Note that as librarians develop their own tool kits with active-learning strategies and instructional technology resources, their assessment of students' experiences can inform them about what to keep, modify, or discard altogether. An important element of this tool kit involves keeping it up to date by staying current with new technology resources that can be applied in the structure of Bloom's Revised Taxonomy (Anderson et al., 2001). To do so, librarians may need to look outside of traditional library literature and consider resources from organizations that focus on technology in teaching. For example, the Association for Educational Communications and Technology (AECT) publishes several resources, including *Educational Technology Research and Development, TechTrends*, and *Educational Media and Technology Yearbook* that may be instructive as to current pedagogical strategies and trends for using technology in teaching, especially in online environments. Multimedia Electronic Resource for Learning Online and Teaching (MERLOT) publishes the *Journal of Online Teaching and Learning* (*JOLT*); this resource allows librarians involved in online or blended education to access additional scholarship on resources for teaching and learning online.

APPENDIX TECHNOLOGY RESOURCES, ORGANIZED AS REFERENCED IN THE CHAPTER

Tool	Alignment with Bloom's revised taxonomy	Intended use
zotero.org	Remembering	Note taking, annotating, sharing resources
diigo.com	Remembering	Note taking, annotating, sharing resources
evernote.com	Remembering	Note taking, annotating, sharing resources
polleverywhere.com	Understanding	Virtual polling
socrative.com	Understanding	Virtual polling and assessment
padlet.com	Understanding	Online wall
screencast-o-matic.com	Applying	Web-based screen casting; downloadable
screenr.com	Applying	Web-based screen casting
techsmith.com/jing	Applying	Screen casting; downloadable
onlinecharttool.com	Analyzing	Chart and graph creation
creately.com	Analyzing	Chart and graph creation
bubbl.us	Analyzing	Concept mapping and mind mapping
text2mindmap.com	Analyzing	Concept mapping and mind mapping
mindmeister.com	Analyzing	Concept mapping and mind mapping
surveymonkey.com	Evaluating	Survey creation
chatzy.com	Evaluating	Back-channel discussion
wikispaces.com	Creating	Writing collaboration
blogger.com	Creating	Writing collaboration
drive.google.com	Evaluating, creating	Collaboration—writing, spreadsheets, forms, presentations

(*Continued*)

Tool	Alignment with Bloom's revised taxonomy	Intended use
sites.google.com	Creating	Web-site creation
weebly.com	Creating	Web-site creation
prezi.com	Analyzing, creating	Presentation
glogster.edu	Creating	Virtual poster creation
animoto.com	Creating	Video creation

REFERENCES

Anderson, L. W., Krathwohl, D. R., & Bloom, B. S. (2001). *A taxonomy for learning, teaching, and assessing: A revision of Bloom's taxonomy of educational objectives.* Abridged edition. New York, NY: Pearson Education.

Andrews, T. M., Leonard, M. J., Colgrove, C. A., & Kalinowski, S. T. (2011). Active learning not associated with student learning in a random sample of college biology courses. *CBE Life Sciences Education,* 10(4), 394−405.

Association of College and Research Libraries. (2000). *Information Literacy Competency Standards for Higher Education.* Retrieved from <http://www.ala.org/acrl/sites/ala.org.acrl/files/content/standards/standards.pdf>.

Association of College and Research Libraries. (2015). *Framework for Information Literacy for Higher Education.* Retrieved from <http://www.ala.org/acrl/sites/ala.org.acrl/files/content/issues/infolit/Framework_ILHE.pdf>.

Bonwell, C. C., & Eison, J. A. (1991). *Active learning: Creating excitement in the classroom. ASHE-Eric higher education report no. 1.* Washington, DC: The George Washington University, School of Education and Human Development.

Braxton, J. M., Jones, W. A., Hirschy, A. S., & Hartley, H. V., III (2008). The role of active learning in college student persistence. *New Directions for Teaching and Learning,* 2008(115), 71−83.

Brown, J.S. (2002). *Growing up digital: How the web changes work, education, and the ways people learn.* Retrieved from <http://www.johnseelybrown.com/Growing_up_digital.pdf>.

Campisi, J., & Finn, K. E. (2011). Does active learning improve students' knowledge of and attitudes toward research methods? *Journal of College Science Teaching,* 40(4), 38−45.

Chickering, A. W., & Gamson, Z. F. (1987). Seven principles for good practice in undergraduate education. *AAHE Bulletin,* 39(7), 3−7.

Dabbour, K. S. (1997). Applying active learning methods to the design of library instruction for a freshman seminar. *College & Research Libraries,* 58(4), 299−308.

Gogus, A. (2012). Active learning. In N. M. Seel (Ed.), *Encyclopedia of the sciences of learning* (pp. 77−80). New York, NY: Springer.

Greer, K. (2014). Cooking up a concept map with Prezi: Conceptualizing the artist statement. In K. Calkins & C. Kvenild (Eds.), *The embedded librarian's cookbook* (pp. 84−85). Chicago, IL: Association of College and Research Libraries.

Hoffman, C., & Goodwin, S. (2006). A clicker for your thoughts: Technology for active learning. *New Library World,* 107(9/10), 422−433.

Hoppenfeld, J. (2012). Keeping students engaged with web-based polling in the library instruction session. *Library Hi Tech,* 30(2), 235−252.

Jacklin, M. L., & Robinson, K. (2013). Evolution of various library instruction strategies: Using student feedback to create and enhance online active learning assignments. *Partnership: The Canadian Journal of Library and Information Practice and Research, 8*(1), 1–21.

Kenney, B. F. (2008). Revitalizing the one-shot instruction session using problem-based learning. *Reference & User Services Quarterly, 47*(4), 386–391.

Obenland, C. A., Munson, A. H., & Hutchinson, J. S. (2012). Silent students' participation in a large active learning science classroom. *Journal of College Science Teaching, 42*, 90–94.

Perrin, A., & Duggan, M. (2015, June 26). *Americans' internet access: 2000-2015*. Pew Research Center. Retrieved from <http://www.pewinternet.org/2015/06/26/americans-internet-access-2000-2015/>.

Powner, L. C., & Allendoerfer, M. G. (2008). Evaluating hypotheses about active learning. *International Studies Perspectives, 9*(1), 75–89.

Roehl, A., Reddy, S. L., & Shannon, G. J. (2013). The flipped classroom: An opportunity to engage millennial students through active learning strategies. *Journal of Family & Consumer Sciences, 105*(2), 44–49.

Ross, A., & Furno, C. (2011). Active learning in the library instruction environment: An exploratory study. *Portal: Libraries and the Academy, 11*(4), 953–970.

Schrock, K. (2015, April 1). *Bloomin' apps*. Retrieved from <http://www.schrockguide.net/bloomin-apps.html>.

Williams, J., & Chinn, S. J. (2009). Using web 2.0 to support the active learning experience. *Journal of Information Systems Education, 20*(2), 165–174.

Zayapragassarazan, Z., & Kumar, S. (2012). Active learning methods. *NTTC Bulletin, 19*(1), 3–5.

SECTION II

Pedagogy

CHAPTER 4

Designing Online Asynchronous Information Literacy Instruction Using the ADDIE Model

4.1 BACKGROUND

The University of South Florida (USF) Tampa Library has always had an active information literacy program. Housed within the Academic Services department, the unit once boasted more than 14 faculty librarians who maintained responsibilities for instruction. Over the past 10 years, budget shortfalls, retirements, and other staff departures has cut the unit in half in terms of faculty and staff. Presently, the department employs eight faculty librarians with information literacy instruction duties, as well as responsibilities for reference and departmental liaisons. Because of the loss of staff members and the realization that these positions would not be restored, Academic Services has found it necessary to reevaluate its instructional activities and make deliberate decisions to cease providing instructional services to certain populations that traditionally visited the library for introductory instruction but did not actually engage in a research assignment. For the most part, these populations were found in introductory courses that emphasized college success strategies such as time management and study skills, but they did not incorporate any kind of assignment that would require students to engage with the library. By working with identified departmental chairs and faculty to inform them of the library's limited staff resources, the library successfully transitioned away from providing face-to-face information literacy instruction for these populations (e.g., reducing one-shot library lecture requests to almost nothing) by creating a series of online tutorial videos housed on YouTube that addressed almost all of the entry-level knowledge and skills that the typical undergraduate student would need. These included information for remote use, basic catalog searches and some of the more popular multidisciplinary databases, avoiding plagiarism, citation basics, and so on.

Distributed Learning.
© 2017 T. Maddison and M. Kumaran.
Published by Elsevier Ltd. All rights reserved.

In 2012, the department was faced with the retirement of key faculty who had continued to provide instructional services for the last remaining large population of students who visited the library out of tradition and not because of research needs. Furthermore, the department was still trying to keep pace with face-to-face instruction request for the university's first-year composition (FYC) program, which incorporated thousands of students in hundreds of course sections who *did* have assignments that required a certain degree of research. In response, the department made the decision to use this departure as a turning point to transition our undergraduate library instructional program to a series of interactive, multimedia online modules housed within Canvas, our course management system (CMS).

This transition would not be simple, however. Librarians in the department needed to ensure that instructional integrity and learning goals were met while also appealing to an audience made primarily of freshmen students who, although familiar with online learning, were difficult to engage in an online environment. In addition, the hope was to improve on the "one-shot" face-to-face model of instruction, leveraging the online instructional setting so the learner could access and review the instructional materials any time and for as many times as desired. This would allow librarians to design and implement instructional materials that could delve deeper into the various components of information literacy while integrating in a more meaningful way with the subject-specific curriculum for which we were designing. Complicating our efforts was the relative inexperience of our instruction team in designing instructional materials and the general knowledge gap when it came to instructional design theory and practices.

Prior to making this transition, the department researched various instructional design models to determine which would suit our needs best.

4.2 INSTRUCTIONAL DESIGN MODELS

The field of instructional design includes a wide variety of models from which to choose that guide the instructional design and development process. Some models, such as that of Dick and Carey (Dick, Carey, & Carey, 2001), or the three-phase design model, are built around a linear, prescriptive set of steps that define the instructional design process from start to finish. Others such as the recursive reflective design and development (R2D2) (Bonk & Zhang, 2006) and four-component instructional design (4C/ID) (Van Merriënboer, Clark, & De Croock, 2002) models emphasize the identification, analysis, and subsequent dissection of learning tasks into simpler subtasks that can then be addressed with instructional materials. Still others such as Gagné's model (Gagné, 1985) concern themselves primarily with learner motivation and interest.

In selecting a model to guide the online instructional design efforts at USF, the instruction team looked closely at all these models to determine the best fit for our instructional team and the project at hand.

4.3 DICK AND CAREY

Perhaps one of the most well-known and often-cited instructional design models is that of Dick and Carey (Dick, Carey, & Carey, 2001). Since the first edition of *The Systematic Design of Instruction* was published in 1978, the Dick and Carey model has influenced instructional design to a great degree. The model describes a recursive and reiterative instructional design process that includes 10 stages of instructional design arranged linearly with feedback loops in key areas that assist the designer in continuous improvement of instruction and assessment (please refer to Dick, 1996, p. 58 for a diagram of the systems approach model for designing instruction).

This model has several advantages in that it clearly illustrates the instructional design process and lays out a clear process for designers to follow. However, by his own admission, Dick cites limitations in the model that make it insufficient as a complete instructional systems design (ISD) model, including lacking procedures for implementing and maintaining instructional content (Dick, 1996). In addition, Dick notes that the model was originally designed for the novice designer at a time when the profession of instructional design was nascent. The most oft-cited criticism of the model, however, was its linear nature and the perception among designers that the stages of design described in the model were fixed and rigid. What's more, Dick suggests that many designers feel the model is outdated on a philosophical level, seeing it as overly behaviorist, positivistic, or objectivist in an ISD environment that is increasingly constructivist in nature, objecting to prespecified learning objectives and criterion-referenced evaluation (Dick, 1996).

While the Dick and Carey model has many factors that make it a valuable design tool to this day, the instruction team opted not to use the model, citing its overly prescriptive nature and perceived lack of flexibility.

4.4 RECURSIVE REFLECTIVE DESIGN AND DEVELOPMENT MODEL

The R2D2 model was designed specifically to meet the challenges of designing instruction for the online environment (Bonk & Zhang, 2006). However, as its creators state, the model differentiates itself from other instructional design models by not specifically addressing the instructional design process but focusing instead on reflection about the different kinds of learner tasks, activities, and resources that the designer may wish to incorporate into online instructional modules that would appeal to varying learning styles. The R2D2 taxonomy, if it could be called that, organizes these tasks into four categories, including reading and listening, reflecting and writing, displaying, and doing.

While the R2D2 model can be helpful with ensuring that online learning activities are varied and interactive, this model presupposes a familiarity with instructional design processes and is therefore inappropriate for faculty engaged in design, but helpful to those with little knowledge or experience with design models and require guidance in the development process.

4.5 FOUR-COMPONENT INSTRUCTIONAL DESIGN

Originally developed in 1992 by van Merrienboer, Jelsma, and Paas, the 4C/ID model attempts to address key perceived deficits in previous instructional design models. The 4C part of the model comprises four components: (1) learning tasks, (2) supportive information, (3) procedural information, and (4) part-task practice. In this model, tasks required for achieving a learning goal are identified and ordered by difficulty (Van Merriënboer et al., 2002). Information is introduced to the learner through task practice with a large amount of scaffolding at the beginning, which tapers off as the learner progresses. By focusing on task-specific skills instead of knowledge types and making distinctions between early supportive information and just-in-time required information, as well as an emphasis on a combination of part-task practice supporting whole-task learning, this model seeks to support the development of instructional curricular goals that accommodate learning for highly complex cognitive skills. Van Merrienboer contends that earlier models that "assume that a large, complex set of interrelated tasks are achievable

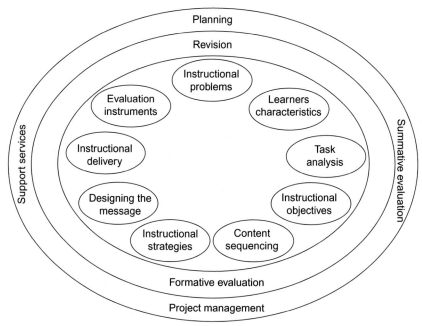

Figure 4.1 Kemp model. *Adapted from Kemp, 1985, p. 304.*

… by sequencing a string of simplified, component task procedures until the complex task is captured" are inaccurate and that complex learning involves mastery in coordinating and integrating separate skills that constitute real-life performance (Van Merriënboer et al., 2002). As an example, van Merrienboer et al. propose the moderately complex skill of "searching for relevant research literature," suggesting that the learning goal is not simply the acquisition of separate constituent skills but the ability to combine each constituent skill "in a coordinated and integrated fashion while doing real-life literature searches" (please refer to Van Merriënboer et al, 2002, p. 41 for further details on the 4C/ID model's skill hierarchy).

This model is attractive for its emphasis on holistic complex learning and its problem-based approach. However, its inherent complexity and the difficulty in implementing this type of model with a team of instructors who are not necessarily well versed in instructional design, educational psychology, or learning theory puts it just out of reach in terms of usability and accessibility.

4.6 KEMP

The Kemp model identifies and attempts to order nine components of the instructional design process in a nonlinear, cyclical arrangement that emphasizes the systemic nature of instructional design (see Fig. 4.1)

(Kemp, 1985). Ideally, a designer could enter the development process at any point in the model and proceed accordingly. Many of the steps share similarities to both the Dick and Carey and 4C/ID models, representing a kind of hybrid approach that emphasizes both the development process and task analysis (Andrews & Goodson, 1980).

The Kemp model is specifically designed for use by multiple groups engaged in the design and development of large-scale programs and was also intended to represent a continuous cycle of development in which planning, revision, evaluation, and management are persistent, concurrent, and continually ongoing throughout the life of the project.

While there is much that can be useful in this model, its emphasis on a cycle without clear starting or ending points makes it needlessly complex for an instruction team whose members are educators, not instructional designers.

4.7 GAGNÉ

Robert Gagné offers more of a theoretical framework than an instructional design model, but nonetheless his work in instructional design is one of the most influential and well-known ideas in the field. Gagné, along with colleague L. J. Briggs, was among the first in the field of educational psychology to advocate for a systems approach to instructional design in which development takes place in a structured environment and all components of instruction can be analyzed "and designed to operate together as an integrated plan for instruction" (Encyclopedia of World Biography, 2004, para. 2). Gagné's theory is composed of three major components that together serve as a blueprint for effective instructional design. These components include a taxonomy of learning outcomes, conditions for learning, and nine events of instruction.

Gagné's taxonomy includes five domains of learning that encompass verbal information, intellectual skills, cognitive strategies, attitudes, and motor skills. His conditions of learning serve as a system for analyzing various levels of learning from simple to complex that are arranged hierarchically; the most basic behavioral skills are at the bottom and complex higher-order thinking is at the top (see Fig. 4.2) (Gagné, 1985).

Rounding out the theory are Gagné's "nine events of instruction," which are intended to correlate with the conditions for learning to promote knowledge acquisition and transfer in the learner (please refer to Gagné, 1985, p. 304 for a visualization of the nine events of instruction).

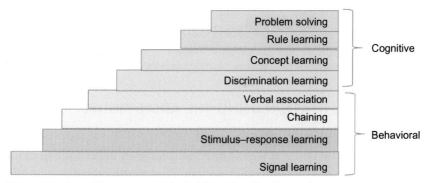

Figure 4.2 Gagné's taxonomy. *Adapted from Gagné, 1985.*

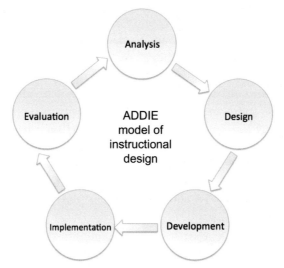

Figure 4.3 ADDIE model. *Adapted from Molenda or refer to Molenda, 2003, p. 35.*

4.8 ADDIE

The ADDIE model—**analysis, design, development, implementation**, and **evaluation**—was developed by the Center for Educational Technology at Florida State University in the mid-1970s to formalize and codify the processes inherent in the development of ISD theory. It represents a more generic framework that serves as a guideline for the instructional design process (Molenda, 2003). In fact, most instructional design models previously cited can be said to adhere in some respect to the ADDIE framework (see Fig. 4.3).

The instruction team at USF chose the ADDIE model for its simplicity, flexibility, and ability to accommodate some of the various bits and pieces from other ISD models that were selected for incorporation into the current customized design process. The following is a description of each phase of the model and how we applied it to the development of online information literacy instructional modules.

4.9 IMPLEMENTATION USING THE ADDIE MODEL

4.9.1 Analysis

The analysis phase of the process focuses on gaining an understanding of the target audience and the instructional goals. This means attempting to gain insight into the existing skill levels and experience that the learners might bring to the lesson. In addition, the team must work out exactly what learners should be able to do after they finish the instructional program.

The challenge at this phase was to determine the learner characteristics of our audience—i.e., freshman students enrolled in FYC courses. Because of the high enrollment and diversity of learners in the program, it was determined that the team would likely not be able to assess the qualities of the entire target population. Librarians in the team therefore decided to select a sample population based on existing relationships with instructional faculty and conduct a skills pretest to get an approximate baseline idea of students' information literacy skills. In addition, certain assumptions were made about the population based on existing empirical evidence informed by previous experience and generalizations made concerning this population.

The pretest was constructed by examining the assignments that students had to complete for their FYC course and linking the research-related items inherent in the assignments to existing ACRL information literacy standards (Association of College and Research Libraries, 2000).

The relevant ACRL standards for the FYC assignments coincided mainly with standards one, two, three, and five; more specifically, they linked to the following outcomes:

Standard One:

1. Develops a thesis statement and formulates questions based on the information need
2. Explores general information sources to increase familiarity with the topic
3. Defines or modifies the information need to achieve a manageable focus
4. Identifies key concepts and terms that describe the information need

5. Identifies the purpose and audience of potential resources (e.g., popular vs scholarly, current vs historical)
6. Reviews the initial information need to clarify, revise, or refine the question
   Standard Two:
1. Identifies keywords, synonyms, and related terms for the information needed
2. Identifies gaps in the information retrieved and determines if the search strategy should be revised
   Standard Three:
1. Identifies verbatim material that can be then appropriately quoted
2. Examines and compares information from various sources to evaluate reliability, validity, accuracy, authority, timeliness, and point of view or bias
   Standard Five:
1. Demonstrates an understanding of what constitutes plagiarism and does not represent work attributable to others as his or her own

Specific learning objectives were formulated based on the standards selected. The task was made easier by the use of a grading rubric developed by the FYC program that provided specific grading criteria for information literacy components (categorized in the rubric under "Evidence") (see Appendix 1) within the assignment. To make the pretest as easy as possible on the students while still obtaining the information needed, the instruction team of librarians limited the test to 10 questions addressing manageable topic selection, the research question, identification of keywords, scholarly versus popular sources, correct use of citations, plagiarism, and source evaluation.

The pretest was sent to students in five sections of Composition I, each with an enrollment of 22–25 students to be completed the semester before module development was to take place. Out of 97 students asked to complete the pretest, 89 completed and submitted it. The results for all sections were collected and compiled using the same survey instrument, so it was not possible to determine if any variation existed between course sections. Nonetheless, the results indicated that more than half the students were not proficient in the areas of recognizing manageable topics, identifying keywords, defining a research question, and properly formatting citations in Modern Language Association style (see Appendix 2). Conversely, the overwhelming majority of students understood the definition of peer review and plagiarism, as well as the concept of source evaluation.

After reviewing these results and consulting with faculty members from the FYC department, the following course modules for information literacy instruction were planned:

- Topic selection
- Topic investigation
- Evaluating sources
- Citing sources and avoiding plagiarism

4.10 DESIGN

The design phase is where much of the work takes place. During this phase, learning goals and objectives are formulated and drafted (but not finalized), learning activities are conceived, media selection is determined, and learning assessments based on objectives are tentatively sketched out. Questions that instruction teams should ask themselves at this phase deal primarily with the availability of resources and how the team feels they should be allocated. Some of these questions include:

- What kinds of media will be used in designing the instructional materials? Will the team have various media resources at its disposal?
- Will the material be interactive? Do the team members have the required skills to create interactive online learning objects?
- How much time will be required to complete the planned lessons?
- How much time does the team have at its disposal to design the materials?
- How will the team determine that the learners have acquired the desired learning outcomes?
- What sort of interface will the team design for? Will the team make use of the CMS or will the material be accessed outside, perhaps from a web page?

In addition, the team decided that all online modules should follow a design template, ensuring a level of continuity to facilitate students' navigation through the materials.

At this phase, a few components from other ID models were also examined. Specifically, the task-analysis feature from the 4C/ID model and the nine events of instruction described by Gagné were found valuable by the team. For the task analysis, FYC research assignment was examined and then analyzed to determine specific targeted skills that would be required to complete the research portion of the assignment.

Students enrolled in Composition I, for example, are asked to complete an annotated bibliography and a corresponding academic essay on a topic of their choosing. Students are graded using a 10-point rubric that assesses the students in five key areas, including focus, evidence, organization, style, and format. The information literacy skills that are assessed can be found mostly within the evidence criterion and include credibility of sources, use of primary and secondary sources, timeliness and relevancy of sources used, and proper citation.

The library instruction team first set about drafting initial learning objectives for each of the four modules in relation to the graded criteria on the students' assignments. These learning objectives were modeled on the three-part learning objective suggested by Mager (1997). Care was taken to include only those learning objectives that directly addressed the skills required to complete the corresponding assignment to ensure that no more than five to seven objectives appeared in any one module. Task analysis (see e.g., Table 4.1) of the assignment's information literacy components directly informed the learning objectives and ensured that our module objectives integrated well with the overall curriculum.

Design of the online instruction template began after a task analysis was completed for all the criteria for evidence within the FYC grading rubric. The rough drafts of learning objectives were written and assigned to their proper module.

The team agreed to follow a standard template for each module, which would contain a module overview, statement of learning objectives, module activities, and a criterion-based assessment at the end. The module template also provided prompts for the instructor developing the material. For example, the division for "Lesson Objectives" includes the phrase "After completing this module, you should be able to ..." along with a brief list of action verbs taken from Bloom's taxonomy. The team hoped this would remind instructors about the necessity for providing clear, measurable learning objectives for the module. Similar prompts were provided for the Module Activities section, where the team hoped instructors would go beyond passive reading to include more interactive components that would ask students to *watch* or *complete*. The template was designed based on reviews of sample course templates provided by Instructure Canvas, the company that hosts our institution's CMS. We ultimately agreed on a standard look and feel for our modules by using a common layout and set of icons to distinguish the various module divisions (see Fig. 4.4).

Table 4.1 Example task analysis for single rubric criteria

FYC assignment rubric criteria for evidence	Identified tasks	Learning objective (After completing the module, learners will be able to …)	Instructional activity	Module placement
Credible and useful sources and supporting details.	1. Select a manageable research topic. 2. Distinguish between sources that are credible and those that are not. 3. Distinguish between sources that are useful (aka *relevant*) and those that are not. 4. Identify and locate resources that provide credible sources. 5. Search identified resources to locate items that are "useful" (aka *relevant*): a. Identify search terms from topic and corresponding keywords. b. Identify the meaning of the term *keyword*.	1. Distinguish among example research topics and categorize those that are manageable, too broad, or too narrow. 2. Apply evaluation criteria to given examples of sources to determine credibility. 3. Apply evaluation criteria to given examples of sources to determine relevancy. 4. Navigate the library's website and locate the course guide for FYC. 5. Identify main ideas from a given research question and gererate corresponding keywords.	1. Interactive learning object: categorize sample topics. 2 & 3. Worksheet: Apply TRAAP test to given example sources. 4. Video tutorial: Locating the FYC course guides. 5. Worksheet: Select main idea words from sample topic; brainstorm keywords. 6. Interactive learning object: Match lists of relevant keywords with sample topics.	1. Topic selection (Module 1) 2. Evaluating sources (Module 3) 3. Evaluating sources (Module 3) 4. Topic investigation (Module 2) 5. Topic investigation (Module 2)

Figure 4.4 Module template in Canvas CMS.

4.11 DEVELOPMENT

The development phase of the ADDIE model is meant to allow for the instruction design team to put the plans from the design phase into action. During this phase, the team works on finalizing learning goals and objectives, drafting the actual instructional materials, and constructing online learning objects and assessments. At this point, the instruction team may want to address any concerns that relate to the time frame allocated for the development of the materials.

In the case of the USF Tampa Library, the team realized that the time frame for development would not allow for the creation of an entire set of original in-house video tutorials and learning objects as originally planned. As a result, members of the team scoured the Internet for relevant tutorial videos from other university libraries for inclusion within the modules in order to cover basic learning activities that addressed low-level objectives such as research topic selection and the identification of topics that are too broad or too narrow. Thankfully, a wide range of possibilities was discovered. The team reviewed several instructional videos from other universities for inclusion within our learning modules, giving credit within the module to the institution where it originated.

For interactive learning objects, the team *did* opt to create them in-house, using e-learning authoring tools such as Captivate and SoftChalk. Although it required additional time and planning, the team determined that interactive learning objects, unlike video tutorials, were not so ubiquitous and freely available on the Internet. Even a thorough search of learning object repositories such as MERLOT yielded little in the way of readily reusable objects. In fact, the amount of time required to modify an existing object acquired from a repository would be equal to if not greater than the amount of time required to author one from scratch.

After gathering together the relevant media and learning objects, the team used the developed module template to begin writing the instructional content (see Fig. 4.5). This content, because of the preliminary planning in the design phase, went quite smoothly. Instruction team members were asked to write a module overview that would start with a question aimed at the target audience about the research assignment they were about to begin. For example, in the Topic Selection module overview, the paragraph is introduced with the question, "Would you believe that the selection of research topic is probably the most important step in completing a successful research assignment?" This question is intended to subtly engage the attention of students by communicating that careful topic selection will result in the successful completion of their assignment. Phrased as a question, the natural response from readers will be to attempt to mentally answer it, pulling learners in a little further. They might then want to read on to see what the reasoning is behind the assertion that topic selection is the most important step. The overview's purpose is to describe in a few sentences a general summary of what the module will cover and why that information will be important to learners as they complete their research assignments.

Next, learners are presented with the module objectives that provide a roadmap to the knowledge and skills they will be expected to acquire by the conclusion of the module. Learners then proceed to the learning activities, which include a combination of written material, video tutorials, and interactive learning objects. Each learning activity must relate directly to one or more of the learning objectives so that the alignment between the objective and the information and hands-on practice opportunities presented in the learning activities are readily apparent. Finally, students are asked to complete a module quiz. The quiz is a criterion-based assessment that relates directly back to the learning objectives and simply provides proof that students engaged with the material (see Fig. 4.5).

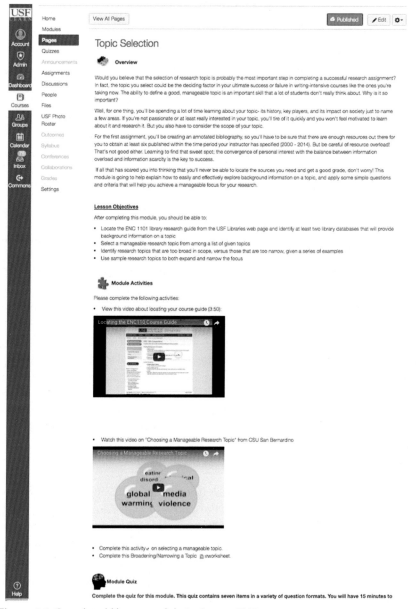

Figure 4.5 Completed library module in Canvas CMS.

4.12 IMPLEMENTATION

The implementation phase of the ADDIE model is intended to serve as a period of training for instructors who may interact with materials they did not design or develop. This phase also allows the design team to review the materials, test them out, and make any necessary modifications before the material goes live.

The USF Library instruction team found this phase to be the most problematic. Pressures from time constraints for rolling out the modules by the start of the academic year combined with the sizable yet largely distributed and inaccessible population of FYC instructors made testing the modules with this population unrealistic. What's more, the modules were designed to be largely self-contained and intuitive, making any kind of extensive training unnecessary. The team did, however, periodically request feedback and review from the FYC advisory team, which was overwhelmingly supportive and provided constructive criticism in the use of language and terminology.

In addition, working with the FYC advisory team during this phase, we were able to work out how the library instructional modules would be integrated into the Canvas course shells for all of the FYC course sections and modules would be placed chronologically within the weekly schedule and syllabus for the program.

4.13 EVALUATION

The last phase of the ADDIE model is evaluation. The phase usually consists of both formative and summative evaluations. A formative evaluation concerns the instructional materials themselves and how well they are facilitating the learning process. Formative evaluation is largely done through feedback from learners and instructors, but it can also be derived from usage reports from the CMS or large-scale observation of student performance on criterion-referenced quizzes. In the latter case, patterns of widespread failure to select correct answers on quizzes can indicate problems with either the quiz or the instructional material. Formative evaluation is often continuous and gets fed back into the design process on a reiterative and recursive cycle of content revision.

After the library instructional modules were integrated into all of the FYC course shells within the Canvas CMS, we received feedback from both students and instructors that prompted us to consider changes.

As one of our adopted "best practices," the team originally included the average amount of time a student could expect to spend on each module. Typically, it was no more than 30 minutes per module. Feedback obtained from instructors via two focus groups held at the end of the semester indicated that students were feeling pressured if it took them longer than the suggested time to complete the module and that they felt they were not learning at the expected pace. Many instructors also suggested that the library did not provide enough instructional material in a few areas, including the definition and function of an annotated bibliography, the definition and formulation of a research question, and the research process. In response to this feedback, the team decided to remove the average time for completion from each module and to develop three new modules that would address the indicated instructional gaps. Overall, however, information from instructors gathered at the focus groups overwhelmingly indicated that both instructors and students had positive experiences with the library modules and felt the information provided was helpful in contributing to students successfully completing their research assignments.

Additional formative feedback was gleaned from students, typically via the email system within the learning management system (LMS), who reported various errors within the modules with respect to broken links and quizzes that were set to the wrong answer, thus marking correct answers wrong. In addition, we were able to obtain formative feedback by examining and analyzing student quiz scores and responses. By paying careful attention to the scores of each individual question, it became apparent which questions were perhaps not well constructed. If the failure rate of any one question was greater than 70% overall, the item was marked for review since it seemed more likely that the assessment item was at fault rather than the students' understanding of the concept supposedly assessed by the item. Because the instructional content was copied from a master course shell into more than 100 separate course sections within the CMS, correcting these kinds of problems required a significant amount of staff time, prompting the team to think of future alternate means of integrating library instructional content into large-scale undergraduate courses.

Summative evaluation refers to the measurement of learners' achievement after they conclude the lesson. Summative evaluation typically takes the form of tests and quizzes or other criterion-referenced assessment where a score is assigned based on learner-supplied evidence that he or she has mastered the desired knowledge or skills. Each library instructional module concluded with a criterion-referenced quiz that was aligned to the module's learning objectives. Students in some sections were required to complete these quizzes, which collectively counted toward a homework grade, while others were not. Students were given two opportunities to complete each quiz, with the highest grade from the two attempts retained. It is no surprise that students performed especially well on the quizzes overall, prompting a discussion about the value and validity of the assessments. Some felt that an overwhelmingly positive performance on the quizzes indicated that the instructional materials were not challenging enough and reflected information and skills that the students probably already had. The pretest information, however, indicated otherwise, perhaps indicating instead that while the information literacy skills required for the successful completion of the students research assignments may not be difficult to acquire, they still represent a gap in learning that the library has been able to address. Future plans for summative assessment will involve a grade comparison between final course rubric scores for students who completed the online library instructional modules versus those who did not.

4.14 CONCLUSION

Using a standard template, the USF Tampa Library instruction team developed online information literacy instructional content using instructional design principles based on the ADDIE model. Teaming up with the FYC program to design these modules and integrating some of the academic department's curriculum materials with our own, we now provide students with online, course-embedded information literacy instruction that meets the needs of the student at the time and place they deem best.

Overall, the effort has produced mostly positive results. Quiz grades for students who completed the online information literacy modules reflect a success rate of more than 90%, and anecdotal feedback from students and instructors seems to indicate that the material provides valuable information for students completing their research assignments. Future plans to include a comparative analysis of students' final grades should provide additional insight regarding the impact that these modules have had on students' success.

The design and development process was also a learning experience for the librarians who participated. While the move to online learning has often been seen as a regrettable but necessary trend to make up for the loss of library teaching faculty, the team has discovered that online learning also has some highly positive aspects. First, the process of preparing to deliver online instructional content forced instructors to think about how that content would be best presented. Thus, instructors were prompted to research and evaluate instructional design models that in turn allowed for thoughtful and reflective process of their our own teaching efforts overall. Designing effective online content within the guidance of an instructional design model also requires instructors to engage with their instructional materials in more depth and intensity, often revealing gaps or confusing and misaligned material within the lessons.

Second, although instructors did not survey students to address it, specifically providing interactive online information literacy instruction within the LMS benefits student learning by providing asynchronous instructional opportunities where students can interact with the content at the time of their choosing and as many times as needed. In contrast to a one-shot traditional library instruction session, it would seem that online instruction has a distinct advantage. Moreover, much more information has been integrated into the course than could have possibly been covered in a single 50- to 75-minute session.

Based on the mostly positive experiences with online instructional design and delivery, online instruction has become and will remain a cornerstone of USF's information literacy program. Not only will our team continue to revise and develop these modules, but our instructors also intend to expand the online information literacy efforts into other academic units, creating new modules as the need arises.

REFERENCES

Association of College and Research Libraries. (2000). Information Literacy Competency Standards for Higher Education. Retrieved from http://www.ala.org/acrl/sites/ala.org.acrl/files/content/standards/standards.pdf.

Andrews, D. H., & Goodson, L. A. (1980). A comparative analysis of models of instructional design. *Journal of Instructional Development, 3*(4), 2−16.

Bonk, C. J., & Zhang, K. (2006). Introducing the R2D2 model: Online learning for the diverse learners of this world. *Distance Education, 27*(2), 249−264.

Dick, W. (1996). The Dick and Carey model: Will it survive the decade? *Educational Technology Research and Development, 44*(3), 55−63.

Dick, W., Carey, L., & Carey, J. O. (2001). *The systematic design of instruction* (5th ed). New York: Longman.

Encyclopedia of world biography. (2004). Retrieved October 7, 2015, from http://www.encyclopedia.com/doc/1G2-3404702355.html

Gagné, R. M. (1985). *The conditions of learning and theory of instruction* (4th ed). New York: Holt, Rinehart and Winston.

Kemp, J. E. (1985). *The instructional design process*. New York: Harper & Row.

Mager, R. F. (1997). *Preparing instructional objectives*. Atlanta, GA: Center for effective performance.

MERLOT II: Multimedia Educational Resource for Learning and Online Teaching. (n.d.). Retrieved October 7, 2015, from https://www.merlot.org/merlot/index.htm

Molenda, M. (2003). In search of the elusive ADDIE model. *Performance Improvement, 42*(5), 34–37.

Van Merriënboer, J. J., Clark, R. E., & De Croock, M. B. (2002). Blueprints for complex learning: The 4C/ID-model. *Educational Technology Research and Development, 50*(2), 39–61.

APPENDIX 1: FYC RUBRIC

University of South Florida Academic Writing Scoring Criteria

CRITERIA		Emerging 0	1	Developing 2	3	Mastering 4
Focus 25%	Basics	• Does not meet assignment requirements		• Partially meets assignment requirements.		• Meets assignment requirements.
	Critical Thinking	• Absent or weak thesis. • Ideas are underdeveloped, vague, or unrelated to thesis. • Poor analysis of ideas relevant to thesis.		• Predictable or unoriginal thesis. • Ideas are partially developed and related to thesis. • Inconsistent analysis of subject relevant to thesis.		• Insightful and/or intriguing thesis. • Ideas are convincing and compelling. • Cogent analysis of subject relevant to thesis.
Evidence 25%	Critical Thinking	• Sources and supporting details lack credibility. • Poor synthesis of primary and secondary sources/evidence relevant to thesis. • Poor synthesis of visuals, personal experience, or anecdotes relevant to thesis. • Rarely distinguishes between writer's ideas and source's ideas.		• Fair selection of credible sources and supporting details. • Unclear relationship between thesis and primary and secondary sources/evidence. • Ineffective synthesis of sources/evidence relevant to thesis. • Occasionally effective synthesis of visuals, personal experience, or anecdotes relevant to thesis. • Inconsistently distinguishes between writer's ideas and source's ideas.		• Credible and useful sources and supporting details. • Cogent synthesis of primary and secondary sources/evidence relevant to thesis. • Clever synthesis of visuals, personal experience, or anecdotes relevant to thesis. • Distinguishes between writer's ideas and source's ideas.
Organization 25%	Basics	• Confusing opening. • Absent, inconsistent, or non-relevant topic sentences. • Few transitions and absent and unsatisfying conclusion.		• Uninteresting or somewhat trite introduction. • Inconsistent use of topic sentences, segues, transitions, and mediocre conclusion.		• Engaging introduction. • Relevant topic sentences, good segues, appropriate transition, and compelling conclusion.
	Critical Thinking	• Illogical progression of supporting points. • Lacks cohesiveness.		• Supporting points follow a somewhat logical progression. • Occasional wandering of ideas. • Some interruption of cohesiveness.		• Logical progression of supporting points. • Very cohesive.
Style 20%	Basics	• Frequent grammar and/or punctuation errors. • Inconsistent point of view.		• Some grammar and/or punctuation errors occur in some places. • Somewhat consistent point of view.		• Correct grammar and punctuation. • Consistent point of view.
	Critical Thinking	• Significant problems with syntax, diction, word choice, and vocabulary.		• Occasional problems with syntax, diction, word choice, and vocabulary.		• Rhetorically-sound syntax, diction, word choice, and vocabulary.
Format 5%	Basics	• Little compliance with accepted documentation style (i.e., MLA, APA) for paper formatting, in-text citations, annotated bibliographies and works cited. • Minimal attention to document design.		• Inconsistent compliance with accepted documentation style (i.e., MLA, APA) for paper formatting, in-text citations, annotated bibliographies and works cited. • Some attention to document design.		• Consistent compliance with accepted documentation style (i.e., MLA, APA) for paper formatting, in-text citations, annotated bibliographies and works cited. • Strong attention to document design.

Notes
This is a "living document", in other words, we revise our scoring criteria in response to teacher and student feedback.
A "0" for "**Does not meet assignment requirements**" results in an overall 0 (F) for the project.

APPENDIX 2: FYC PRETEST SURVEY SUMMARY

	A	B	C	D	E	F
1	FYC InfoLit Pretest					
2	1. Examine the drop-down menus below. For each drop-down, select the topic that matches the heading (I.e. Manageable, Too Broad, Too Narrow)					
3	Manageable					
4	Answer Options		1. Does food from genetically modified crops pose a health threat to consumers?	2. Are vaccinations harmful?	3. What percentage of female students at USF have been victims of domestic violence?	Response Count
5	Research Topic		7	63	19	89
6						
7	Too Broad					
8	Answer Options		1. Does food from genetically modified crops pose a health threat to consumers?	2. Are vaccinations harmful?	3. What percentage of female students at USF have been victims of domestic violence?	Response Count
9	Research Topic		31	31	2	89
10						
11	Too Narrow					
12	Answer Options		1. Does food from genetically modified crops pose a health threat to consumers?	2. Are vaccinations harmful?	3. What percentage of female students at USF have been victims of domestic violence?	Response Count
13	Research Topic		47	7	35	89
14						
15						Question Totals
16					answered question	89
17					skipped question	0
18						
19						
20	2. Which of the definitions below best describes the term "keywords" within the research context?					
21	Answer Options		Response Percent	Response Count		
22	Special terms used to link search words together to define a		71.4%	60		
23	Synonyms derived from the "main idea" words or initial terms		16.6%	14		
24	Words that sound the same as your search terms, but have		11.9%	10		
25	A list of words that have the opposite meaning of the "main		0.0%	0		
26			answered question	84		
27			skipped question	5		
28						
29						
30	3. What questions should you ask when evaluating sources for use in your research? (Select all that apply)					
31	Answer Options		Response Percent	Response Count		
32	Who published the article?		100.0%	89		
33	What are the qualifications (authority) of the author or		100.0%	89		
34	What evidence is used to back up arguments?		100.0%	89		
35	How recently was the article published?		100.0%	89		
36			answered question	89		
37			skipped question	0		
38						
39						

4. Of the choices listed below, which one reflects the best list of possible keywords for the search term "poverty"?

Answer Options	Response Percent	Response Count
Lazy, begger, loser	17.4%	15
Destitute, indigent, poor	46.5%	40
Prosperous, rich, comfortable	1.1%	1
Depression, homeless, bankrupt	34.9%	30
	answered question	86
	skipped question	3

5. What does the term "peer reviewed" mean when applied to a journal article?

Answer Options	Response Percent	Response Count
That the article has been endorsed by the editor of the journal	6.0%	5
That the article has been read and evaluated by a group of	90.3%	75
That the authors friends have provided testimonials stating	3.6%	3
That everyone who subscribes to the journal has voted the	0.0%	0
	answered question	83
	skipped question	6

6. It is only necessary to provide a citation if you use a direct quote from a source.

Answer Options	Response Percent	Response Count
True	78.7%	70
False	21.3%	19
	answered question	89
	skipped question	0

7. Examine the definitions below and select the one that best defines the term "Plagiarism".

Answer Options	Response Percent	Response Count
Summarizing the ideas from the work of another person,	0.0%	0
Taking someone else's work or ideas and passing them off as	97.8%	87
Writing in the style of a well-known author for the purposes of	1.1%	1
Recording a potentially damaging conversation between two	1.1%	1
	answered question	89
	skipped question	0

8. Which of the following citations is the the correct one using MLA format for the electronic copy of the journal article "Stalking in the Twilight Zone: Extent of Cyberstalking Victimization and Offending Among College Students"?

Answer Options	Response Percent	Response Count
Reyns B, Henson B, Fisher B. Stalking in the Twilight Zone:	2.4%	2
Reyns, Bradford W., Billy Henson, and Bonnie S. Fisher.	54.1%	46
Reyns, Bradford W., Billy Henson, and Bonnie S. Fisher.	29.4%	25
Reyns, Bradford W., Billy Henson, and Bonnie S. Fisher.	14.1%	12
	answered question	85
	skipped question	4

9. What is the definition of a Research Question?

Answer Options	Response Percent	Response Count
Something you wonder about after making an observation.	7.9%	7
A question with no clear answer.	65.2%	58
A brief, clear, and testable question that serves as the point	27.0%	24
A long, detailed, and complex question that requires decades	0.0%	0
	answered question	89
	skipped question	0

10. What is a Primary Source?

Answer Options	Response Percent	Response Count
A source aimed at children in grades K - 6	0.0%	0
A source in which the author has self-published his or her	4.5%	4
A source that comments on, summarizes, or evaluates	19.3%	17
A source that has been written by the person who performed	76.1%	67
	answered question	88
	skipped question	1

CHAPTER 5

Enhancing Kuhlthau's Guided Inquiry Model Using Moodle and LibGuides to Strengthen Graduate Students' Research Skills

5.1 INTRODUCTION

Many faculty at the graduate level assume students have acquired the necessary research abilities to be successful in their coursework. Recent studies, however, illustrate that graduate students lack awareness of library resources and services (George et al., 2006; Hoffmann, Antwi-Nsiah, Feng, & Stanley, 2008). The need for academic libraries and librarians to provide opportunities for graduate students to improve their information literacy abilities is well documented (Baruzzi & Calcagno, 2015; Monroe-Gulick & Petr, 2012). Information literacy instruction (ILI) programs are generally not mandatory at the graduate level, and students require fewer course credits than undergraduate programs. This translates into even fewer opportunities for graduate ILI. Graduate enrollments in online distance education courses are rapidly growing, making it more difficult for librarians to provide face-to-face interactions that directly connect students with librarians or course-integrated IL instruction (Ismail, 2013).

The focus of the ethics, law, and technology course at Adelphi University was designed to give aspiring administrators a working knowledge of education law to support the rights of teachers, students, and parents and to prevent legal difficulties in the school environment. To increase students' legal knowledge, a research paper addressing a law topic in education was assigned as a major component of the course. To that end, a faculty—librarian collaboration was initiated, and a 50- to 75-minute library

Distributed Learning.
© 2017 T. Maddison and M. Kumaran.
Published by Elsevier Ltd. All rights reserved.

instructional session (commonly referred to as a "one-shot") was added to the course curriculum early in the semester to assist students with this assignment.

The faculty—librarian observations and other formative assessments during their own instructional session validated that most students had only a rudimentary knowledge of academic library resources and searching techniques. To address this issue, the authors investigated strategies to progressively develop and extend IL instruction to better meet students' needs. Recent studies however, indicate that a one-shot, librarian-led instructional session integrated into a course does not provide adequate time for students to understand, reflect, or explore resources effectively (Mery, Newby, & Peng, 2012; Porter, 2014). Using tools and resources in an online learning environment would offer the librarian opportunities to deliver and build information literacy abilities beyond the traditional classroom. The rationale for the redesign of the course is to demonstrate the application of a distributed learning environment in a graduate educational administration program employing technology to facilitate and extend learning beyond the classroom without place or time constraints.

5.2 REDESIGN RATIONALE

5.2.1 Limitations of a One-Shot

The 50- to 75-minute instructional collaboration between librarian and teaching faculty continues to be the method most used at Adelphi and receives the most attention in the literature (Blummer, 2009; Critz et al., 2012; Harkins, Rodrigues, & Orlov, 2011). However, several studies indicate that the knowledge, methods, and research skills necessary for information literacy to be viable cannot sufficiently be covered in a single library research class (Johnston & Webber, 2003; Martin, 2008; Reece, 2005).

Students entering graduate programs come with diverse levels of information literacy abilities, which pose their own challenges for course-integrated, one-shot instruction (Masuchika & Boldt, 2012). Instructors preparing a one-shot instruction session can never assume baseline competencies, which is the major disadvantage of this type of instruction. General fundamental skills must be assessed and reinforced before advanced research techniques can be introduced. The time required to cover the fundamentals leaves little if any time to teach advanced bibliographic search skills and strategies that are so necessary for graduate-level research (Catalano, 2010). In addition, the repetition and redundancy in these classes reinforces the misconception that the basic skills they possess are all they need to succeed and may contribute to overestimations of their research abilities. Students then may perceive library instruction as irrelevant to their specific research need (Gandhi, 2004).

The original one-shot for the ethics, law, and technology course was scheduled at the onset of the project to provide instruction and resources for students for their assignments. The session included several interdisciplinary reference tools for researchers and educators not familiar with the law and a variety of relevant article databases. Because the time to present and discuss tool content and other pedagogy is limited, as corroborated by Reece (2005) and Gandhi (2004), students were encouraged to contact the librarian after the session for additional assistance. Some student comments submitted during and after the first one-shot session indicated that the material covered was overwhelming. About one-third of the class contacted the librarian for additional one-on-one help in person, by phone, or by email. When their final papers were submitted, some student works cited lists that included reference sources and texts introduced in the early part of the session and a few included relevant journal articles. The overall lack of scholarly sources and articles in the reference lists demonstrated the students' inability to absorb all the instruction in one sitting. Martin's (2008) study suggests there is no difference in students' research behavior if there is one session or no session. Only multiple library sessions may prove more effective in changing students' information-seeking behavior than the one-shot approach. The results of a recent study conducted by Mery et al. (2012) comparing one-shot to an online credit bearing course "... indicate that [ILI] skills are complex, cognitively challenging skills that need repeated application and practice, ideally from a formal course" (p. 373). In addition, if coverage is interdisciplinary as it is in this case, it would prove even more difficult to cover with one-shot instruction.

Masuchika and Boldt's (2012) study exploring Cambourne's learning theory in relation to one-shot sessions discusses some drawbacks associated with two pedagogical approaches most frequently used with one-shot course integrated instruction. The critical mass approach emphasizes basic skills mastery for a variety of resources with little time for student searching. The user-centered approach introduces a few subject-related resources to provide time for students to practice and work with the librarian during the class. In the critical mass approach, the obvious disadvantage is that the time period allotted does not allow for in-depth coverage of resources, so there is insufficient time to focus on the complex research concepts or convey the necessary knowledge associated with graduate-level research (Bean & Thomas, 2010; Gandhi, 2004). Another disadvantage of the critical mass approach is information overload. Librarians try to convey as much as possible in this short time, and students become overwhelmed retaining little of what is presented (Bean & Thomas, 2010; Gandhi, 2004; Reece, 2005).

In the user-centered approach librarians try to devote enough time for active hands-on learning exercises. A disadvantage of this approach is that students often become more focused on the mechanics of searching than the results themselves, thus reducing information literacy to result-oriented search-and-find sessions that ignore the critical thinking skills required throughout the research process. The pedagogy attached to research methodology is often devalued in the quest to learn the tool or get materials. Masuchika and Boldt's (2012) article argues that the user-centered model is the better approach, although they concede that both approaches are insufficient for assessing students need, skill levels, or learning styles.

Cambourne's research suggests that the librarian must build time into the class to allow the librarian to better understand and address the needs of individual students by providing time for the students to try out the demonstration for themselves while the librarian observes and then recognizes and addresses difficulties as they arise in the students' efforts. Even for more seasoned researchers, it is important that the information provided addresses the needs of that researcher as specifically as possible (p. 294).

The Association of College Reference Librarians Framework for Information Literacy for Higher Education (2015) recognizes research as a complex endeavor in which students must repeatedly ask questions to deepen their knowledge base. Instruction at the graduate level needs to go well beyond the one-shot paradigm and requires more in-depth recursive instruction. The new framework also emphasizes reflection and critical thinking as a major tenet with relation to research. Instruction in relation to academic resources must move away from a hunting and gathering mode to a more reflective pedagogy of knowledge while guiding students through the research process. The authors recognized the complexity of research and the time needed to reflect during the research process. They sought to develop a program that extended the classroom environment and provide venues for asynchronous learning. Kuhlthau's information search process (ISP) supports a curriculum framework for a distributed learning environment.

5.3 CURRICULUM DEVELOPMENT FOR DISTRIBUTED LEARNING ENVIRONMENT

5.3.1 Kuhlthau's ISP Model

This case study uses Kuhlthau's ISP as a model for student inquiry. Kuhlthau's ISP provides instructors with an understanding of the learning stages individuals must experience to accomplish their information needs (Kuhlthau & Maniotes, 2010). Kuhlthau makes a strong case for faculty—librarian collaborations fostering positive and enduring research practices by acknowledging and confirming the thoughts, feelings, and actions that students are likely to experience during their research process. The usefulness of Kuhlthau's information-seeking model has been validated in numerous research studies that have included diverse library users and a variety of information environments (Kuhlthau, Heinström, & Todd, 2008). Kuhlthau's ISP supports information seeking as a process with different stages that take place over time. It is a comprehensive view of information-seeking behavior that identifies six stages of knowledge construction in the information-search process from the user's perspective. Each stage incorporates three areas of experience: cognitive (thoughts), affective (feelings), and physical (actions).

- Initiation, is when a person first becomes aware of a lack of knowledge or understanding, and feelings of uncertainty and apprehension are common.
- Selection, is when a general area, topic, or problem is identified, and initial uncertainty often gives way to a brief sense of optimism and a readiness to begin the search.
- Exploration is when background information is explored. In this stage, students must take a topic and focus the research. This transition often produces inconsistent, incompatible information, and uncertainty, confusion, and doubt frequently increase. Students find themselves in the "dip" of confidence.
- Formulation is when a focused perspective is formed and uncertainty diminishes as confidence begins to increase. Information they have gathered begins to make sense and they can begin to organize their ideas and research needs cogently.
- Collection is when information pertinent to the focused perspective is gathered and uncertainty subsides as interest and involvement deepens.
- Presentation is when the search is completed with a new understanding, enabling the person to explain his or her learning to others or in some way put the learning to use. (Kuhlthau, 2008).

5.4 KUHLTHAU'S GUIDED INQUIRY PROGRAM FOR INFORMATION LITERACY

Kuhlthau has also created a program for developing information literacy called *guided inquiry*, which is based on the ISP model. Guided inquiry engages students in the information-seeking process by using a concepts approach to information literacy, and it introduces students to information-seeking strategies that will assist them through a variety of sources. "Guided inquiry involves students in critically thinking about the research process. It encourages students to think of inquiry as a journey and to find a trail through the information" (Kuhlthau, 2008, p. 72).

By understanding these basic concepts students develop the lifelong information literacy skills needed for searching and using information wisely throughout their academic careers and personal lives.

The curriculum for educational law and technology was redesigned from the standard one-shot and extended to three sessions. Based on Kuhlthau's Guided Inquiry Model the faculty—librarian developed a new curriculum for ILI utilizing a distributive learning environment. The distributed learning environment offers the optimum setting for student research and inquiry by providing room for multiple sessions, learning and instruction asynchronously and by employing different modes of delivery beyond the classroom. Kuhlthau's Guided Inquiry Model focuses on successfully developing the ILI skills to critically think and reflect on the different information needs and motivations for creating information.

Kuhlthau's model contradicts the assumption that confidence steadily increases throughout the research process; central to her model is strategic intervention at times of uncertainty in the search process. Applied in an educational setting, guided inquiry provides the format for faculty and librarians "to recognize critical moments when instructional interventions are essential in students' information-to-knowledge experiences" (Kuhlthau et al., 2008; application in educational contexts, paragraph 7).

The faculty—librarian's instructional pedagogy supports the ISP in the Guided Inquiry Model, a program created to develop students' information literacy skills. Each instructional session was designed around various research concepts that students could apply in their own research. The standard one-shot library session was expanded to three face-to-face sessions (initial, midcycle, and near end). The three sessions integrated into the cycle were developed to correspond to a critical "dip of confidence" identified with ISP stages when students would benefit most from support and intervention. The librarian and ILI support was now available throughout the course cycle and research project.

Based on the literature, student feedback, and other formative assessments, the faculty—librarian's collaborative redesign was created to address one-shot issues to better meet students' needs. In addition to the multiple sessions, a distributed learning environment was created to supplement the classroom sessions and allow students to engage in an asynchronous online environment in which they could have access to the librarian, resources, or instructional materials as their needs arose.

5.5 ONLINE ZONES OF INTERVENTION

In a recent article, Ina Fourie (2013) encourages librarians to employ innovative methods for incorporating Kuhlthau's zones of intervention using information technology to align the information-seeking process with students. The zone of intervention in information seeking is that area in which information users can do with advice and assistance with what they cannot do alone or do only with difficulty. Intervention within this zone enables individuals to progress in the accomplishment of their task. Fourie (2013) suggests interventions described by Kuhlthau, including support with locating information, identifying a group of relevant sources, advice on a pattern or particular order for use with relation to resources, and counseling. The Moodle course management system (CMS) and LibGuides provided the venues for the faculty—librarian to provide some of Kuhlthau's zones of intervention.

5.6 TECHNOLOGY

5.6.1 Moodle CMS

CMSs offer unique online opportunities for the librarian to enhance and extend one-shot instructional sessions. Collaboration between faculty members and librarians for face- to-face one-shots—even if they are assignment specific—is often minimal because "the librarian is an invited guest with little input in designing the assignment so it is difficult for librarians to understand how the assignment fits into the course" (Rabine & Cardwell, 2000, p. 326). In a CMS such as Moodle, which is used at Adelphi, librarians can be invited as participants (like students, they are included into the course and can post online) or they can have a more permanent online presence as an instructor (which allows them to add content). In both cases, librarians can have more access to course content, thus allowing them to become more familiar with faculty goals and student needs, which lead to librarians being more involved in the course. The faculty member in this case study included the librarian as a co-instructor in Moodle so that she had full access to course content, and the ability to provide support at critical times throughout the cycle was available.

Discussion forums are an integral part of CMS. Creating a library research forum offers students a venue to ask questions at point of need during their research project so they can receive guidance and support outside class through the flexible online system. Students could now learn from peer postings and librarian responses. In addition, as a co-instructor the librarian has access to the valuable discourse in other discussion forums as well, which were an integral part of the hybrid class. The 24/7 availability of these forums allows students to ask questions and review materials at their own pace and in their own time. Students have more time to reflect after each session and build a deeper understanding and knowledge of the research process, which leads to independent learning. Recent articles in the library field have explored the positive impact associated with librarians integrated into online delivery using a CMS such as Moodle and Blackboard (Bowen, 2012; Schulte, 2012).

5.7 LibGuides

The librarian can also extend the face-to-face session for the students by posting guides with additional handouts, tutorials, external links, and so on within the CMS course materials. Library research guides that connect patrons to resources have been a mainstay in libraries for years. When academic libraries created websites on the Internet, many libraries digitized their guides as well. Since its inception in 2007, LibGuides has become the dominant platform in most academic libraries (Bowen, 2012). The guides allow librarians to organize and share materials virtually with their patrons. While LibGuides can take a variety of forms, the approach the librarian—faculty chose to use for educational law and technology was to build a customized guide. A customized guide includes hyperlinks and persistent links to numerous resources specific to the subject matter relevant to the course, including, catalog entries, ebooks, databases, information literacy tutorials, citation guides, and many other useful tools for research. The librarian can offer access to the guide through their library website or through the course CMS. The LibGuide for this course reinforces concepts covered in instruction sessions to guide students' inquiry progressively through the research process. The guide includes relevant resources and online tutorials for additional research help. Students in the education law class could link to this LibGuide from their CMS or go directly to the guide through the library website. The LibGuide contains seven pages accessible through tabs (home, background information, books, databases, journals, web evaluation, and citations). Each tab refers students to a dynamic page useful for their research. The instruction in the face-to-face classroom session is reinforced by repetition in an online format.

5.8 INFORMATION LITERACY SESSIONS OVERVIEW

Session One: Overview and Research Topic Selection. This coincides with the ISP initiation and selection phase.

Session Two: Web Evaluation for Law-Related Websites. This website evaluation skills sheet was included to integrate the website portfolio project, which was a developmental step for the students' research assignment. This coincides with the exploration and formulation phase.

Session Three: Database Searching. This coincides with the collection and presentation phase.

5.9 SESSION ONE: TOPIC SELECTION

The first and second stages of Kuhlthau's ISP—initiation and topic selection and early exploration—are coordinated to coincide with the content in the first module. The faculty member distributes the research assignment, including a list of suggested topics and the option to choose a topic, via Moodle before the first face-to-face class. Students are encouraged to discuss their topic with the librarian via a discussion forum prior to the library session.

Students' topic selection may vary from too broad or too specific to having no idea what they want to research. This interaction affords the librarian an opportunity to brainstorm topic ideas with students and acknowledge their apprehension. By the time they arrive at the first session, a general topic is settled on and students largely feel encouraged and optimistic about starting their search. The librarian also has an idea of the topics and context of students' research. This formative assessment enables the librarian to direct specific resources to individual students' research and establishes the librarian's relationship within the course from the onset.

In the first session, the students are introduced to the LibGuide created for this course and how to use it for their assignments. The first session focuses on the home page, background information, and books tabs within the guide. Critical to a legal research paper was explaining primary (case law, statutory, and regulatory law) versus secondary materials (law review articles, legal treatises, and other interdisciplinary journal articles) in education and legal research and how they apply to this assignment. Concepts covered in this session include review of the assignment and discussion about student and faculty perceptions and expectations. Both faculty member and librarian stress the benefits of using the selected library resources in the guide for academic research. The librarian provides an introduction to sources such as subject encyclopedias and other resources supplied under background information.

Using some of the preliminary topics suggested by students, the librarian then demonstrates how to use the reference tools discussed for background information, topic exploration, and identifying subject-specific vocabulary and concepts relevant to the topic. Students are then encouraged to use the LibGuide in class to explore topics for their research. To scaffold this session, the librarian inserted various video clips on how to define a topic and developing a research question that reinforces the concepts covered. In addition, a worksheet was inserted into the LibGuide to assist students with their ISP. Students are given a worksheet to develop their own research focus before the next session. This worksheet is also available on the LibGuide. The students are expected to complete the worksheet and contact the librarian with any questions or problems before the second session.

5.10 SESSION TWO: WEBSITE EVALUATION AND SCHOLARLY SOURCES

The second session was designed to coincide with the second and third stages of the ISP model (exploration and formulation). During the previous semester's one-shot sessions, the librarian and faculty both observed that students seemed to suffer from information overload. They also observed that students had difficulty with the formation of a research focussed question and the development of workable search queries. According to Kuhlthau, this proves to be the most challenging task for students and is noted as a pivotal point in much of the literature on ISP. "It is at this stage that the individuals are at the greatest risk of losing momentum and reverting to bad habits" (e.g., relying only on Wikipedia or other non—peer-reviewed sources) (Mortimore, 2010, p. 5). Before the second session, the faculty member reminds students through Moodle to continue to work with the librarian on their projects through a research forum established online. Both the librarian and faculty member continue to stress possible pitfalls and problems students may encounter at this stage. The students are encouraged to complete the worksheet from session one, and the librarian continues to brainstorm ideas and explain and develop search strategies.

Both faculty and librarian answer questions and acknowledge students' feelings related to their projects, offering support where needed during the course of their individual research. The faculty—librarian realized that students often lack the critical skills necessary to extrapolate useful information when they find it. For example, they try to find information exactly on point and get discouraged when they do not find the perfect article or find information that may seem incompatible or inconsistent with their focus. Often they discard important information or related articles in search of an elusive source they believe to have the answer to their research question. Students are again referred to the LibGuide and the Journals tab, which allows them to browse articles from relevant journals and learn how to locate articles from citations.

The latter half of this second session is scheduled to coincide with an upcoming website assignment related to their research papers, so content for the session was designed to emphasize credibility and reliability of sources in all formats. Previous papers submitted by students indicated they did not clearly understand how to evaluate their resources effectively, especially websites. Concepts covered for this part of the session include popular versus scholarly sources, free web versus library resources, scholarly publishing cycles, authorship, bias, audience, domain and publication origin, documentation, and annotated versions of primary sources in legal research. To scaffold this session, the students are guided through the Web Evaluation tab on the LibGuide, and the worksheet with Criteria for Evaluating Law-Related Websites is distributed as a guideline for this project. The discussion forum for the research project was also continued.

5.11 SESSION THREE: DATABASE SEARCHING

The third module was designed to correspond to stages four and five of the ISP model: formulation and collection. To complete their research assignments, students must not only have a topic but also a well-formed aspect, idea, point of view, or perspective associated with the topic and construct their understanding of the information they encounter around that concept. Prior to this class and through the discussion forum, students are encouraged to move forward with their research. They are given guidance on search strategies and database selection.

In the third session, the librarian reviews the content of the Databases tab in the LibGuide. Students work on the research worksheet as they move forward and continue looking for research materials. The librarian and faculty member repeat information from previous sessions online and in class. At this point in their research, students are again encouraged to locate and use law reviews or other scholarly articles relevant to their research. Both the faculty member and librarian communicate to students that scholarly articles can be found using the library databases listed in the LibGuide.

This final session is more individualized and student-centered. The librarian elicits questions from students regarding their own research focus and asks them about their database search techniques and their progress. Using students' questions and answers, the librarian demonstrates database searches and points out commonalities that can be employed by all. The librarian and faculty member found students to be highly engaged at this stage. Students asked specific questions about their searching needs and about information they found. They also asked about citing sources they were using. This was extremely gratifying because many of the questions posed dealt with several misconceptions the faculty—librarian had observed in previous sessions. This session proved to be a real teaching moment for the faculty and librarian.

5.12 ASSESSMENT MEASURES

5.12.1 Postsession Survey

A pre- and post course questionnaire was created to survey students' perceptions regarding the library research sessions and online tools. An additional space was added for student comments regarding the sessions. Comments from the postsurvey with student's permissions are listed below:

> I think the course offered help like no other. It would be completely our fault if we didn't know how to research, since they were available all the time. They would also always reach out and always tell us they were there waiting for us to reach out. I felt supported.

> Very helpful! I suggest having a law librarian embedded in all Ethics law and technology classes.

> We got hands on individual attention on exactly what our topics were, and how to find information. The one on one attention was very helpful and made it less stressful on the students.

> I found the listing of resources in the research worksheet to be helpful but did not really use the questions.

> The LibGuide helped with citations, provided databases, and most importantly provided school law books where a lot of worthwhile information was found.

> The LibGuide helped in every way-gave me ideas of what to look for and helped with entering search words.

> The LibGuide had a multitude of sources available at any time, and gave strategies and tutorials on all things necessary for the research project.

The comments indicated that students employed the technology and felt more comfortable and better prepared to pursue their legal research.

5.13 MOODLE AND LibGuide ANALYTICS

Moodle and the LibGuide provided data that supported a librarian—faculty scaffolding approach for Kuhlthau's Guided Inquiry Model developed for this program. Through Moodle's Ask the Librarian discussion forum, the 6 students enrolled in this last class contacted the librarian with research inquiries 14 times. The dates for the inquiries occurred throughout the course cycle. This demonstrates that students used the technology at various times during their research. Four inquiries using the Moodle discussion forum took place before the second session. The majority of questions coincide with the Kuhlthau's research on students' ISP. It is during the exploration and formulation phase that students might experience a dip in their confidence when researching. The majority of questions posed through the discussion forum took place after session two when students sought more direction with their research. In addition, the analytics provide statistics on the specific tools or resources that users clicked on within a given page.

The LibGuide's platform enabled the authors to view a usage chart that lists how frequently students viewed various tabs and resources. The usage chart statistics show the number of hits for each tab and how many times resources available within the tab were accessed. The relevant books, databases, and other resources tabs introduced in class sessions were clearly used by the students for their research papers.

The administrative tools also creates graphs that show when users access pages within the guide. From the graphs, it is evident that students used the LibGuide throughout the cycle as needed, especially during and immediately after instructional sessions.

5.14 CONCLUSION

This case study demonstrates the benefits of a faculty—librarian partnership using a collaborative model within the course curriculum to benefit students' research skills. Because of their association with this course, the librarian and faculty member have developed a strong partnership that has been invaluable in assisting the faculty member develop assignments. The librarian is no longer a "guest lecturer" but an integral part of the coursework. By including legal research in an education law, ethics, and technology course, aspiring administrators who have limited familiarity with scholarly and academic resources learned how interdisciplinary resources can be used.

Before this study, students in the ethics, law, and technology course (with a varying population of 6—10 graduate students) would depend on popular search engines for research sources in their legal research papers. Because of the distributed learning environment implemented in this case study, the additional meetings supplemented with online interventions gives the librarian more opportunities to work with students and develop an in-depth knowledge of students' research strengths and weaknesses that was not afforded in one-shot sessions.

The authors recommend that some additional activities be included during the collection and presentation phase of the ISP model to develop and elicit higher-order critical thinking skills. One activity assigns students to create their own outlines for their papers based on a sample research paper. Next, the topic of research and database worksheets should be required rather than be voluntary and used in conjunction with student outlines. As the students' research progresses, the faculty member and librarian instruct students to review and revise the outlines and worksheets. This would provide a research and writing log that would enable the librarian and faculty member to assess students' progress. The outline and worksheets help illustrate to the students the connection between searching and making meaningful connections as they research.

Based on an examination of the students' final research papers, there was improvement in the quality of the resources used. Requiring students to use scholarly sources increased the use of scholarly journal articles in their papers. The research paper content also integrated appropriate legal cases. Based on data collected from the LibGuide, there was an increased use of the LibGuide following all sessions. Students used the technology available and accessed the librarian and resources online. For any modality, the distributive learning environment offers users avenues for independent learning by connecting the library, librarian, faculty members, and students whenever needed.

REFERENCES

Baruzzi, A., & Calcagno, T. (2015). Academic librarians and graduate students: An exploratory study. *Portal: Libraries & the Academy, 15*(3), 393−407. Retrieved from <http://www.press.jhu.edu/>.

Bean, T. M., & Thomas, S. N. (2010). Being like both: Library instruction methods that outshine the one-shot. *Public Services Quarterly, 6*(2), 237−249. Available from <http://dx.doi.org/10.1080/15228959.2010.497746>.

Blummer, B. (2009). Providing library instruction to graduate students: A review of the literature. *Public Services Quarterly, 5*(1), 15−39. Available from <http://dx.doi.org/10.1080/15228950802507525>.

Bowen, A. (2012). A LibGuides presence in a blackboard environment. *Reference Services Review, 40*(3), 449−468. Available from <http://dx.doi.org/10.1108/00907321211254698>.

Catalano, A. J. (2010). Using ACRL standards to assess the information literacy of graduate education students in an education program. *Evidence Based Library and Information Practice, 5*(4). Retrieved from <http://ejournals.library.ualberta.ca/index.php/EBLIP>.

Critz, L., Axford, M., Baer, W. M., Doty, C., Lowe, H., & Renfro, C. (2012). *Development of the graduate library user education series*. Available from <http://dx.doi.org/10.1108/00907321211277341>

Fourie, I. (2013). Twenty-first century librarians: Time for zones of intervention and zones of proximal development? *Library Hi Tech, 31*(1), 171−181. Available from <http://dx.doi.org/10.1108/07378831311303994>.

Gandhi, S. (2004). Faculty-librarian collaboration to assess the effectiveness of a five-session library instruction model. *Community & Junior College Libraries, 12*(4), 15–48. Available from <http://dx.doi.org/10.1300/J107v12n04_05>.

George, C., Bright, A., Hurlbert, T., Linke, E. C., St.Clair, G., & Stein, J. (2006). Scholarly use of information: Graduate students' information seeking behaviour. *Information Research, 11*(4). Retrieved from <http://informationr.net/ir/11-4/paper272.html>.

Harkins, M. J., Rodrigues, D. B., & Orlov, S. (2011). "Where to start?". Considerations for faculty and librarians in delivering information literacy instruction for graduate students. *SLA Practical Academic Librarianship: The International Journal of the SLA Academic Division, 1*(1). Retrieved from <http://journals.tdl.org/pal>.

Hoffmann, K., Antwi-Nsiah, F., Feng, V., & Stanley, M. (2008). Library research skills: A needs assessment for graduate student workshops. *Issues in Science and Technology Librarianship, 53*, 1.

Ismail, L. (2013). Closing the gap. *Reference & User Services Quarterly, 53*(2), 164–173. Retrieved from <https://journals.ala.org/rusq>.

Johnston, B., & Webber, S. (2003). Information literacy in higher education: A review and case study. *Studies in Higher Education, 28*(3), 335–351. Available from <http://dx.doi.org/10.1080/03075070309295>.

Kuhlthau, C. C. (2008). From information to meaning: Confronting challenges of the twenty-first century. *Libri, 58*(2), 66–73. Retrieved from: <http://cissl.scils.rutgers.edu/research/imls/> http://citeseerx.ist.psu.edu/viewdoc/download?doi=10.1.1.597.3068&rep=rep1&type=pdf>.

Kuhlthau, C. C., HeinstrÖm, J., & Todd, R. J. (2008). The 'information search process' revisited: is the model still useful? *Information Research, 13*(4). paper 355. Available at <http://InformationR.net/ir/13-4/paper355.html>.

Kuhlthau, C. C., & Maniotes, L. K. (2010). Building guided inquiry teams for 21st-century learners. *School Library Monthly, 26*(5), 18–21.

Martin, J. (2008). The information seeking behavior of undergraduate education majors: Does library instruction play a role? *Evidence Based Library and Information Practice, 3*(4), 4–17. Retrieved from <https://ejournals.library.ualberta.ca/index.php/EBLIP/index>.

Masuchika, G. N., & Boldt, G. (2012). One-shot library instruction and cambourne's theory of learning. *Public Services Quarterly, 8*(4), 277–296. Available from <http://dx.doi.org/10.1080/15228959.2012.730394>.

Mery, Y., Newby, J., & Peng, K. (2012). Why one-shot information literacy sessions are not the future of instruction: A case for online credit courses. *College & Research Libraries, 73*(4), 366–377. Retrieved from <http://www.ala.org/nmrt/>.

Monroe-Gulick, A., & Petr, J. (2012). Incoming graduate students in the social sciences: How much do they really know about library research?. *Portal: Libraries & the Academy, 12*(3), 315–335. Retrieved from <http://www.press.jhu.edu/>.

Mortimore, J. M. (2010). Making research make sense: Guiding college students into information literacy through the information search process. *The Southeastern Librarian, 58*(3), 3–13. Retrieved from <http://www.selaonline.org/sela/publications/SEln/issues.html>.

Porter, B. (2014). Designing a library information literacy program using threshold concepts, student learning theory, and millennial research in the development of information literacy sessions. *Internet Reference Services Quarterly, 19*(3), 233–244. Available from <http://dx.doi.org/10.1080/10875301.2014.978928>.

Rabine, J., & Cardwell, C. (2000). Start making sense: Practical approaches to outcomes assessment for libraries. *Research Strategies, 17*(4), 319—335. Available from <http://dx.doi.org/10.1016/S0734-3310(01)00051-9>.

Reece, G. J. (2005). Critical thinking and cognitive transfer: Implications for the development of online information literacy tutorials. *Research Strategies, 20*(4), 482—493. Available from <http://dx.doi.org/10.1016/j.resstr.2006.12.018>.

Schulte, S. J. (2012). Embedded academic librarianship: A review of the literature. *Evidence Based Library & Information Practice, 7*(4), 122—138. Retrieved from <http://ejournals.library.ualberta.ca/index.php/EBLIP>.

CHAPTER 6

A Model for Teaching Information Literacy in a Required Credit-Bearing Online Course

6.1 INTRODUCTION

Excelsior College (EC) and the Entrepreneurial Library Program (ELP) of the Sheridan Libraries at Johns Hopkins University have been partners since 2000. The EC Library is a full service, online library managed by the ELP librarians. The importance of information literacy instruction was recognized early on; as such, a customized information literacy course (INL102) was created at EC's request in 2001. At this time, the librarians worked in collaboration with an administrator at EC to discuss how to tailor INL102 for adult learners in an online environment while addressing the needs of students with different learning styles. In the initial creation of INL102, many of the learning exercises and videos were based on the Texas Information Literacy Tutorial (TILT), "an educational Web site designed to introduce first-year students to research sources and skills" (TILT, 2015). The quiz and final exam questions, which are both multiple choice, were created by the library team using the guidelines set forth by the assessment unit at EC.

The mission and the strategic plan of the Library are designed to fully support the College's mission and strategic plan, which helps with faculty and administrative support. EC's mission is to "provide[] educational opportunities to adult learners with an emphasis on those historically underrepresented in higher education. The college meets students where they are" (Excelsior College, 2015).

The library team supports EC's mission by meeting the students where they are—both geographically and academically. Through the years, the team has enhanced and modified content in INL102 to include more

Published by Elsevier Ltd. All rights reserved.

multimedia, content created by the librarians, transcripts for all videos, and updates that reflect the changing landscape of information literacy.

In addition, a free library-created information literacy tutorial is available to all members of the EC community on the library website; the tutorial is designed to supplement learning before or after the course. It covers the same main concepts that are taught in the online course so that learners can prepare for INL102 by referring to content before signing up. Conversely, once students complete INL102, they can return to the tutorial for a refresher as needed. Faculty members also make use of the information literacy tutorial by linking to it at a point of need in their courses.

Taking and passing an information literacy course is one of EC's general education requirements, and INL102 is the recommended option for fulfilling this one-credit requirement. Mayer and Bowles-Terry (2013) describe how a credit-bearing information literacy course is unique, but making a course credit bearing is an essential factor needed to make a course mandatory. In their article, Mayer and Bowles-Terry explain how completing a course of this nature is part of improving a student's overall grade point average (GPA).

INL102 is designed to be completed independently with a self-paced study module structure within the Blackboard learning management system (LMS). The course provides a broad overview of information literacy concepts organized into five distinct modules. These modules introduce skills for locating, using, and evaluating resources, as well as the legal and ethical uses of information.

To provide a layer of customization to this course, the librarians aligned the college's information literacy goals and outcomes to coincide with the EC curriculum and the Middle States Commission on Higher Education's accreditation requirements, as published in its online report (Middle States Commission on Higher Education, 2016). Additional customization is found in module outcomes that link directly to EC's academic honesty and electronic use policies. Wording in each module outcome is designed for the EC student population; the librarians work closely with EC staff to craft a succinct set of outcomes. Lastly, the course adheres to the current ACRL Framework for Information Literacy for Higher Education.

One of the best ways to ensure that INL102's outcomes are met is by collaborating closely with an EC team of educators, including instructional designers, faculty members, and administrators. These relationships are essential in maintaining the support that ensures that the scaffolding of information literacy concepts occurs in assignments across all disciplines and levels. Many librarians often experience some level of resistance to deep collaboration, but headway is often made over time with marketing and by gaining the trust of the faculty. As noted in the article by Jennifer Murphy (2013), the librarian's role in teaching information literacy should be considered on par with other faculty on campus. With a strong working partnership, outside team members will encourage students to take INL102 early in their college career, which has been found to help students achieve a higher GPA as evidenced and referenced in the section of this chapter titled "Measure of INL102's Success."

Some of these best practices incorporated into the model were gleaned from the literature, but most of them were perfected based on student feedback and personal firsthand experience over the years. This firsthand trial-and-error experience was a key factor in tailoring INL102 to meet the specific needs of EC students. Integral parts of the model that make this online self-paced course a success are integrating technology into INL102, the availability of customized library resources, how student success is measured, and finally the schedule followed for updating the course. While reading about this model, please note that as a part of the partnership, the authors have agreed to share only certain information, which precludes specific details about the work being done surrounding assessment, statistics, and evaluation tools from being included in this chapter.

6.2 MEASURE OF INL102's SUCCESS

Through the assessment unit at EC, a five-year pattern analysis was conducted to determine the impact of INL102 on graduates' GPAs. The study compared students who completed and passed the required INL102 information literacy course to students who had not taken INL102 before or who transferred in the information literacy credit from another institution. With a 5% GPA increase, it was clear that those students who completed and passed the customized course were better prepared to complete the rest of their EC coursework (Ostrowski, 2014). In addition to this concrete data, the library team receives a significant amount of anecdotal comments via phone and email as well as through course evaluations that confirm how this class has helped improve success in other courses completed at EC.

The significance of the difference in scores among these groups proves both the value of requiring an information literacy course and the value of a course customized to a specific school's need. As noted by Sanborn (2015), online education allows "for a flexible and responsive digital education that adjusts to student learning ... and it allows for a truly customized, individualized, supportive educational experience" (p. 97). The student population of EC is nontraditional, working full time and often juggling careers, family life, and school life. Adult learners often need extra assistance with technology and guidance through general information literacy concepts, and they often must have information placed strategically at the point of need throughout the course (Birdsong & Freitas, 2012). The combination of this in-house study, scholarly articles, and firsthand experience working with students shows that the course is providing a valuable service to students at EC.

6.3 INTEGRATING TECHNOLOGY

Research indicates that students learn best in a variety of ways. After completing a study on learning styles, Dr. Darren C. Wu concluded, "educators should saturate [distance education] courses with as much variance as feasible given technology and cost limitations to account for possible learning style differences" (2014, p. 123). For this reason, information is delivered in a variety of manners in INL102, a top priority for the EC library team. In addition, much of what is done to meet the needs of different learning styles coincides with supporting the needs of students with disabilities. INL102 includes audio recordings of module outcomes, videos showing learning outcomes in action (along with transcripts), and visuals that are all intended to make sure that all students have the chance to learn the material in a way that suits their needs. As Mestre's research suggests, "[students] prefer that the learning objects include both images and sound, are visually engaging, and are available at point-of-need" (2010, p. 827).

6.3.1 Audio

Audio is created for certain sections of the course so that students can listen to what they are required to read. This audio can be played while reading along with the content, or it can be played and listened to on its own as an alternative to reading the content. As updates are made to INL102, the library team aims to have audio for every section of the course, but for now it was determined that it is most important to have audio for each module's introduction. Historically, Jing was used for recording the audio so that it would be paired with a video of the text being read. These files were hosted on Screencast and allowed students to submit comments. The librarians received an enormous amount of positive feedback on this feature from students, so it was determined that the time investment is worthwhile.

With current updates, the library team has started using SoundCloud to host audio, which is recorded using Audacity. The librarians found this to be an ideal setup because Audacity is open-source software, and SoundCloud allows for a free subscription with unlimited listening bandwidth. Such unlimited bandwidth is of particular importance to the course because hundreds of students could potentially be listening simultaneously during any term. In addition, SoundCloud allows users to make audio private unless shared with a distinctive URL, which is useful for controlling who has access to the information in a course. Now, as updates are done module by module, the SoundCloud player will be embedded into each section so the student can easily listen and read along in the LMS.

As Drouin, Hile, Vartanian, and Webb (2013) found, another benefit to including audio in INL102 is that it gives a human connection to online students who often crave this. The audio lends itself to creating a sense of community that is essential to student success and persistence.

6.3.2 Video

Videos are a particularly useful tool for students in an online asynchronous self-paced course. INL102 has no instructor or discussion boards, and students work on the course at different times during the term. While videos are vital for auditory and visual learners, they are also helpful for any student who may not be comfortable with self-directed learning. Videos are used to help explain more complex topics that lend themselves to additional explanations and visual illustrations. For example, when working through the section on Internet basics, students are provided with a video that explains how the Internet works, which can be an abstract topic when first introduced. Use of the video format allows for greater comprehension of the topic when explained with visuals, comparisons, and examples.

The vast majority of videos are sourced from the library's collection. The Excelsior College Library subscribes to multiple video databases, which allows for embedding materials at the point of need in the course. All of the videos are closed captioned, which is useful for all students but essential for students with disabilities. Videos also can be produced if one does not exist that meets the learning outcome. The librarians use a basic Captivate template that has been customized with the EC branding. Occasionally, a video found on the free web will be included, but generally it is best not to rely on these sources as they can disappear without any warning. This method of including videos is advantageous for some librarians because it requires little to no production skills and minimal additional cost but has a positive impact for the learner.

6.3.3 Visuals

In the context of this chapter, the term *visuals* encompasses all still images. For INL102, these include charts, photos, infographics, and any other type of visual explainer that enhances a topic. Just like with videos, one should first try to find something suitable from the library's collection and then create what is still needed to support learning outcomes. Also, it is possible to find exceptional visuals from the web to incorporate into a course. One example of a visual that was created for INL102 is a chart comparing three major search engines, which highlights their similarities, differences, and tips for performing searches.

6.4 LIBRARY STAFF AND COLLECTION AS A RESOURCE

Typically, more than 1000 students enroll in this self-paced course per term (six terms a year). Students can get support in this course using the synchronous and asynchronous options of contacting library staff, accessing the library's knowledge base, and referencing the INL102 course guide.

6.4.1 Library Staff as a Synchronous and Asynchronous Resource

As the managers of INL102, the EC Library staff answers questions related to the course throughout the day. These questions run a broad spectrum ranging from simple factual answers to talking students through topics they are struggling to comprehend. Sometimes students are looking for encouragement that assures them they are on the right track. Having a person they can turn to with specific questions and concerns makes them feel less isolated and helps with persistence; as noted by

Sutton (2014), "the greater the level of academic and social integration, the greater the student's chances of persisting until graduation" (p. 6).

Whatever the students' needs, the library staff is there to support them. Through experience, the library team has learned that students in a self-paced online course must have a person they can turn to when they have questions or concerns. Students can reach library staff members via email, fax, phone, chat, and Springshare's FindAnswers 24/7 interface. Most questions come in over the phone or via email. During business hours, the majority of phone calls are answered immediately or within a few hours at worst. Library staff members have a commitment to respond to written questions within 24 hours (excluding weekends). This commitment to timely turnaround gives students confidence in the library's assistance and helps them continue through the course in a timely manner. As one can see from this model, librarians do not have to reinvent the wheel when branching out into managing an online course; instead, they can provide support to students taking a self-paced credit-bearing information literacy course via their everyday reference services.

6.4.2 Library Collections as a Synchronous Resource

Because EC is an online college, the majority of students are adult learners who are completing their coursework in addition to many other responsibilities. EC students' lifestyles necessitate the need for flexibility in how we offer assistance. With EC students' schedules in mind, the librarians create as many materials as possible to assist students when the library is not open. The EC Library uses Springshare's LibAnswers knowledge base and LibGuides software to create resources that students can access at the point of need. According to Springshare, 4800 libraries are utilizing its services, which shows how pervasive these tools are in the library community and suggests that many libraries could comfortably copy this model.

When an applicable guide or frequently asked question (FAQ) resource is created, it is linked to the course to ensure that it is easily accessible by students. These resources are also discoverable via the library's webpage, which not only makes them more visible but also allows them to be a resource for students preparing for INL102 before enrolling. Currently, the library's knowledge base contains 15 FAQs created in response to the needs of students taking this specific course. The most popular FAQs with thousands of views is the set of questions specific to the INL102 course.

Examples of FAQs include "How do I access the syllabus?" and "Can you give me tips to help me for prepare for the final exam?" These FAQs also help guarantee consistency in how library staff members answer inquiries because they act as a standardized response. A persistent link to this FAQ set is on the homepage of INL102.

In addition to the library's knowledge base, the librarians have used LibGuides to create an INL102 course guide. This guide serves as a "one-stop-shop" for information about the course. It includes testimonials from previous students explaining how the INL102 helped them, a course outline, the syllabus, and a link to the FAQ set in the knowledge base. Even though this guide was created less than a year ago, it is already one of the most visited course guides in the Excelsior College Library's collection.

6.5 ASSESSMENTS

EC has a published set of assessment guidelines that all courses are recommended to follow when creating multiple-choice exam questions. For example, the words *never* and *always* should not be included as choice options. In addition, each answer selection must be plausible but not possible. This design makes it clear that there is only one correct answer, which is obvious to the student who has comprehended the material. The library team works collaboratively with the assessment unit to ensure that the standard protocol is followed. In addition, INL102's content is evaluated using voluntary student evaluations at the end of each term.

6.5.1 Design of Assessments

The first step of the assessment process was to design a 10-question multiple-choice quiz per module based on each module's learning outcomes. Then three versions of the final exam were created. Each the three sections of INL102 that run simultaneously has its own final exam, a 50-question multiple-choice exam that encompasses all module concepts without duplicating any of the quiz questions. Finally, each module has at least one ungraded question where students can check in and test their knowledge as they work through the content. Multiple-choice questions were chosen as the style of assessment not only because this is a self-paced course but also because the LMS allows for instantaneous grading of all exams. In fiscal year 2015, this course had a range of approximately 1100—1700 students per term, so this grading method allows for this large number of students to get immediate feedback on their progress throughout the course.

This one-credit course is pass—fail and requires a score of 70% or above on the final exam to be considered passing. The module quizzes must be taken sequentially, and the student must score a 70% or above to be able to move on to the next quiz. The quizzes can be retaken as many times as the student chooses to help practice with the content. The quiz score totals are not factored into the student's final grade for INL102, but they do provide students with an idea about what they can expect for the final exam. Another aspect of the quizzes that helps the student learn the material is the quiz feedback provided after submission. Any incorrect answer on the quiz generates a pop-up text box that provides feedback and hints to the correct answer.

6.5.2 Managing Assessments

While working with the director of the assessment unit, the library team learned best practices for managing the course's assessments. Using the LMS reporting feature allows for psychometric reports to be downloaded and mined for data. Librarians and EC staff can use these reports to review the reliability of each final exam question. Each question is given a percentage weight based on how many of the students in that section got it correct. For example, if all students in a section get the same final exam question correct, then that question is considered to be too easy and must be rewritten. Consistently, these reports speak to the validity of assessments in that they reflect the assessment unit's standards such as falling within the unit's acceptable bell–curve range.

6.5.3 Mapping Assessments

For the final exam and module quizzes, a map of the main concepts from INL102 was created using an Excel spreadsheet. The spreadsheet outlines the question and the concept it refers to in each assessment item. This map ensures all concepts are covered, and it clearly communicates to internal staff where to find specific assessment questions if changes need to be made in the future (see Fig. 6.1).

6.5.4 Course Evaluations

An optional course evaluation is available for students after they complete INL102, and at the end of each term all student evaluations are downloaded

Figure 6.1 Concept assessment map of Module 1.

to review the responses from the 20-question set. The evaluation questions review various aspects of INL102 such as how the student values the course content and suggestions for improvement. The librarians compile the responses and group them according to themes so that notes can be made about areas of improvement for future edits. For example, over the past few years one overarching theme has been a request for more audio to be added, so it became a priority to improve this aspect of the course. These evaluations reflect the informal feedback received by phone and email, which shows that the majority of students feel INL102 has helped them learn concepts central to success in their studies and careers.

6.6 SCHEDULE FOR UPDATES

Keeping courses current is essential, and it is best to have a plan in place that addresses how an updated course schedule can be maintained. The course consists of five modules, and each module contains large amounts of information in various formats. While changes are made as needed for immediate updates such as broken links or typos, the vast majority of changes are saved for scheduled revisions throughout the fiscal year. Waiting to complete major revisions is necessitated by the design of assessments and the fact that the course runs every term. The library team is conscientious about how any change to the content could affect learning outcomes, how it relates to a quiz or exam question, or how it will effect students currently taking the course.

Initially, the whole team undertook updating the entire course in a fiscal year, which was a challenge for a small library team due to the sheer volume of content in the course. Maintaining a course makes up a fraction of a librarian's overall schedule, but an entire course revision is a time-intensive process. After learning from this experience, the team decided to try updating the course one module at a time with a small working group for one fiscal year. This idea was tested by revising module one of the course with two to three librarians during 2015, and it proved to be a more manageable method. It also allowed for more than one librarian to offer expertise on a module, which led to a higher-quality finished product.

Through trial and error, the team discovered the best revision schedule for our need. Where the team had been updating the entire course once every 5 years, the librarians now do one module a year. A recommendation is to keep in mind the importance of having a revision schedule in place for efficiency and continuous updates. Otherwise, students could be getting outdated information that not only would be detrimental to student learning but also could potentially diminish the library's credibility.

6.7 CONCLUSION

The team's experience in developing and maintaining a one-credit information literacy course for EC has given the librarians an opportunity to test various methods of revisions and determine in-house best practices over many years. The library team is fortunate to have EC administrators' support in creating and maintaining INL102. However, many librarians will have to be the catalysts in getting an information literacy course integrated into the curriculum. In addition, the GPA study conducted by EC's assessment unit establishes the positive long-term learning effects INL102 has on students, which confirms EC administrators' thinking and can be used as evidence for other librarians and administrators looking to create an information literacy course.

This experience shows that many elements are involved in creating a successful online information literacy course. Adding various types of multimedia formats—including library-branded audio, video, and graphics enhances course development and can be done over time with a well thought out plan. Then embedding existing library resources to round out supplemental learning concepts is beneficial to comprehensive course design and student success. Lastly, developing sustainable assessments mapped to the main information literacy concepts is essential to creating a final exam that reflects what students have learned. Many variables must be considered for overall design and to successfully implement these ideas, but with collaborative and strategic work between a team of librarians and administrators one can create and manage a successful credit-bearing information literacy course in an online environment.

REFERENCES

Birdsong, L., & Freitas, J. (2012). Helping the non-scholar scholar: Information literacy for lifelong learners. *Library Trends, 60*(3), 588—610. Retrieved from <http://www.press.jhu.edu/>.

Drouin, M., Hile, R. E., Vartanian, L. R., & Webb, J. (2013). Student preferences for online lecture formats: Does prior experience matter? *Quarterly Review of Distance Education, 14*(3), 151—162. Retrieved from <http://www.infoagepub.com/>.

Excelsior College. (2015). *About excelsior college.* Retrieved from <http://www.excelsior.edu/about>.

Mayer, J., & Bowles-Terry, M. (2013). Engagement and assessment in a credit-bearing information literacy course. *Reference Services Review, 41*(1), 62—79. Available from: <http://dx.doi.org/10.1108/00907321311300884>.

Mestre, L. S. (2010). Matching up learning styles with learning objects: What's effective? *Journal of Library Administration, 50*(7—8), 808—829. Available from: <http://dx.doi.org/10.1080/01930826.2010.488975>.

Middle States Commission on Higher Education. (2016). *Characteristics of excellence in higher education: Requirements for affiliation and standards for accreditation.* Retrieved from <http://www.msche.org/publications/CHX-2011-WEB.pdf>.

Murphy, J. (2013). Marketing the library in an on-line university to help achieve information literacy. *Education Libraries (Online), 36*(2), 17—24. Retrieved from <http://education.sla.org/wp-content/uploads/2012/12/36-2-17.pdf>.

Ostrowski, A. (2014). *Analysis of information literacy patterns of FY2008-2012 graduates.*

Sanborn, L. (2015). The future of academic librarianship: MOOCs and the robot revolution. *Reference & User Services Quarterly, 55*(2), 97—101. Available from: <http://dx.doi.org/10.5860/rusq.55n2.97>.

Sutton, R. (2014). Unlearning the past: New foundations for online student retention. *Journal of Educators Online, 11*(3), 1—30. Retrieved from <http://www.thejeo.com/>.

TILT. (2015). Texas information literacy tutorial. Retrieved from <http://library.utb.edu/tilt/nf/intro/internet.htm>.

Wu, D. C. (2014). Learning styles and satisfaction in distance education. *Turkish Online Journal of Distance Education, 15*(4), 112—129. Available from: <http://dx.doi.org/10.17718/tojde.31724>.

CHAPTER 7

Engaging Learners Online: Using Instructional Design Practices to Create Interactive Tutorials

7.1 INTRODUCTION

Increasing student enrollment coupled with limited resources in the library requires creative solutions to deliver information literacy instruction, which often takes the shape of online tutorials. Working closely with faculty members, librarians at the University of California, San Diego (UCSD), created a required online interactive tutorial. The tutorial replaced a library-led face-to-face information literacy workshop for students enrolled in an undergraduate microbiology course. To make an effective online tutorial that engages students in active learning without the benefit of live instruction, evidence-based instructional design practices were used to create a student-centered learning experience that makes learning more effective and efficient. Using instructional design pedagogy, this chapter presents the decision-making process for the tutorial, focusing on design and evaluation, including writing practical learning outcomes, assessing student understanding, and selecting a technology to support student engagement.

7.2 DESIGN

The design process began with the instructional design librarian discussing the requirements of the tutorial with the teaching faculty. The purpose of the tutorial was to replace the in-person library workshop as much as possible. The tutorial needed to be self-paced, delivered asynchronously, and include opportunities for students to practice what they were learning. Since the tutorial was a course assignment, student completion needed to be captured in the campus course management system's grade book. After examining the learning outcomes and content from the existing

Distributed Learning.
Published by Elsevier Ltd. All rights reserved.
© 2017 T. Maddison and M. Kumaran.

workshop, the learning goal of the tutorial was determined: Undergraduate biology students will be able to find, obtain, and use journal articles for their lab reports.

Once the learning goal had been decided, learning outcomes were developed using the ABCD objectives model (Heinich, Molenda, & Russell, 1989). This well-known instructional design model suggests that each outcome specify:

- the learners (or **audience**),
- the **behavior** to be observed as a result of learning,
- under what **conditions** the learning will occur, and
- the **degree** to which correct behavior must be observed for the outcome to be "achieved" (if applicable).

The information literacy outcome for this module written in the ABCD format looks like this:

> Given description and examples of primary and secondary sources, students will be able to categorize examples of sources used for science research.

In this example, the *audience* is "students," the *behavior* is "categorize examples of sources," the *condition* is "given a description and examples of primary and secondary sources," and there is no degree included. (A condition statement is more commonly used when one is able to capture assessment data.)

Using Ruth Clark's Content Performance Matrix as illustrated in Table 7.1 (2008, p. 50), learning outcomes were mapped accordingly. Clark's matrix is adapted from David Merrill's Component Display Theory (CDT). Turnbow (2012) describes the CDT approach as one in which instructors are encouraged "to focus on content types (facts, concepts, process, procedures, principles) separately from performance (find, use, remember), to create meaningful learning activities and design effective assessment" (p. 1).

Following Clark's model, the authors first determined whether each objective was "apply" or "remember." For example, the objective above would be "apply" because students are asked to categorize which is an

Table 7.1 Clark's content performance matrix

Apply Remember					
	Facts	Concepts	Process	Procedure	Principle

application of learning. Next, it was determined if the objective was a "fact," "concept," "process," "procedure," or "principle." This exercise, while challenging, is invaluable to facilitating the process of how to teach and assess each outcome. Appendix 1 includes outcomes for this tutorial and how they mapped to Clark's matrix. It also illustrates where each learning outcome appears in the tutorial modules.

Once the learning outcomes were determined, the authors needed to think about how to present the content. Drawing from instructional design approaches, the flow of each module was designed according to Gagné et al nine events of instruction (Gagné, Wagner, Golas, & Keller, 2005).

1. Gain attention: audio is used throughout the modules and the narration addresses the individual learner in the first person while also using colloquial language.
2. Describe the goal: the learning outcomes are presented at the beginning of each module.
3. Stimulate recall of prior knowledge: whenever possible, students were pointed to examples that had been presented in prior modules or ones that they might already be familiar with.
4. Present the material to be learned: new material is presented in each of the modules.
5. Provide guidance for learning: guidance is provided by way of practice exercises and activities.
6. Practice: practice questions/activities are provided and can be mapped directly to module outcomes.
7. Provide feedback: real time feedback is provided to learners for all activities.
8. Assess: assessment is captured in the form of students completing the tutorial. While it is self-paced they must complete all activities in order to move forward. Students do not receive credit for the tutorial unless they complete the tutorial in its entirety.
9. Enhance transfer of knowledge: opportunities to practice each skill are provided within the tutorial

Each module of the tutorial begins with a slide that describes what the student will know and be able to do at the end of it. While these are based on the learning outcomes for the module, they are communicated in plain language. Throughout the tutorial, recall of prior knowledge was stimulated by connecting information literacy concepts with experiences students are likely to have had in the past. Each learning outcome can be mapped to an interactive exercise within the tutorial that students

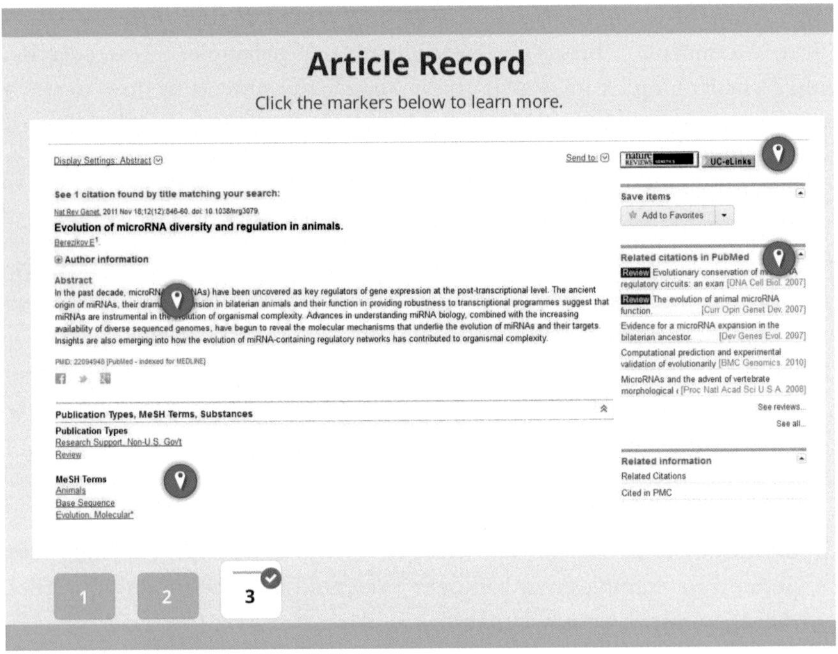

Figure 7.1 Worked example.

complete in order to practice what they have learned. Whenever possible, the tutorial used what Clark describes as a worked example (Clark & Mayer, 2011). In doing so, students were provided an illustration and description of how to decipher a search results screen (Fig. 7.1).

When each marker is clicked, a description of that part of the record appears.

Next, there is a partial example that the student completes as an activity (Fig. 7.2).

Finally, the answers were provided on a third screen so learners can compare their work to the example (Fig. 7.3).

When each marker is clicked, a description of that part of the record appears.

When students answer questions incorrectly or do not complete interactions accurately, they receive feedback that provides clues to the right answers. Students have two chances to complete an interaction correctly before they automatically move forward in the tutorial. The assessment and transfer of knowledge takes place in the classroom after the tutorial has been completed where students have another opportunity to practice their search skills with the guidance of their teaching assistant or

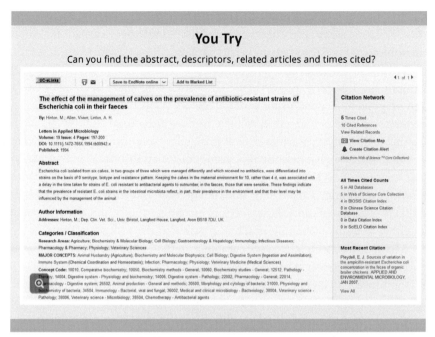

Figure 7.2 Worked example learner practice.

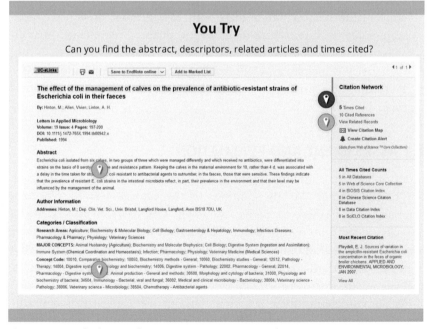

Figure 7.3 Worked example answers.

faculty member. This practice also facilitates the transfer of knowledge, thus ensuring they can apply the research strategies learned for their microbiology class in other university courses.

Finally, throughout the tutorial, Richard Mayer's Cognitive Theory of Multimedia Learning was used when considering the placement of text and graphics and when using audio elements. Mayer argues, "People learn better from words and pictures than from words alone" (2009, p. 223). To this end, Mayer describes several principles based on his argument that people have two channels for processing information (visual and auditory) that are limited in their capacity, therefore learning is an active process of filtering, selecting, organizing, and integrating information. While it is unlikely a designer can follow every principle perfectly, the authors made an effort to employ them as much as possible in the following ways:

- Multimedia principle: used graphics rather than text to represent content whenever possible.
- Contiguity principle: provided instructions to activities on the same screen as the activity, included feedback to activities near the questions, and timed audio to correspond with affiliated graphics.
- Modality principles: used audio to define and describe content rather than words on a screen whenever possible.
- Redundancy principle: used either spoken words or written words to convey information rather than both at the same time (exceptions were made for cases where there was no pictorial representation available, which is supported by Mayer).
- Coherence principle: avoided extraneous audio, graphics, and words.
- Personalization principle: narration is presented in a conversational tone.

7.3 ASSESSMENT AND EVALUATION

In an online asynchronous learning environment, assessing learning goals can be challenging. As the purpose of the tutorial was to replace a librarian-led workshop, formative assessment was provided to students as they progressed from start to finish. Throughout the tutorial, students were provided with opportunities to apply their learning through activities. For example, after learning about primary, secondary, and tertiary sources, students are asked to categorize different types of sources (Fig. 7.4).

While completing these activities, students were given two opportunities to check their understanding of a given learning outcome via an

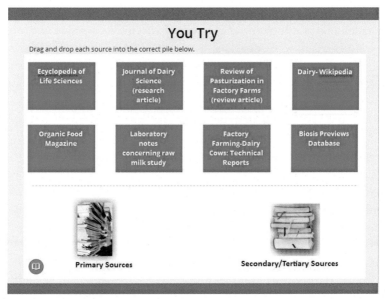

You Try

Drag and drop each source into the correct pile below.

| Ecyclopedia of Life Sciences | Journal of Dairy Science (research article) | Review of Pasturization in Factory Farms (review article) | Dairy- Wikipedia |
| Organic Food Magazine | Laboratory notes concerning raw milk study | Factory Farming-Dairy Cows: Technical Reports | Biosis Previews Database |

Primary Sources   Secondary/Tertiary Sources

Figure 7.4 Example activity.

interactive activity with feedback before moving on to the next section of the tutorial. For example, if students completed an activity successfully on the first try, they were given positive reinforcement on a job well done. On the other hand, if students incorrectly completed the activity on the first attempt, they were given feedback that would help lead them to a correct answer. If an additional failed attempt was made, the feedback provided the correct answer and a gentle suggestion to review the previous material. This approach provides targeted formative feedback based on a particular response. Unfortunately, individual feedback was not possible so that the tutorial could accommodate hundreds of students completing it at the same time. However, part of the design was to require students to complete or at least attempt each activity twice before moving on to a new section or completing the tutorial. This directed learning approach enabled us to provide feedback to each learner based on his or her understanding of presented material.

Kirkpatrick's Four-Level Evaluation Model (n.d.) suggests that assessment and evaluation should be approached in four levels. Level 1 (reaction) is focused on satisfaction with the tutorial, especially with regard to engagement and relevance. Although in the past, Google forms were used to collect satisfaction data as part of capturing student completion, one of the technological requirements of this tutorial was to have it connected to the gradebook of a learning management system. As a result, the authors now rely on faculty feedback and anecdotal comments with regard to student satisfaction when previously feedback was gained from a survey. Level 2 (learning) examines what students have learned as a result of the instruction. While the authors were unable to observe each student complete the tutorial, it was designed in a way that required the student to apply the learning before proceeding with the tutorial. Level 3 (behavior) is concerned with whether or not learners change their behavior as a result of their learning. In this case, are students using the search strategies presented in the tutorial (e.g., truncation, related articles, following cited records) or are they relying on the same ones they used previously? Changing behaviors take time and cannot adequately be assessed after one tutorial or even after one course. The students complete a lab report at the end of the quarter, and informal feedback from faculty has indicated that students are in fact using search strategies presented in the tutorial. Faculty feedback is the only means of assessing this level for a variety of reasons, including the fact that the authors do not have access to final lab reports. Finally, Level 4 (results) is concerned with the return on investment. In library instruction, completing Level 4 would mean students can demonstrate they are information literate as a result of the instruction. Obviously, this is not something that can be assessed as a result of any single library instruction event and is beyond the scope of this tutorial.

7.4 SELECTING TECHNOLOGY

Selecting technology for tutorial creation can be complicated; even the initial price of the software can be layered with considerations. Here is a short checklist of cost considerations when trying to determine if a particular technology fits within a budget:

- Does the software have a one-time cost or a license agreement?
- How many people will need access to the software, and do all persons need their own access?
- Is there an education discount?
- What version is needed?
- If the software is expensive, does it have multiple uses?

At UCSD, the rapid-authoring tool Articulate Storyline is used in creating tutorials because it offers a good balance between cost and ease of use.

7.4.1 Cost

The software is relatively inexpensive considering how much it is used by the authors at UCSD. While Articulate requires one license per user, the library only required two licenses, so the cost was not prohibitive. In addition, Articulate offers a significant educational discount that increases based on how many user licenses are purchased.

7.4.2 Ease of Use

Beyond price, when selecting technology one should consider the tool's ease of use. This is important not just for the person who is going to be using the software but also for those who may come after. It is important to think about the learning curve. Does the tool have a short or long learning curve? Can it be used immediately to accomplish a task or is expertise needed right away to complete a project? One reason why Articulate was selected over other rapid-authoring tools is because the base functionality and layout is similar to Microsoft PowerPoint. This allows new Articulate users to recognize common functionalities and transfer skills they may already have to the new software, which makes Articulate user friendly and reduces training time.

Software tools that allow an individual to complete projects as experience and expertise are gained are ideal for individuals who are just starting out with tutorial creation. Few librarians have the time to become master software users before a project needs to be completed. One way to determine if a software tool is easy to use is to look at its support and help resources. Does the tool have easy to locate and use "how to" resources? Is there a community of users who share their experience with others? Is the functionality built from preexisting skills? Help resources such as community blogs and how-to videos are extremely useful. Along with being a training resource, they are often a place that sparks inspiration and creativity.

Engaged learners are individuals who are able to work with and act on concepts presented in a tutorial's content. The idea is to go beyond the typical read and quiz format. To help achieve interactivity within an online tutorial, the selected software needs to offer tutorial creators the ability to build interactive activities. Some common interactions built into rapid authoring tools include:

- Matching drag and drop: this activity allows students to match items together by using a mouse to drag a content item and drop it on another. This feature is useful for learning definitions, putting like items in categories, and as a replacement for traditional quiz questions.

- Sequence drag and drop: this activity is similar to the functionality just described. It is most useful in helping students put content items in order for step sequences or timelines.
- Quizzes: traditional quizzes and grading can be used to test student recall.
- Hotspots: hotspots use a click indicator to highlight an aspect within an image or text. This allows students to click highlighted sections of information to learn more about a given topic; hotspots also can be used for identification purposes.
- Fill in the blank: this activity mirrors traditional fill-in-the-blank quiz functionalities.
- Games: rapid-authoring tools often allow the designer to use built-in games to present information. A well-known example of this type of presentation is the Jeopardy board.
- Time-line interactions: a time-line interaction allows students to select a specific content item along a time line and learn more about it. This feature can also be used to replace tab functionality in order to present information from several different subcategories.

With a little creativity, these built-in activities help students develop a deeper level of understanding of the lesson's material because it requires them to model behavior that was identified in the tutorial's learning outcomes and assessment.

In a world of multiple devices, it is important to determine the output of the tool or how the tutorial is published. Can a tutorial be embedded into a learning management system or is it accessible through a link on a website? If there is an interest in embedding a tutorial into a learning management system, a software solution that enables Sharable Content Object Reference Model (SCORM) data will likely be needed. Another consideration when it comes to publishing is responsive design. Many software tools now offer responsive design publishing options so that the same tutorial can be viewed on a desktop, as well as a smartphone. Often outputs that allow for responsive design limit the ability to create interactive activities. Before deciding for or against responsive design, think about the instructional need. Is it more important for students to interact and engage with the material or is it more important for them to be able to access material in a menulike structure on a smartphone? Remember that the goal is to create effective and engaging learning. Consider how the method of output may influence the way in which students engage and interact with tutorial content.

7.5 CONCLUSION

Instructional design practices provide an evidence-based framework that facilitates the design, development, and implementation of student-centered online tutorials. While working through the various instructional design practices such as creating learning outcomes and working with Ruth Clark's Content Performance Matrix (2008) as well as Gagné's nine events of instruction (2005), a blueprint for how to integrate content, active learning, and assessment emerges. With this blueprint in place, one can begin to explore technology options that enable the incorporation of multimedia learning. By using instructional design practices, one can feel confident that the learning object that is created is based on sound evidence-based decision making designed to engage student learning and provide an enriching instructional experience in the absence of in-person instruction.

REFERENCES

Clark, R. C. (2008). *Developing technical training: A structured approach for developing classroom and computer-based instructional materials* (3rd ed.). Hoboken, NJ: John Wiley & Sons, Inc.

Clark, R. C., & Mayer, R. E. (2011). *E-Learning and the science of instruction: Proven guidelines for consumers and designers of multimedia learning* (3rd ed.). San Francisco, CA: Pfeiffer.

Gagné, R. M., Wagner, W. W., Golas, K. C., & Keller, J. M. (2005). *Principles of instructional design* (5th ed.). Belmont, CA: Wadsworth.

Heinich, R., Molenda, M., & Russell, J. D. (1989). *Instructional media and the new technologies of instruction* (3rd ed.). New York, NY: Macmillan Publishing Company.

Kirkpatrick Partners. (n.d.) *The new world Kirkpatrick model.* Retrieved from <http://www.kirkpatrickpartners.com/OurPhilosophy/TheNewWorldKirkpatrickModel/tabid/303/Default.aspx>.

Mayer, R. (2009). *Cognitive theory of multimedia learning.* Cambridge, NY: Cambridge University Press.

Turnbow, D. (2012). Incorporating instructional design approaches into library instruction. In *Proceedings from California Academic & Research Libraries conference 2012: Creativity and sustainability: Fostering user-centered innovation in difficult times.* San Diego, CA.

APPENDIX 1 CLARK'S CONTENT PERFORMANCE MATRIX FOR MICROBIOLOGY TUTORIAL

Instructional Goal

Undergraduate biology students will be able to find, obtain, and use journal articles for their research.

	Outcome	Clark's content performance matrix
Module 1: Prepare	Given example sources, student will be able to identify if the source is primary, secondary, or tertiary.	Apply concept
	Given a topic, student will be able to select the appropriate source for background information.	Apply concept
	Given examples of a resource (e.g., PubMed, library catalog), student will be able to determine if it is a catalog or database.	Apply concept
Module 2: Search	Given a list of databases and topics, select the best article database.	Apply concept
	Given a screenshot of a database, student will be able to identify where they would find features such as abstract, descriptors, related articles, and times cited.	Remember fact
	After watching a video about UC-eLinks, student will be able to identify the features of UC-eLinks.	Remember fact
Module 3: Read	Given the "Anatomy of a Journal Article," student will be able to identify parts of an article.	Remember fact
Module 4: Write	Given an overview of quoting, paraphrasing, and summarizing, student will be able to identify if an excerpt is plagiarized.	Apply concept
	Given a citation, student will be able to identify the parts.	Remember fact

CHAPTER 8

Developing Best Practices for Creating an Authentic Learning Experience in an Online Learning Environment: Lessons Learned

8.1 INTRODUCTION

Distance education statistics clearly show the popularity of online courses. About 63% (3322) of all Title IV institutions—i.e., those institutions whose students are eligible to receive federal financial aid—offer distance education (National Science Foundation, 2014). Advantages such as flexibility, availability, and variety drew nearly 14% of the 20.6 million students enrolled in 2012 to at least one distance education course as part of their program (National Center for Education Statistics, 2013). Professionals who provide online instruction are under a myriad of pressures to address emerging user needs and look ahead to new directions and approaches. Librarians are being challenged to adapt traditional, face-to-face instructional techniques to work in an online setting while still creating a meaningful learning experience. Libraries are seeking to develop best practices and approaches to effectively meet the information literacy needs of their online students and faculty. This chapter will offer insights and suggestions on how to develop best practices for libraries serving online students. These insights and suggestions are based on the experiences of a mid-sized regional comprehensive university as part of its own approach to providing quality online library services and resources to its online students and faculty and is supported by the research literature.

8.2 SAGE, GUIDE, MEDDLER

Two educational concepts are poised to replace the "sage on the stage" teaching method and become best practice in online learning: the "guide

Distributed Learning.
© 2017 T. Maddison and M. Kumaran.
Published by Elsevier Ltd. All rights reserved.

on the side" and "meddler in the middle." While initially developed for use in traditional classrooms, these approaches show much promise in enhancing active, authentic learning in the online environment.

The traditional sage on the stage paradigm features the learned instructor imparting his or her wisdom to a classroom of pupils. This approach has been the standard of education for hundreds of years but is being deposed by the explosion of knowledge in the 20th and 21st century.

> There is a powerful myth lurking behind our habitual thinking that the teacher is the Knower who ought to be providing all the maps in the learning process. . .. While teachers need knowledge, this should not be confused with a powerful memory or the capacity to seem all-knowing. It is much more important to model how to be usefully ignorant, and to assist students who fear not having all the answers all the time
>
> **(McWilliam, 2009, p. 287)**

Most librarians have realized it is more important to guide students through the discovery process than to present themselves as the ones with the all the answers. This is even truer in the world of online education because students often need even more hands-on assistance than usual. Finding information is not that difficult today, but students will need guidance on evaluation of the almost infinite sources of information available to them and how to effectively incorporate the most appropriate sources into their work.

This realization has led many educators to embrace the guide on the side mentality in which the educator facilitates learning rather than disseminating information. This shift in responsibility releases the educator from the cycle of lecture and then test and allows students to direct their own learning, occasionally "guided" by their teacher when needed. The guide approach functions well in the online environment when some amount of structure is needed to direct learning while allowing some level of autonomy on the part of the students. Examples of this include guided discussion board threads, online interactive lectures, and supervised group work via meeting applications such as Skype or Adobe Connect.

While the guide still has the implication of authority, the meddler in the middle sets aside any such pretense and approaches the learner as a colleague and an equal. The meddler provides support and structured activities that challenge students to take ownership of the process as they work toward an outcome and *does not rush to save students from the struggle that higher order thinking involves, by giving them either the answer or the template for finding it. They allow their students to experience the risks and confusion of authentic learning by allowing their students to experiment with possibilities in ways that put their ignorance to work*

(McWilliam, 2009, p. 291)

Though the role of a meddler requires more work and in many ways more effort than the sage or the guide approach, the meddler encourages learning in the moment allowing individuals to create their own learning experience.

Librarians are natural meddlers because many of us are accustomed to being in the middle on a regular basis. Working toward a goal with a student or group of students is an everyday situation for many librarians as they help students with research or other projects. Librarians are rarely the ones giving the assignments and are most often in the trenches with the student, trying to figure out how to meet the standards set by teaching faculty. This creates a natural camaraderie that can be exploited to encourage students to buy in to the meddler in the middle approach.

Beyond the reasons already mentioned, shifting to the meddler in the middle approach will help libraries maintain relevance into the future. As McWilliam (2007) states,

The message from industry is that increasing numbers of university graduates will be working in digitally enhanced environments where there are few transportable blueprints for project design and management. To develop the sorts of learning dispositions that are appropriate in such contexts, academic educators will need to spend less time explaining through instruction and more time in experimental and error-welcoming modes of engagement (pp. 1–2).

Eastern Kentucky University (EKU) Libraries mirror the transition from sage to meddler but have not officially adopted any of these three teaching approaches. Most EKU librarians would be considered guides on the side, with several embracing the meddler in the middle approach and beginning to develop best practices using that mindset on their own. Several new approaches are being experimented with and implemented in this transition—e.g., EKU librarians are offering open online office hours or virtual office hours. These open forums have proved to be a collaborative and democratizing tool when used to facilitate sharing of ideas between students and librarians. Librarian-created open discussion threads within the course management system (CMS), allowing the students to lead the conversation, have also been an effective meddling strategy.

The use of online whiteboards is another strategy being explored. Online whiteboards offer a level of anonymity and can be a valuable tool in leveling the playing field and encouraging student experimentation and interaction. Librarians at EKU have experimented with the online white-board feature in Adobe Connect as well as Padlet (which will be discussed further in this chapter.) Evaluations of the two tools showed that while students in the Adobe Connect session did contribute ideas on the white-board, participation using the Padlet application was stronger and more effective.

Establishing best practices, or setting standards based on effective prac-tices proven through personal trial and error and augmented by others' experiences, is an important component of any libraries' information lit-eracy efforts. Once approved, these principles should be communicated to the organization as clearly as possible. While it varies from library to library, some of the most important components that EKU Libraries have identified as part of a set of best practices governing online library services are discussed below.

8.3 BEFRIEND THE GATEKEEPERS

Many of the services and resources offered by libraries are constrained or enabled by various outside groups. Recruiting and befriending these gate-keepers is one of the most important things a librarian can do to ensure quality services and resources for their online students and faculty. In their book *Academic Librarianship by Design: A Blended Librarian's Guide to the Tools and Techniques* (2007), Bell and Shank agree that "[we] need to part-ner and form learning communities with our faculty, instructional designers and technologists, and other staff" (p. xii). The number of out-side groups that impinge on librarians' ability to deliver content and ser-vices forms a daunting list. Of course, these gatekeepers will vary by institution, but some of the most common are course developers, infor-mation technology (IT) administrators and technologists, both library and college-level administrators, learning management system (LMS) adminis-trators and technicians, regional or satellite campus administrators and staff, in-house library administrators, staff librarians, and IT personnel.

Cultivating positive one-on-one interactions is an excellent strategy for building strong relationships with gatekeepers. According to Bell and Shank (2004), "collaborating and engaging in dialogue with instructional technologists and designers is vital to the development of programs, services and resources needed to facilitate the instructional mission of academic libraries" (p. 374). Having as many of these individuals and groups kindly disposed toward a librarian and, by extension, the library in general can open up a world of opportunities for collaboration and provide librarians access to online courses in ways that were never thought possible just a few years ago. As a case in point, at EKU Libraries, an exceptionally strong partnership with Blackboard administrators developed by the distance learning librarian has resulted in the librarian being granted administrator rights, including the ability to add librarians to online courses. This is a huge advantage to librarians interested in working with online courses and saves much time and reduces the number of requests to Blackboard administrators. It is good practice in this sort of situation to be careful to have the faculty's permission and to not disturb anything in the course without prior notification and approval. Faculty can be understandably nervous about granting access to their courses, but in EKU Libraries' experience, once instructors have worked with a librarian they are much more comfortable with inviting them back. Instructors tend to be more worried about the integrity of their grade book being disturbed when considering whether to allow an outsider into their course. To address this concern, unless there is a specific request by the faculty member teaching the course, the library administrator never grants librarians more than course-builder status in a course, which prevents access to the grades section of Blackboard. In negotiating with the faculty member, librarians make this clear up front, which often eliminates any resistance they might encounter.

There is no one approach to winning over these groups, and there are exhaustive examinations of collaborative techniques in the literature that offer in-depth advice on how to recruit and win over possible collaborative partners, but several suggestions illuminated by EKU experiences present themselves and deserve mention in the following.

8.4 POSSIBLE APPROACHES

- Look for individuals who are high energy and willing to collaborate. Almost every office has at least one person who is creative and open to new opportunities. Sometimes it is extremely easy to identify these individuals, but sometimes it can require patience and multiple attempts to find the right person.

- New hires are often an excellent target audience because they are not rooted in traditional practices or invested in institutional culture, are often full of new ideas, and are looking for ways to excel.
- Look for opportunities to assist the individual gatekeeper or group. Find ways to demonstrate your value and expertise.
- Be specific in your proposals and make certain the gatekeeper understands your commitment to student success and that you are willing to act on it. Make sure they know and believe that you are sincere and there to help.
- Look for committees and other campus work that will enable you to identify, work with, and get to know the gatekeepers of your institution.
- Embrace individual expertise. Identify those individuals whose personal interests or talents might help you. For example, a librarian who enjoys video editing might be a perfect partner for a collaboration to create an online video tutorial.
- Be persistent. It is good practice to continually look for ways you can collaborate, especially in ways that will make things easier for the prospective partner or exploits common interests. If one proposal does not elicit interest, try something else after a suitable interval. Try to avoid overdoing it, as being perceived as overeager can damage your collaborative opportunities.

Some of the strategies previously mentioned have been extremely successful for EKU Libraries, their collaborators, and, most importantly, EKU's students. Here are a few of their most noteworthy successes:

- The library has become part of the online course design process, and librarians are included in initial consultations as faculty members begin developing their courses. Getting in on the front end of course design and becoming part of the course development process and not a tacked-on afterthought has been incredibly rewarding for both EKU Libraries and their partners. Being able to establish contact, educate faculty members about library services and resources up front, and collaborate on how best to integrate these resources into their online courses has had a transformative effect on the level of library involvement and the impact librarians can have on student success. More than 90% of EKU faculty members who develop online courses incorporate library services and resources as part of their courses.
- The need to consult with library representatives before developing online courses has been inserted in the contract that faculty members must sign

to be paid for teaching their courses. This lends legitimacy to collaborative efforts by librarians and sets an expectation that library services (and librarians) are an important part of online teaching at EKU.

- As a result of the good working relationship developed between the distance learning librarian and local LMS administrators, as mentioned previously, the librarian was granted administrator privileges to enable better access to online courses. This has turned into a great time-saver and has streamlined many day-to-day operations that once took much longer as librarians were forced to wait for the changes to be made by third parties. This situation is a perfect exemplar of the principle of persistence. It took several years to establish the trust necessary to achieve this level of collaboration.

- The distance learning librarian was asked to sit on the search committee for the new director of the Instructional Development Center, who is in charge of all online course development and online course designers at EKU. After years of resistance to the libraries' collaborative efforts by the previous administration, EKU librarians were able to have a voice in choosing collaborative partners and were able to forge a close working relationship with the incumbent leadership; this has led to many excellent collaborations benefiting online students and faculty.

These are good examples of the sort of benefits that can be garnered by identifying and befriending the gatekeepers who will either help or hinder a library in its efforts to provide quality services and resources to its users. Situations will vary from institution to institution, but hopefully these strategies will provide guidance and spur ideas that will help librarians reach out to these groups and individuals and form long-term and productive partnerships.

8.5 TEACHING IN THE TRENCHES

In EKU librarians' experience, when a course is offered online for the first time, among other challenges, the instructor must rework the delivery of instruction, address different learning styles in an online environment, consider how to best communicate with students, and adjust delivery of content that typically works best in a face-to-face environment. For the librarian, reorienting instruction for an online course will almost certainly require more preparation time than a traditional one-shot, face-to-face session. Librarians will need to allow for more planning time to consider options such as using the flipped-classroom approach in which learning tools such as screen casts or videos are used to offer introductory information prior to the actual library meeting. Assignments and activities incorporating active learning are also vital to creating an effective learning experience and should be included. Library instructors should initiate clear communication with the professor as soon as the instruction request is made in order to establish goals of the appointment, seek suggestions on teaching strategies via this type of format, and set expectations on what can be achieved. The librarian should not offer unrealistic hopes for what can be accomplished but should discuss what is reasonable within the time allotted and in the delivery format being used.

Using technology can often seem to be the most cost- and time-effective means of presenting information literacy concepts, but it can lack the personal touch that a librarian can provide students, especially those who need more guidance. Hoffman (2011) suggests, "while tutorials or how-to-videos may suffice, embedding a librarian in an online course offers a more interactive, direct substitute for the hands–on, customized library instruction provided in many face-to-face courses" (p. 445). To provide students with a positive and effective connection with the library, an embedded librarian should employ a variety of tools and communication methods to create an experience equivalent but not necessarily the same as the face-to-face experience of a traditional student. In referring to the 2008 *ACRL Standards for Distance Learning Library Services*, Hoffman explains that "ACRL acknowledges that online faculty and students 'face distinct and different challenges' that cannot be met by simply linking to the library web site from the course software" (Hoffman, 2011, p. 445). The most important factors to consider when developing tools and preparing content for an online class are the goals of the course and the needs of the students juxtaposed against the realistic estimation of what the librarian can reasonably accomplish within the constraints of time and resources because "it is important that the design of the embedded library content, including instructional materials and interactions, be tied to the needs of the learners and the course" (Edwards & Black, 2012, p. 301). The librarian should keep in mind several other factors when creating content, including different learning styles, reusability, longevity of the software being used, cost, and access, among others. Zapalska and Brozik (2006) assert that, "it is necessary to provide a number of different learning options that take into account different learning styles. Combining a mixture of approaches and teaching methods allows online students to choose the instructional style that best fits their individual learning styles" (p. 332).

In an article on embedded librarianship, Morasch (2013) refers to the online course management environment as a *courseroom*. Most, if not all, of the communication between faculty and students occurs inside of the courseroom, providing one avenue by which a librarian becomes an active participant in the class using discussion boards, email, and announcements. Whether the students are fully online or in a hybrid course, librarians who seek to be integral parts of the course can follow the lead of the instructor and look for opportunities to introduce themselves, share links to resources, invite feedback, and answer questions as they arise. Research supports the idea that making a connection with students helps to build trust (Morasch, 2013). Librarians who want to provide an equivalent service to the face-to-face reference desk interaction can offer virtual office hours, post a list of those available time slots in the courseroom, share the times via an embedded Google Calendar, or even use the "Schedule an Appointment" option from Springshare's LibCal program (http://springshare.com/libcal). Regardless of which tools the librarian uses, the goal is to provide multiple avenues of communication so the librarian can create a bridge between the library and online users.

8.6 ONE GOAL, MANY TOOLS

Librarians can use many tools to create a more personalized experience for students in an online environment. Since instructors need to be sensitive to students' multiple learning styles, as well as the varying needs of students, it is wise to take advantage of different types of tools to provide research help. One common teaching application used by instructors to connect with students is web-conferencing software. This type of software allows faculty to create a classroom-like experience by bringing together students in a live, online format with the benefit of screen sharing, a virtual whiteboard, assessment tools, interactivity, and multiplatform delivery. Historically, EKU has utilized several web-conferencing software programs, including Wimba, Elluminate, WebEx, and more recently Adobe Connect to provide online instruction for an entire class, as well as for individual research consultations. All of these programs allow the librarian to sign in to an online session, use a webcam or microphone to communicate with students, incorporate a virtual whiteboard to share ideas and seek feedback, and share his or her screen to demonstrate the various library resources available for students. These software programs allow the instructor to record the session and share it with faculty members and students later, thereby addressing one of the major concerns involved in working with students geographically removed from the originating institution by providing asynchronous materials for student use. Using online meeting software, including free services such as Skype or Google Hangouts, librarians can offer flexible meeting times and formats to connect with students, share ideas, demonstrate research strategies, and make a personalized connection between the library and the student.

Online tutorials are another tool librarians can use to connect the library with the online student. Librarians and instructional designers are using tools such as TechSmith's Camtasia (https://www.techsmith.com/camtasia.html), Adobe Captivate (http://www.adobe.com/products/captivate.html), Screencast-O-Matic (http://www.screencast-o-matic.com/), and TechSmith's Jing (https://www.techsmith.com/jing.html) to develop video tutorials that have many advantages. In their article, "Best Practices for Online Video Tutorials," Bowles-Terry, Hensley, and Hinchcliffe (2010) propose that "video tutorials provide asynchronous library assistance, and students can view them on [their] own time at any hour of the day" (p. 19). In the same article, the authors point out other benefits, including that videos "can be viewed as many times as necessary," "can facilitate a teachable moment," and can "engage visual and auditory learners" (p. 19). General tutorials are useful for introducing and demonstrating basic library research tasks, including the use of resources such as article databases, the online catalog, and the library's interlibrary loan service along with other helpful exercises such as distinguishing between peer-reviewed sources and popular magazines and newspapers. They are designed to meet the needs of a broad audience of users who are in the early stages of their research. Program- and course-specific tutorials can take more time to create, but they demonstrate the librarian's commitment to point-of-need, specialized attention. Librarians can also build modules that incorporate a scaffolded approach to instruction, including tutorials that address basic needs, as well as more advanced research help such as citation formatting and management, tracking a source if a citation is known, specialized database searching, and avoiding plagiarism. Librarians who are embedded in courses may have the opportunity to follow the course syllabus, monitor discussion boards, and communicate directly with students to determine information needs, address learning outcomes, and create assignment-specific tutorials.

Developing tutorials also allows for a flipped classroom approach to teaching in which instruction is delivered outside of class or via an online format and class time is spent on the activities usually given as homework. A librarian can embed tutorials in the CMS or on a website or LibGuide and, with the support of faculty members, require students to view or complete the tutorials prior to an instruction session or online meeting. For example, one of the education librarians at EKU has used this method with positive results. After discussing the course syllabus and assignments with the professor, the education librarian created tutorials explaining library resources that students would need to use for the various assignments throughout the semester and posted them in Blackboard, the CMS currently used by EKU. Before the Adobe Connect online instruction session, the librarian required students to view the tutorials and informed them that the online meeting time would be spent talking about how to use the tools to complete the assignment along with questions the students had rather than using the short time she had to introduce and teach the tools. She also embedded a quiz that counted for participation points to help her gauge their participation, as well as their understanding. Using a flipped approach allows librarians to spend their class time with students for more advanced instruction or for one-on-one help while students practice the skills learned from material presented prior to the class session.

A proactive librarian who incorporates interactive learning tools into online courses can help improve student learning "by engaging in meaningful and fun activities the students are doing and thinking about library research and ultimately developing information literacy" (Campbell, Matthews, & Lempinen-Leedy, 2015, p. 583). One tool that librarians can use to solicit interaction from students is Padlet. According to its homepage, Padlet is "possibly the easiest way to create and collaborate in the world" ("Padlet," n.d.). Among its many features, Padlet can be used for gathering feedback and input from people using a wall that easily allows users to post comments; upload video, audio, or images; paste links to any number of file types; and manipulate those images or documents to appear anywhere on the Padlet wall ("What can I do with Padlet?" n.d.). Walls can be customized, and the link to the posts shared or exported. A Padlet wall can be a more flexible alternative to the discussion board within a CMS, and can be embedded in a webpage or even within the CMS to be used throughout the course. If a librarian has an opportunity to provide an online instruction session via a web-conferencing program, Padlet could be a great tool to gauge student participation and as a place for students to share answers in real time. They can provide live feedback to their peers (or to the librarian) and the wall on which they post can be shared later for those who want to go back and review the content. Google Drive also offers tools such as Google Docs and Google Forms, which can give the librarian a powerful tool in which to engage students in real time or in an asynchronous format (https://www.google.com/drive/). Google Forms, which can be used as a survey tool or quiz, among other functions, are embeddable and shareable just like Padlet walls. Web-conferencing tools such as WebEx and Adobe Connect also offer simpler methods of encouraging interaction through the use of the raise-a-hand feature, the chat window, microphone, webcam, and screen sharing.

One struggle EKU Libraries have faced is the seemingly simple act of just getting started. Many librarians keep abreast of trends and are familiar with the technology tools used to create tutorials and screen casts, interactive webinar features, and other digital learning objects, but having all that knowledge can create the problem of feeling the need to perfect a digital learning object or teaching tool before presenting it to a faculty member to use in a course. It is easy to get caught up in the planning and researching of the next best software and online programs used to create instructional tools. Examining all the options for creating video tutorials, for example, can be overwhelming. There are numerous free programs such as Jing, Screencast-O-Matic, and Screenr along with more costly programs such as Camtasia Studio, Adobe Captivate, Snagit, and Apple QuickTime player. One can easily get pulled in to spending hours watching videos on how to use each program, reading articles that review the pros and cons of each tool, and installing and testing tools that seem most useful for the instructional need. Librarians should remember the aphorism paraphrased by Voltaire as "Perfect is the enemy of good" (Ratcliffe, 2011, p. 389). Librarians can translate this to our own situation by remembering that a finished product does not have to be perfect; getting caught up in trying to perfect an already good enough product can waste time, effort, and resources that could be better used elsewhere.

Librarians need to invest appropriate amounts of time to create learning objects. For example, one might spend more time on a tutorial that will appear on the library's website and perhaps less time on a demonstration screen cast of a single concept that will be used for a narrow audience or only for one instruction session via Adobe Connect, WebEx, Wimba, or some other webinar software. At some point the librarian has to pick a program, create a quality tool that meets the needs of the faculty and student, and present it via the CMS, LibGuides, websites, or other electronic communication to be used by the students. The librarian should find a way to solicit feedback either formally or informally from students and faculty members to advise on changes that might need to be made in the future. Regular maintenance of online tools is a must; this includes replacing screenshots of database interfaces or website pages that have changed, amending broken URLs, and updating new features of a highlighted resource. The librarian should consider sustainability when determining the development of a tool because of the time commitment involved in the entire process. For EKU Libraries, offering something that may seem simple or sometimes even low-tech is better than creating unsustainable products and services or even not offering any sort of assistance to guide students in their research process.

8.7 MAINTAIN AUTHORITY

During their interactions with online faculty and courses, many librarians will presumably create or use a multitude of digital learning objects. Whenever possible, those digital learning objects should be curated by librarians rather than faculty in order to maintain appropriate resources. This allows librarians to keep the materials up to date and ensure they are used correctly and maintain relevance. It is important to continually collaborate with course instructors so that the librarian is aware of any embedded library content in their courses or use library materials such as handouts to ensure their currency and accuracy.

Not allowing faculty to keep copies of digital learning objects need not be a negative and can be used as a reason to maintain communication with faculty and a springboard to collaboration. Setting up a schedule to consult with faculty and provide them with updated links is a great opportunity to perform outreach on other topics.

8.8 ASSESSMENT

Academic libraries are under a variety of financial pressures—most notably the decreasing state support of colleges and universities—to demonstrate relevance and value. As a result, libraries are putting more emphasis on assessment for many library services, including information literacy instruction. Megan Oakleaf (2009) explains that, "if libraries intend to remain relevant on campus, they must demonstrate their contributions to the mission of the institution by becoming involved in assessment, the process of understanding and improving student learning" (p. 539). Providing accountability as to the effectiveness of online instruction on student learning requires more than the creation of library tutorials and assigning pre- and post-tests, although those types of tools can certainly be included in an assessment toolbox. Libraries need to establish an assessment philosophy and culture, deciding what services to evaluate and why those areas need to be targeted.

One strategic direction implemented by EKU Libraries as part of their 2011–15 strategic plan cycle states that the library will "use assessment to demonstrate value and make decisions" (Eastern Kentucky University Libraries, n.d., Goal 3: Strategic Direction 1.2). To address this goal, several instruction librarians created a document titled the *Information Literacy Matrix* to guide the instruction program. This document contains various competencies aligned with the Association of College and Research Libraries' (ACRL's) *Information Literacy Competencies for Higher Education*. Instruction librarians use this document to assess whether they are meeting specified information literacy standards, including the creation of rubrics designed to measure the quality of student work using predetermined criteria and performance quality indicators. EKU Libraries' approach is to scaffold the competencies to meet students' information needs at each level: first-year courses, core courses, and capstone courses. EKU's instruction librarians are currently revising the matrix they have used over the last several years to align more closely with the new ACRL information literacy framework. Marketing the guiding standards that librarians use to deliver instruction and assess student learning lets course instructors know what goals the library strives to meet and can help guide conversations in relation to how to meet those goals in both a face-to-face environment and an online setting. When librarians establish a relationship with the faculty member and embed course-related library resources in the class, including assessment instruments, students can see the librarian as a partner in the educational process rather than a separate entity.

A prime example of the value of assessment in a distributed learning environment is the assessment efforts of an EKU librarian who works almost exclusively with online health science programs. This librarian has employed a variety of assessment techniques to discover students' prior experience with library tools and services, gauge their understanding of a topic, and determine their self-efficacy in regard to locating and evaluating resources. The health sciences librarian communicates regularly with students via email and the CMS discussion board and gives students assignments that ask them to demonstrate understanding of information literacy topics such as how to obtain the full text of an article not available in our library's databases, how to identify and navigate the primary databases for health science research, and how to evaluate articles for quality and credibility. Key to the success the health sciences librarian has experienced is clear and regular communication with faculty and students, a commitment to offering a variety of tools through which students can access the research resources that they need, and the routine evaluation of the effectiveness of her efforts using input from faculty and students. She has experienced overwhelmingly positive feedback from faculty and students because of her commitment to bringing the reference desk to the students, so to speak, in their online environment.

Before creating any type of assessment tool, librarians should consider teaching and learning goals for classroom faculty, which provide the perfect opportunity for discussion with instructors for whom they offer instruction and in whose classes they are embedded. Librarians involved with blended learning should work closely with the course instructor to determine learning outcomes and how the library can support those outcomes. Librarians, who are generally technology leaders on their campuses, can then share ideas for delivering instruction and assessing the students' learning. This can involve using free tools such as Socrative, Poll Everywhere, Padlet, and Google Docs, among many others. Sometimes, communicating with faculty members about assignments can even lead to a co-teaching relationship. It is not enough, though, to just inform teaching faculty what a librarian can do for them. Librarians should share sample work from other partnerships, demonstrate familiarity with the expectations based on the syllabus and previous conversations with the faculty member, offer specific examples of how she or he can meet the needs of the students in that professor's course, and offer customized content to share with students via an online web chat session, the CMS, or both.

8.9 FUTURE WORK: ONLINE INFORMATION LITERACY LAB

Librarians at EKU provide library instruction for all ENG 102 classes, which is mandated by a collaborative agreement between EKU Libraries and EKU's first-year writing coordinator. Until recently, all of these courses were delivered on campus in a face-to-face format. In the last year or so, a few online sections of this course have been introduced as part of a pilot program. Providing instruction to online students is common for most of EKU's instruction librarians, but taking this first-year course content and delivering it in an online environment presents a challenge. The ENG 102 course has more library-related work than any other class at EKU due to the assignments' close alignment to information literacy skills throughout the semester and requires a significant investment of time by the librarian. This makes it difficult to provide an equivalent experience for online students, and as such the course continues to be developed as EKU's librarians search for the most effective combination of services and tools.

The first introduction to the library for most EKU students is through the ENG 102 instruction session, so initiating a personalized and rich connection with students is a goal of that meeting. Re-creating this in an online environment would require some creativity and intentional efforts to build trust and help students see librarians as a partners in their learning process. A group of EKU librarians has been tasked to develop an instructional strategy to help these students acquire information literacy skills, especially those outlined in the libraries' information literacy matrix, a scaffold approach to teaching core information literacy standards. A few of the librarians who work closest with the professors and students in the ENG 102 course researched what other libraries have done to meet the demands of similar online composition courses for whom librarians provide frequent instruction and guidance through information literacy activities tied to specific research assignments. The librarians at EKU discovered a model developed by the University of Arizona (UA) Library that matched the needs and goals of EKU library's instructional partnership with EKU's first-year writing coordinator. An article coauthored by Sult and Mills (2006) offered insight into the history of the UA instruction program that led to the development of an online information literacy instruction approach.

The UA Library's approach is a blended model that involves using prior knowledge of assignments in the ENG 102 course, for which information literacy instruction is required throughout the various stages of writing and research, and combining that with activities that can assist an experienced ENG 102 professor to lead the instruction rather than have a librarian teach the class in a face-to-face manner (Sult & Mills, 2006). The librarian serves as a partner in the development of the weekly activities prior to the beginning of the course but remains more of a guide in the background throughout the semester. Developed by UA librarians and posted in an online guide, the activities are divided into

12 weeks and aligned with the library's information literacy goals. Sult and Mills (2006) explain that "in this 'invisible' role, librarians could help experienced instructors organize and pace their classes and suggest ways to teach the information literacy skills that complimented the composition course activities for each week" (p. 375).

The blended instructional approach developed by the University of Arizona presents EKU's librarians with a strong model for distributed learning in the ENG 102 course. Librarians at EKU hope to develop an online information literacy lab for both online students and possibly even as an alternative to some of the face-to-face sessions offered to current ENG 102 classes using a model similar to the one used at the University of Arizona. The EKU librarians working on this project have discussed creating information literacy activities that happen every week or two weeks and blend with the course assignments so that they do not appear to be separate, library-specific exercises. This strategy is designed to help students see the librarian as a co-teacher who is part of the educational process. Building this online information literacy lab can also help "teach the teacher." In these days of budget constraints and demand for more online instruction, which generally takes more time to prepare and deliver, it is important to get course instructors to share the responsibility of integrating information literacy skills, especially since they are seen as the primary classroom authority (Sult & Mills, 2006).

Several options have been identified for delivery of the information, including a webpage created for the online ENG 102 courses, a LibGuide with ENG 102 modules, and posts within Blackboard. Topic-based activities or modules could include tutorials, worksheets, quizzes, interactive whiteboard activities, video recordings, document sharing via Google Drive, or other methods of providing interactivity, as well as assessment of learning. The first step to implementation is to discuss ENG 102 learning goals with the first-year coordinator followed by meeting with some of the gatekeepers, such as one of EKU's instructional designers and a faculty member or two who are teaching the course. This is a work in progress, and as of today the ideas are still under development and have not been implemented.

8.10 CONCLUSION

Experience is the most valuable and hard-won guidance any library can obtain and should be included in any framework or collection of best practices. An important aspect of librarianship is effective, consistent, and long-term outreach and collaborative efforts by librarians and libraries targeting and involving the gatekeepers that can help or hinder provision of quality resources and services. These efforts must be approached with a practical plan and realistic expectations. Online librarianship best practices have moved beyond the sage on the stage and guide on the side models of teaching and are transitioning to the more egalitarian meddler in the middle approach that brings the librarian into student interaction as an active and equal participant. This reflection on one institution's successes and failures, augmented by supporting documentation from the literature, can serve as a guide to other institutions seeking to create their own best practices to better serve their own online communities.

REFERENCES

Bell, S. J., & Shank, J. (2004). The blended librarian: A blueprint for redefining the teaching and learning role of academic librarians. *College & Research Libraries News*, *65*(7), 372–375. <http://doi.org/doi:10.5860/crl.67.6.514>.

Bell, S. J., & Shank, J. D. (2007). *Academic librarianship by design: A blended librarian's guide to the tools and techniques*. Chicago, IL: American Library Association.

Bowles-Terry, M., Hensley, M., & Hinchliffe, L. J. (2010). Best practices for online video tutorials: A study of student preferences and understanding. *Communications in Information Literacy*, *4*(1), 17–28. <http://doi.org/10.7548/cil.v4i1.83>.

Campbell, L., Matthews, D., & Lempinen-Leedy, N. (2015). Wake up information literacy instruction: Ideas for student engagement. *Journal of Library Administration*, *55*(7), 577–586. <http://doi.org/10.1080/01930826.2015.1076313>.

Eastern Kentucky Unviersity Libraries. (n.d.) *Strategic plan, 2011–2015*. Retrieved from <http://library.eku.edu/strategic-plan>.

Edwards, M. E., & Black, E. W. (2012). Contemporary instructor-librarian collaboration: A case study of an online embedded librarian implementation. *Journal of Library & Information Services in Distance Learning*, *6*(3–4), 284–311. <http://doi.org/10.1080/1533290X.2012.705690>.

Hoffman, S. (2011). Embedded academic librarian experiences in online courses: Roles, faculty collaboration, and opinion. *Library Management*, *32*(6/7), 444–456. <http://doi.org/10.1108/01435121111158583>.

McWilliam, E. (2007). Unlearning how to teach. *Paper presented at the creativity or conformity conference*. Cardiff, Wales, UK.

McWilliam, E. (2009). Teaching for creativity: From sage to guide to meddler. *Asia Pacific Journal of Education*, *29*(3), 281–293.

Morasch, M. J. (2013). Embedded reference: Providing research guidance within the education courseroom. *Journal of Library & Information Services in Distance Learning*, *7*(3), 297–312. <http://doi.org/10.1080/1533290X.2013.783524>.

National Center for Education Statistics, U.S. Department of Education. (2013). *Table 311.15: Number and percentage of students enrolled in degree-granting postsecondary institutions, by distance education participation, location of student, level of enrollment, and control and level of institution: Fall 2012*. Retrieved October 28, 2015, from the National Center for Education Statistics: <https://nces.ed.gov/programs/digest/d13/tables/dt13_311.15.asp?referrer=report>.

National Science Foundation. (2014). *STEM education and data trends: What percentage of postsecondary institutions offer distance education?* Retrieved from the National Science Foundation: <https://www.nsf.gov/nsb/sei/edTool/data/college-03.html>.

Oakleaf, M. (2009). The information literacy instruction assessment cycle: A guide for increasing student learning and improving librarian instructional skills. *Journal of Documentation, 65*(4), 539–560. <http://doi.org/10.1108/00220410910970249>.

Padlet. (n.d.). Retrieved October 10, 2015, from <https://padlet.com/>.

Ratcliffe, S. (2011). *Concise Oxford dictionary of quotations* (6th ed.Oxford: Oxford University Press.

Sult, L., & Mills, V. (2006). A blended method for integrating information literacy instruction into English composition classes. *Reference Services Review, 34*(3), 368–388. <http://doi.org/10.1108/00907320610685328>.

"*What can I do with Padlet?*" (n.d.). Retrieved from <http://jn.padlet.com/article/55-what-can-i-do-with-padlet>.

Zapalska, A., & Brozik, D. (2006). Learning styles and online education. *Campus-Wide Information Systems, 23*(5), 325–335. Retrieved from <http://www.emeraldinsight.com/doi/abs/10.1108/10650740610714080>.

SECTION III

Technology

CHAPTER 9

Delivering Synchronous Online Library Instruction at a Large-Scale Academic Institution: Practical Tips and Lessons Learned

9.1 INTRODUCTION

Like those at many large academic libraries, librarians at the University of Toronto Libraries (UTL) strive to provide meaningful and engaging information literacy instruction (ILI) for a student population that continues to grow in size and diversity. To accommodate the increasing instruction load that accompanies a growing student population and a desire to reach more undergraduates at the beginning of their academic careers, a group of University of Toronto (UT) librarians[1] and instructional technologists[2] came together in the spring of 2013 to explore alternatives to traditional face-to-face ILI. The resulting working group designed and delivered a session titled "Getting Started at U of T Libraries," a 30-minute library orientation session based on a UTL research guide[3] by the same name. The working group delivered the session nine times over the course of a 3-week period in the fall of 2013: three times in person and six times synchronously online via Collaborate, Blackboard's web-conferencing tool.

---

[1] Patricia Ayala, Maria Buda, Heather Buchansky, Hyun-Duck Chung, Monique Flaccavento, Eveline Houtman, Judith Logan, Vincci Lui, Jeff Newman, Michelle Spence, and Jenaya Webb.
[2] Mariana Jardim and Derek Hunt.
[3] An archived version of the guide as it existed in 2013 can be viewed here: http://wayback.archive-it.org/279/20151008162520/http://guides.library.utoronto.ca/orientationguide/. A current version of the guide is available at http://guides.library.utoronto.ca/orientationguide.

Distributed Learning.
© 2017 T. Maddison and M. Kumaran.
Published by Elsevier Ltd. All rights reserved.

This chapter reports on the UT project. Authors provide a rationale and background for the project and then an overview of the development and delivery of the session. They share the results of the project and some of their successes, failures, and surprises. Finally, they reflect on the experiences of the librarians involved in the project and discuss the possibilities and limitations of synchronous online instruction for incoming undergraduates.

9.2 BACKGROUND

The UT has 83,000 students: more than 67,000 undergraduate students and nearly 16,000 graduate students.[4] Sixty librarians provide ILI to this population, but only a handful have instruction as one of their core job functions. The ratio of students to teaching librarians—approximately 1400:1—combined with the lack of a core first-year writing and research course, led the working group to conclude that providing face-to-face library instruction for all incoming undergraduates would not be feasible. In addition, a large number of UT students live off campus and would likely have limited opportunities to attend in-person library sessions.[5]

UT librarians had already created a vast array of asynchronous online learning objects such as screen casts, interactive guide-on-the-side tutorials, and research guides to support students, but they were hopeful that a larger number of UT's incoming 14,637 undergraduates[6] could be engaged through the delivery of synchronous online ILI.

At the library administration level, there was also a desire to see that a greater number of undergraduate students receive library instruction. The most recent library strategic plan promised to "amplify the University's global reach by developing educational initiatives that can be delivered at scale to a large and diverse student body" (University of Toronto Libraries, 2013).

---

[4] See Part D of the Office of the Assistant Vice-President's 2013 statistics for more detailed information about UT's student body (Office of the Assistant Vice-President, Government, Institutional and Community Relations, University of Toronto, 2013).

[5] An internal report summarizing a survey completed by staff at the University of Toronto's Office of Student Life found that 49% of respondents commuted over 40 minutes to campus (Ashraf et al., 2012).

[6] See Part D of the Office of the Assistant Vice-President's 2013 statistics for more detailed information about UT's student body (Office of the Assistant Vice-President, Government, Institutional and Community Relations, University of Toronto, 2013).

9.3 LITERATURE REVIEW

Whether they are working to support online courses (Kontos & Henkel, 2008; Kumar & Ochoa, 2012) or finding new ways to engage growing numbers of distance users (Bonnand & Hansen, 2012; Lietzau & Mann, 2009; Nicholson & Eva, 2011; Riedel & Betty, 2013), librarians are continuously developing and evaluating methods for delivering instruction outside of the traditional classroom setting.

Online initiatives, whether synchronous, asynchronous, or hybrids, are increasingly accepted as effective methods for delivering ILI sessions, with many studies indicating that online learning can be as successful as face-to-face instruction. In 2007, Zhang, Watson, and Banfield conducted a systematic review and analyzed 10 studies that compared face-to-face and online instruction. Based on those studies, they concluded that computer-assisted instruction and face-to-face instruction were "equally effective for delivery of basic library skills" (p. 483). Subsequent case studies have supported this conclusion by indicating that in a variety of contexts and using different online approaches, online instruction initiatives were as effective as traditional face-to-face instruction (Anderson & May, 2010; Brettle & Raynor, 2013; Greer, Nichols Hess, & Kraemer, 2016 pre-print) and, in some cases, have shown to be students' preferred method of instruction (Silver & Nickel, 2007).

There is also growing representation in the literature of research and case studies that specifically examine synchronous online instruction. The bulk of these studies focus on the use of synchronous delivery in course- and program-based contexts such as librarians embedded in an online doctoral program (Kumar & Ochoa, 2012), using institutionally licensed software to deliver synchronous sessions to online graduate students (Barnhart & Stanfield, 2011), and using Skype to provide information literacy sessions for an academic writing course for students on satellite campuses (Nicholson & Eva, 2011). Although many of these case studies reported technical difficulties, most intend to improve and expand additional sessions.

At the time of the pilot project, little had been reported on efforts to offer online synchronous library instruction outside of a course-based context. One example comes from Lietzau and Mann (2009), who describe various asynchronous instruction initiatives at the University of Maryland University College, a large provider of distance education in the United States. They make mention of a series of six synchronous online RefWorks sessions offered for graduate faculty. These sessions seem to have been offered widely and were well attended by faculty members. The authors report that the response to these sessions was positive and that overall the library found web conferencing to be a valuable service that it would continue to market and offer in the future (Lietzau & Mann, 2009).

More recently, several studies have been published that report on online synchronous instruction outside of a specific course- or program-based environment. In their 2014 article, Jarrell and Wilhoite-Mathews provide a brief example. Building on requests from both graduate students and faculty members, librarians at Ball State University experimented with a synchronous version of one of their most popular graduate workshops. However, despite spreading the word, they had no attendees for the online session. They have offered other workshops as synchronous online sessions but had little success overall and wrote that they decided to abandon the method for non—course-based offerings (Jarrell & Wilhoite-Mathews, 2014).

Another example comes from the University of Alabama at Birmingham's health sciences library, where librarians sought to connect with users not commonly reached by their traditional face-to-face instruction program by reinventing a popular in-person training series "Express Training" as a synchronous online session (Smith & O'Hagan, 2014). The sessions were aimed at distance students and nonstudent users, including clinicians, researchers, and faculty members. Session evaluations and follow-up surveys indicated that the sessions were effective and that participants had an overall positive experience with the online training. Given the popularity of the sessions for both on-campus and remote users, the authors wrote that they plan to continue to expand the offering (Smith & O'Hagan, 2014).

In their 2015 article, Moorefield-Lang and Hall took a different approach to examining synchronous instruction in non—course-based contexts at Virginia Tech's Newman Library. They conducted one-on-one interviews with five graduate students who had participated in at least one library webinar to better understand the students' experiences with and perceptions of online learning more generally. Analysis of the interviews suggested that overall students liked the webinars and found them convenient. However, the authors write that "there is still much to learn and avenues to explore as library webinars are offered" (p. 66). Moorefield-Lang and Hall still had questions about the degree of buy-in for online synchronous sessions not tied to a specific courses, but they indicated that Virginia Tech's libraries would be expanding and offering more webinars going forward (Moorefield-Lang & Hall, 2015).

Despite reporting challenges (including technical difficulties and instances of low or no attendance), most of the preceding studies support further exploration of synchronous online delivery for library instruction in various contexts. This chapter will contribute to this growing body of literature and begin to address a gap in reporting on synchronous initiatives for new incoming undergraduates.

9.4 THE PILOT

With the aim of reaching more of UT's incoming undergraduates, the working group of librarians and instructional technologists decided to offer in-person and online sessions. A synchronous session, it was hoped, would allow for a more personal user experience for students just arriving at UT and give students the opportunity to connect with a librarian as they would in a face-to-face session. In addition, this approach offered an opportunity for librarians to connect with a larger number of undergraduates, including the population of commuter students who might not be on campus every day.

UT had acquired Blackboard Collaborate as its institutional web-conferencing solution in Dec. 2012 as an add-on to Blackboard, UT's learning management system. The question was whether or not Collaborate could provide an effective platform for the delivery of large-scale synchronous online ILI. To answer this question, the working group tasked itself with:

- conducting an environmental scan to learn about similar initiatives at UT and other institutions,
- exploring the functionality and feasibility of Collaborate to deliver online ILI sessions,
- developing at least one online pilot ILI session for first-year undergraduates for the fall of 2013,
- reviewing course evaluations and using these to suggest improvements to future online ILI sessions, and
- evaluating the pilot project—documenting lesson learned and making recommendations regarding next steps.

Members of the working group met in person in May and June and online via Collaborate in July and August. Throughout August, members paired up to practice delivering the session. Following extensive

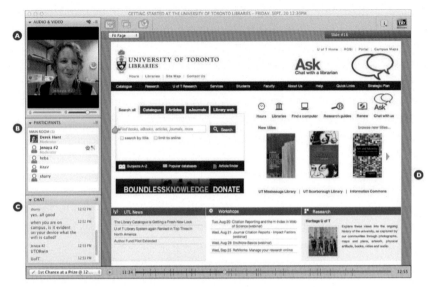

Figure 9.1 Participant view of Collaborate interface showing (A) a live video window, (B) active participants, (C) a chat box, and (D) a slide presentation window.

explorations of the Collaborate platform and its features, the working group determined that it would be best to deliver a synchronous session in Collaborate with a live feed of the teaching librarian accompanied by a PowerPoint slideshow, interactive polls, and a chat window (see Fig. 9.1) rather than rely on live searches that were technologically unreliable. In addition to a slide deck, the working group developed:

- a checklist to help teaching librarians prepare, deliver, and follow up after a Collaborate session;
- a script for teaching librarians;
- an online session evaluation;
- a precourse email to students with instructions for testing system requirements, selecting their Internet connection speed, and logging into the session;[7]
- a postcourse email to students, including links to a recording of the session, the "Getting Started at the U of T Libraries" guide, and the session evaluation; and
- an extensive marketing plan using a variety of materials and channels.

---

[7] As a number of Collaborate features were not yet available for tablets and smartphones, the working group recommended that students use laptops or desktop computers.

The session outlined 10 important tips for new students' first few weeks at UT:

1. Get your TCard: Explains that the UT student card functions as a library card.
2. Study at the library: Promotes the wide variety of study spaces available.
3. Get online: Details Wi-Fi and library computer access.
4. Print and photocopy: Gives basic cost and instructions for printing and photocopying.
5. Borrow books: Summarizes loan privileges and procedures.
6. Be a good (library) neighbor: Discusses basic conduct regulations and community norms.
7. Find your course textbooks: Introduces course reserves.
9. Ask for help: Promotes research help services.
9. Take advantage of the library's collection: Points to access points for discovering the library's collections.
10. Learn more: Suggests upcoming library tours and workshops.

The session was delivered nine times over the course of a 3-week period at the end of September and the beginning of October: six times online and three times in-person. Sessions were offered at different times of day and days of the week[8] to accommodate the schedules of as many students as possible. Enrollment for in-person sessions was capped at 40, but there was no limit placed on enrollment for online sessions.

9.5 EVALUATING THE PILOT

9.5.1 Registration and Attendance

Student participation in sessions was entirely voluntary. Sessions were not associated with any academic program or formal orientation event. Students learned about library sessions through Tweets,[9] Facebook posts, promotional images on UTL's digital wall signs and library websites; posters, buttons, and print tags; word of mouth; and a direct email sent to

---

[8] Sessions were offered Mondays, Wednesdays, Thursdays, and Fridays. Starting times ranged between 10 am and 8 pm.

[9] Tweets were posted to @uoftlibraries (>4000 followers), @GersteinLibrary (>1000 followers), and @OISELibrary (>2000 followers)

Table 9.1 Comparison of registration and attendance for online and in-person sessions with average attendance and attendance rates calculated

	Number of sessions	Number of students registered	Average number of students registered per session	Number of students who attended sessions	Attendance rates (%)
In-person sessions	3	66	22	43	65
Online sessions	6	51	9	16	31
All sessions	9	117	13	59	50

approximately 2800 incoming first-year students via the personal librarian program.[10]

A total of 117 students registered for 9 workshops, with 59 attending the sessions, making an overall attendance rate of 50%. This is just under the average rate of attendance—56%—for other University of Toronto Libraries open registration workshops aimed at undergraduate students (Fratarcangeli, 2015).

In-person "Getting Started at the U of T Libraries" sessions received more registrants and attendees on average than online sessions. On average, 22 students registered for each in-person session and 14 attended. This is higher than the average nine students who registered for each online session with only three attending. Table 9.1 shows that the overall attendance rate for in-person sessions was 65%, while the overall attendance rate for online sessions was less than half: 31%.

9.5.2 Session Evaluation

The working group provided a session evaluation to attendees of both the online and in-person sessions to compare learning and affective outcomes. It was administered as an online questionnaire. At the end of each session, librarians provided a link to the session evaluation and asked attendees to take a moment to provide feedback. Submissions did not include personally identifiable information. Of the 59 students who attended a session (in-person or online), 29 responded to the session evaluation. Of these 29 respondents, 12 had attended an online session and 17 an in-person session.

[10] UT Personal Librarian Program: https://personal.library.utoronto.ca/index.php/content/about

Table 9.2 Summary of academic status of evaluation respondents separated by in-person and online sessions

Status of session evaluation respondent	Number of respondents who attended in-person sessions	Number of respondents who attended online sessions	Total number of respondents
Undergraduate student, new to UT	13	8	21
Undergraduate student, transferred from another postsecondary institution	1	2	3
Graduate student, new to UT	1	1	2
Graduate student, returning to UT	1	1	2
Undergraduate student, returning to UT	1	0	1
Total number of respondents	**17**	**12**	**29**

The working group first wanted to know if it had reached its target audience of new first-year undergraduates. Table 9.2 shows that roughly 72% of the attendees fit this category; approximately 10% of respondents were transfer students and 7% were new graduate students. The working group was surprised to find that a small number of returning undergraduate and graduate students were also attracted to the session.

The working group also wanted to find out if the sessions had helped the attendees adjust to UT. Online and in-person respondents both reported levels of increased comfort with the library (see Table 9.3 for a summary). Eighty-two percent of in-person respondents reported that they "definitely" felt more comfortable using the library; 83% of online respondents reported the same. The remaining 18% of in-person respondents and 17% of online respondents were only "sort of" more comfortable with the library after the session. No respondent reported being "not at all" or "not really" more comfortable with the library after the session.

Online respondents did not report many issues with using Collaborate. Several respondents stated that the technical instructions provided for connecting to the session were clear. One respondent also noted that she or he appreciated the reminder sent 15 minutes prior to the session because his or her Java plugin had not been updated.

Table 9.3 Summary of responses the session evaluation question: "After taking this class, do you feel more comfortable using the library?" by in-person and online attendees

After taking this class, do you feel more comfortable using the library?

	Not at all	Not really	Sort of	Yes, definitely	Total
Respondents who attended in-person sessions	0 (0%)	0 (0%)	3 (18%)	14 (82%)	**17 (100%)**
Respondent who attended online sessions	0 (0%)	0 (0%)	2 (17%)	10 (83%)	**12 (100%)**
All respondents	**0 (0%)**	**0 (0%)**	**5 (17%)**	**24 (83%)**	**29 (100%)**

The final questions in the evaluation allowed respondents to share anything not already covered. Most responses were positive and appreciative; one respondent added that she or he "loved" the convenience of attending the session online and hoped more classes would be delivered that way.

9.5.3 Technical Issues and Teaching Librarians' Experiences

Many studies have reflected on the technical difficulties experienced by attendees and instructors in online synchronous sessions (e.g., see Barnhart & Stanfield, 2011; Bonnand & Hansen, 2012; Kontos & Henkel, 2008). To address anticipated technical issues, the UT working group decided that teams of two would deliver each online session. In addition, a technical support staff member would attend to assist with technical issues encountered during the session. Indeed, while delivering the online sessions, the working group experienced problems uploading presentation slides and adjusting audio settings, glitches with audio and video, and managing Internet connectivity.

During test runs of the online sessions, teams who sat in the same room realized that two active microphones in close proximity caused unpleasant audio feedback. To circumvent this problem during the live sessions, both librarians wore headphones and only one microphone was enabled at a time. One librarian monitored the participants' chat messages while the other led the session. In other sessions, some teams chose not to be in the same location during delivery to avoid audio feedback issues.

Internet connectivity posed serious challenges for members of the working group, as well as participants. Attendees and librarians were advised to connect to the session via a network cable or a strong wireless connection. For those who connected wirelessly, there were sometimes issues with maintaining the connection. In several cases, this prevented a librarian from joining the session, necessitating a backup librarian to join at the last minute. In the end, despite the documentation, training, and testing, the working group still required significant assistance from the University of Toronto's Blackboard support team to set up and run the sessions.

In a follow-up interview, member of the working group stated that the technical aspect of delivering the sessions was more challenging than in-person instruction: "It took up a lot of more time, technically, to prepare for things. Whereas when you are face-to-face in a classroom, you don't have a lot of technical issues to deal with. If something goes wrong you can always talk to the person." The experience of the working group closely reflects what is reported in the literature with respect to the main technical difficulties encountered, as well as the importance of experience, proper training, and documentation for teaching librarians (Bonnand & Hansen, 2012; Lietzau & Mann, 2009; Smith & O'Hagan, 2014). A positive outcome of the pilot was that the experience of developing and delivering synchronous sessions allowed the working group to gain familiarity with Collaborate and to generate a checklist for delivering synchronous online library instruction.

9.6 OUTCOMES

Although enrollment and attendance rates were much lower than anticipated for the online sessions, the pilot was nonetheless successful in the following ways. The working group:
- developed a strong understanding of the possibilities and limitations of synchronous online instruction,
- took a lead role in testing and sharing its experience with UT's institutional web-conferencing software, and
- developed a checklist for delivering future synchronous online library sessions.

One of the working group's goals was to reach a larger number of incoming students with ILI initiatives. The pilot project was an opportunity to closely examine the challenges of delivering large-scale instruction initiatives while still focusing on good pedagogical practices. Reflecting on the pilot, several team members recalled a focus on the technology rather than the pedagogy. One stated, "I remember the sessions ... but what really sticks out is the technical . . . how are we going to deal with the technology, rather than the content itself." Indeed, one of the most important lessons of the pilot was how much technical support was required to run the sessions.

The limitations of the software also shaped the content of the sessions themselves. For instance, leading up to the pilot, members of the working group repeatedly experienced lengthy delays between what was displayed on the teaching librarian's screen and the student's screen, making "live" searching problematic. As a result of this limitation, the group opted to deliver a traditional lecture-style session supported by PowerPoint slides rather than take students on a live tour of the library website as was initially hoped.

Following the pilot, the working group prepared a comprehensive report that documented the successes, challenges, and technical issues experienced over the course of preparing and delivering the sessions. The report opened discussions with technical and instructional support staff across the university about the specific types of functionalities librarians need to deliver online instruction. Top among these was the importance of live searching and screen sharing. Ultimately, these discussions helped technical and instructional support staff advocate for changes to Collaborate's functionality and implement available upgrades across UT. This improved the likelihood that teacher librarians would employ the software again and benefited instructors, students, and others at the university using the software.

9.7 CHECKLIST FOR DELIVERING SYNCHRONOUS ILI

One key outcome of the pilot project was a checklist of practical tips and considerations for librarians planning to deliver synchronous online ILI sessions. A condensed version of the checklist follows.

Checklist for Teaching Librarians
Before Your Session
- Learn about the accessibility features and limitations of the software. See, e.g., Blackboard Collaborate's "Accessibility Guide for Moderators."[11]

(Continued)

[11] www.blackboard.com/docs/documentation.htm?DocID=611001PDF

cont'd

- Send a reminder email to students registered in the session. Include instructions about system checks or other information they'll need to join the session.
- If possible, arrange to have an IT staff member available on call during your session in case technical issues arise.
- Consider what will appear behind you in the room: remove any unnecessary clutter because this can create visual distractions; post something meaningful like a library workshops poster or a "We're Here to Help" sign.
- Ensure that the software works properly on the computer you will be using.
- Conduct audio and video checks.
- Connect with an ethernet cable rather than a wireless network. Wireless can be unreliable and may cause delays in audio and video or even drop the session.
- Upload slides at least a half hour before the session starts in case there are problems with the upload.
- Set up the system so that participants can join the session at least 15 minutes in advance. This will give you time to help students troubleshoot any technical issues that arise.
- Include a "Welcome" slide that lets participants know that the session will be starting shortly. Consider including the start time for the session.
- Close your door to reduce noise; post a sign notifying others that an online class is in progress.
- Mute your office phone.

During Your Session

- Welcome participants as they join the session orally or using the chat window.
- Remember to inform participants that the session will be recorded.
- Wear a headset with a built-in microphone to ensure clear audio.
- It is challenging to lead a session and monitor or respond to questions in the chat window at the same time. If possible, invite someone to co-teach the session with you.
- Remember to wear headphones to prevent an audio feedback loop or echo if both instructors are in the same space and each is using a computer.
- Begin the session with an audio and video check. Ask participants to give you a thumbs-up if they can hear and see you.
- Conduct audio and video checks periodically throughout the session.

(Continued)

cont'd
- Explain that participants can ask questions by typing in the chat window or by using the "raise your hand" feature if you have asked them to mute their microphones.
- Make sure to show students where to find (and how to use) features such as raising and lowering their hands, answering yes or no for polls, and so on.
- Look straight into the camera.
- Remember to smile. It is important to communicate approachability.

After Your Session
- Stop recording the session.
- Remember to send any follow-up materials, a link to the session recording, a feedback survey, and so on to participants.

9.8 DISCUSSION

One main goal of the pilot project was to explore ways to reach out to more of UT's incoming undergraduate students. The working group ran an extensive marketing campaign during orientation week that targeted incoming students via social media, print, and an email via the personal librarian program that reached approximately 2800 students directly. Despite this outreach, registration and attendance numbers for the online orientation sessions were much lower than anticipated and failed to connect with the large number of participants the team had hoped to reach. This was one of the key surprises of the pilot project.[12]

Furthermore, registration numbers for the online sessions were lower than registration numbers for the in-person sessions. The working group contacted all online registrants (including those who did not attend the sessions) to follow up with the slides, as well as provided a link to the workshop evaluation. However, the evaluation did not include a question for those who did not attend the session to explain *why* they did not attend; in retrospect, this question and respondents' answers may have provided valuable feedback.

[12] Other examples of low uptake appear in the literature. For instance, Barnhart and Stanfield (2011) and Bonnand and Hansen (2012) make reference to limited uptake to their synchronous initiatives. Other studies (Jarrell & Wilhoite-Mathews, 2014; Kontos & Henkel, 2008) point to low to no student attendance for the synchronous sessions they offered.

The low uptake for the sessions (and the online sessions in particular) points to a gap in the working group's understanding of the perspectives and preferences of incoming undergraduates at UT. Were the low online registration and attendance rates a result of technological barriers, timing and marketing, or the appeal of the course content? The following three key questions emerged from the working group's reflections.

9.9 WERE THERE TECHNOLOGICAL BARRIERS TO STUDENT ATTENDANCE?

The online sessions had low registration rates and even lower attendance rates. Based on the working group's experiences and technical difficulties, it seems possible that potential participants encountered technological barriers that stopped them from logging on for the sessions.

Students may have had difficulty following the log-in instructions or been intimidated by their complexity. It is also possible that students were not able to complete Java updates or compatibility tests, or that their system requirements did not accommodate Collaborate's needs. Or perhaps they made it through the setup steps but, for any number of reasons, were not in a location with a good Internet connection at the time of the session. The sessions required students log in to Collaborate without necessarily having had any previous experience with Blackboard or any other LMS. None of the students who attended online sessions reported technological issues.

9.10 WAS THE TIMING APPROPRIATE?

Could the time of day, day of week, timing in the semester, or even timing within a student's academic career have been factors in the low registration and attendance numbers? Registration in the online sessions was especially low overall, with the exception of the one evening session (8 pm), where registration numbers were comparable to those of the in-person daytime sessions (11:30 am, 12:30 pm, and 4 pm). However, even the most well-attended sessions were well below capacity. In addition, given the limited number of sessions, it is not possible to say conclusively whether or not the day of week played an important role in registration and attendance numbers.

In terms of the time of semester when library orientation sessions were offered, registration and attendance numbers for in-person sessions were higher earlier in the semester and dropped off later in the term. This pattern is consistent with other open-registration library sessions offered at UTL. Despite this pattern, it's important to remember that all of the "Getting Started at U of T Libraries" sessions underperformed in terms of attendance. Could this be because incoming undergraduate students, the target audience for these sessions, were just too busy or overwhelmed in the first weeks of their academic careers to be interested in attending a library orientation session? Researchers in the United Kingdom (Hughes & Smail, 2015) recently conducted a large qualitative study of new university students to gain a better understanding of which aspects of university life are priorities during their first weeks and months. Key themes in the responses suggest that social integration is a much more urgent concern to students than academics. With so many social opportunities available during incoming students' first few weeks on campus, it seems likely that a library orientation session would be a low priority.

9.11 IS SYNCHRONOUS ONLINE DELIVERY RIGHT FOR THE TARGET AUDIENCE?

Given that the in-person sessions received more registrants and far more attendees than online sessions, the working group wondered whether students might simply lack familiarity with synchronous online learning. In the fall of both years following the pilot project, only the face-to-face version of the "Getting Started at U of T Libraries" session was offered. Table 9.4 provides an overview of all the in-person offerings of the session. Scheduling, registration, and attendance rates were comparable between 2013—14 in-person and 2014—15 sessions. The 2015—16 sessions were not promoted as heavily and consequently received fewer

Table 9.4 Comparison of registration and attendance for all in-person "Getting started at U of T Libraries" sessions

	Total number of in-person sessions	Total number of students registered	Average number of students registered per session	Total number of students who attended sessions	Attendance rate (%)
2013—14	3	66	22	43	65
2014—15	4	61	15	37	61
2015—16	3	21	7	19	90
All sessions	10	148	15	99	72

registrations. The attendance rate for these sessions was high, however. When compared with Table 9.1, this suggests that the low attendance rate for online sessions (31%) was not a result of the content of the sessions.

In addition, the group hypothesized that it might be hard for students to find a quiet space and the equipment needed to participate in an online session. Reflecting on the target audience, the working group recognized that students entering their first year at university may never have participated in an online course. If this is the case, a completely new learning environment with a steep technological learning curve might be daunting. Perhaps it might be more fruitful to explore learning environments that are more familiar to students such as Skype and Google Hangouts.

While much of the literature points to the effectiveness of online delivery methods, synchronous online delivery for non−course-based ILI sessions specifically is still in the experimental phase. Indeed, in their literature review, Greer, Nichols Hess, and (Kraemer 2016, pre-print) remind us that there is substantial evidence showing that library instruction of any kind is most effective when tied to a course or a particular learning need. Moorefield-Lang and Hall (2015) also recognized enrollment as a problem, particularly for webinars: "When webinars are publicized students and faculty are very excited to sign up. However, when it is time for the actual sessions to air online, the percentage of participants who follow through to attend the session is low" (p. 67).

Perhaps new students might prefer in-person sessions and the opportunity to meet other students, make friends, and get to know their campus. UT research corroborates this hypothesis. A major focus group project, sponsored by the University of Toronto's Council on Student Experience, found that "feeling community in the classroom setting is very important to students.... They report feeling more motivated and encouraged to study when they experience community" (Office of the Vice-Provost, 2013, p. 22). In light of this, the online sessions may not have been attractive as a means of meeting other new students than the in-person sessions.

9.12 CONCLUSION

In striving to reach more incoming students, librarians at UT will continue to build on what they have learned to explore options for delivering synchronous online ILI in an engaging and meaningful way. Although enrollment and attendance rates for the "Getting Started at U of T Libraries" sessions were much lower than anticipated, the project was still successful in developing a strong understanding among the working group of the possibilities and limitations of synchronous online instruction. However, librarians will still need to address significant gaps in their knowledge about incoming first-year undergraduate students at the University of Toronto.

Questions raised in the discussion section of this paper stem from a lack of knowledge about the experiences, expectations, and priorities of the target audience. In addition, the working group made a number of assumptions about what would be more convenient, accessible, and appealing to students that might not be reflective of what they actually prefer. The authors believe that further research is needed to deepen understanding of this diverse group of students to help ensure the library's instructional objectives match more closely with the experiences and expectations of incoming undergraduates.

REFERENCES

Anderson, K., & May, F. A. (2010). Does the method of instruction matter? An experimental examination of information literacy instruction in the online, blended, and face-to-face classrooms. *The Journal of Academic Librarianship, 36*(6), 495–500. <http://doi.org/10.1016/j.acalib.2010.08.005>.

Ashraf, A., Getchell, L., Jardim, M. E., Kim, D., Minella, K., & Saad, S. (2012). *Survey of commuter students*. Toronto: University of Toronto Office of Student Life.

Barnhart, A. C., & Stanfield, A. G. (2011). When coming to campus is not an option: Using web conferencing to deliver library instruction. *Reference Services Review, 39*(1), 58–65. Available from: <http://dx.doi.org/10.1108/00907321111108114>.

Bonnand, S., & Hansen, M. A. (2012). Chapter 15: Embedded librarians: Delivering synchronous library instruction and research assistance to meet needs of distance students and faculty. In H. Wang (Ed.), *Interactivity in e-learning: Case studies and frameworks*. (pp. 326–339). Hershey, PA: IGI Global. Available from: <http://dx.doi.org/10.4018/978-1-61350-441-3.ch015>.

Brettle, A., & Raynor, M. (2013). Developing information literacy skills in pre-registration nurses: An experimental study of teaching methods. *Nurse Education Today, 33*(2), 103–109. Available from: <http://dx.doi.org/10.1016/j.nedt.2011.12.003>.

Fratarcangeli, B. (2015). Open registration attendance rates 2014–2015 *[data file]*. Toronto: University of Toronto Libraries, Unpublished dataset, cited with permission.

Greer, K., Nichols Hess, A., & Kraemer, E. W. (2016 pre-print). The librarian leading the machine: A reassessment of library instruction methods. *College & Research Libraries*, Accepted: April 17, 2015; Anticipated Publication Date: May 1, 2016

Hughes, G., & Smail, O. (2015). Which aspects of university life are most and least helpful in the transition to HE? A qualitative snapshot of student perceptions. *Journal of Further and Higher Education, 39*(4), 466–480. Available from: <http://dx.doi.org/10.1080/0309877X.2014.971109>.

Jarrell, L., & Wilhoite-Mathews, S. (2014). Making connections: Reaching online learners at the Ball State University Libraries. *Indiana Libraries, 33*(2), 33–36. Retrieved from <https://journals.iupui.edu/index.php/IndianaLibraries/article/view/16506>.

Kontos, F., & Henkel, H. (2008). Live instruction for distance students: Development of synchronous online workshops. *Public Services Quarterly, 4*(1), 1—14. Available from: <http://dx.doi.org/10.1080/15228950802135657>.

Kumar, S., & Ochoa, M. (2012). Program-integrated information literacy instruction for online graduate students. *Journal of Library & Information Services in Distance Learning, 6* (2), 67—78. Available from: <http://dx.doi.org/10.1080/1533290X.2012.684430>.

Lietzau, J. A., & Mann, B. J. (2009). Breaking out of the asynchronous box: Using web conferencing in distance learning. *Journal of Library & Information Services in Distance Learning, 3*(3—4), 108—119. Available from: <http://dx.doi.org/10.1080/15332900903375291>.

Moorefield-Lang, H., & Hall, T. (2015). Instruction on the go: Reaching out to students from the academic library. *Journal of Library & Information Services in Distance Learning, 9* (1—2), 57—68. Available from: <http://dx.doi.org/10.1080/1533290X.2014.946347>.

Nicholson, H., & Eva, N. (2011). Information literacy instruction for satellite university students. *Reference Services Review, 39*(3), 497—513. <http://doi.org/10.1108/00907321111161458>.

Office of the Assistant Vice-President, Government, Institutional and Community Relations, University of Toronto. (2013). *Facts and figures 2013.* <http://www.utoronto.ca/sites/default/files/about/Facts_Figures_2013/Part_D_Students5085.pdf> Accessed 26.10.15.

Office of the Vice-Provost, University of Toronto. (2013). *A sense of community.* <http://www.viceprovoststudents.utoronto.ca/councils-committees/CSE/senseofcommunity.htm> Accessed 30.10.15.

Riedel, T., & Betty, P. (2013). Real time with the librarian: Using web conferencing software to connect to distance students. *Journal of Library & Information Services in Distance Learning, 7*(1—2), 98—110. <http://doi.org/10.1080/1533290X.2012.705616>.

Silver, S. L., & Nickel, L. T. (2007). Are online tutorials effective? A comparison of online and classroom library instruction methods. *Research Strategies, 20*(4), 389—396. Available from: <http://dx.doi.org/10.1016/j.resstr.2006.12.012>.

Smith, S. C., & O'Hagan, E. C. (2014). Taking library instruction into the online environment: One health sciences library's experience. *Journal of the Medical Library Association, 102*(3), 196—200. <http://dx.doi.org/10.3163/1536-5050.102.3.010>.

University of Toronto Libraries. (2013). *Charting our future: University of Toronto Libraries' strategic plan 2013—2018.* <https://onesearch.library.utoronto.ca/sites/default/files/strategic_planning/UTL-Strategic-Plan-2013-18.pdf#overlay-context = strategic-plan/strategic-plan-2013-2018> Accessed 30.10.15.

Zhang, L., Watson, E. M., & Banfield, L. (2007). The efficacy of computer-assisted instruction versus face-to-face instruction in academic libraries: A systematic review. *The Journal of Academic Librarianship, 33*(4), 478—484. Available from: <http://dx.doi.org/10.1016/j.acalib.2007.03.006>.

CHAPTER 10

Making Library Research Real in the Digital Classroom: A Professor–Librarian Partnership

10.1 INTRODUCTION

Online learning has become prevalent in higher education. According to the Alfred P. Sloan Foundation's latest report (Allen & Seaman, 2016), the number of students in the United States who take at least one online course tripled over the last 13 years (p. 3); and in the fall of 2014, 28% of students took some of their courses online (p. 12). To address this need to take online courses, library instruction professionals are designing new methods to introduce information literacy skills to their students. This study introduces an integrated course design technique borrowed from frequently used education practices that provides research instruction to first-year college students.

Methods of integrating research instruction into online courses have been documented since 1999, when a team of librarians at the University of California–Riverside offered library instruction using the university's Web-Course-in-a-Box software (Getty, Burd, Burns, & Piel, 2000). Librarians at Central Missouri State University began collaborating with professors who were teaching online courses in 1999 (Lillard & Dinwiddie, 2004, p. 136). Articles from the first decade of this century suggest that a librarian's presence in the online course's learning management system (LMS) creates opportunities for students to engage with a research professional via discussion boards, blog posts, and video tutorials (Lillard & Dinwiddie, 2004; Kraemer, Lombardo, & Lepkowski, 2007). Studies indicate that bibliographic instruction via the Blackboard interface encourages students to connect with the library when gathering research (Daly, 2010; Stewart, 2007).

© 2017 T. Maddison and M. Kumaran.
Published by Elsevier Ltd. All rights reserved.

Frederiksen and Phelps's review of literature in the book *Virtually Embedded: The Librarian in an Online Environment* (Leonard & McCaffrey, 2014) cites many research articles that describe the services that librarians delivered online in recent years: personal librarian outreach (Ismail, 2011; Kadavy & Chuppa-Cornell, 2011), original web tutorials (Held, 2010; Hemming & Montet, 2010), research guides (Daly, 2010), web conferencing (Montgomery, 2010), discussion boards (Herring, Burkhardt, & Wolfe, 2009), research advice (Knight & Loftis, 2012; York & Vance, 2009), and library assignments built into professors' online syllabi (Bozeman & Owens, 2008). Librarians are frequently present on other platforms beside the LMS, notably Facebook (Haycock & Howe, 2011) and Twitter (Filgo, 2011). Several recent books have covered best practices in online research instruction, marketing of services, and assessment of the efficacy of librarians' interventions (Leonard & McCaffrey, 2014; Shumaker, 2012; Tumbleson & Burke, 2014). Our project builds on previous models by making information literacy an essential part of the students' online course experience and includes assessments of the efficacy of the project.

10.2 FILLING THE INFORMATION LITERACY GAP IN ONLINE COURSES

Students who habitually use Google, Wikipedia, and YouTube to research their topics are generally not well equipped to assess the websites they use. According to Love and Norwood (2007), e-learning students who use online sources frequently choose websites of questionable authority instead of more credible library materials. Traditional one-shot library instruction classes do not offer enough time for instructors to teach everything or students to practice research skills. Although the increasingly complex process of retrieving academic information demands more library instruction—preferably early in a student's academic career—online students rarely have the opportunity to attend a library instruction class (Stewart, 2007). Also, as Hatfield and Brahmi (2004) point out, "library resources are not well represented in most Blackboard sites" (p. 455).

Experienced teaching librarians know that a single library instruction class does not adequately train students to become savvy researchers or to connect the complex relationship between research and writing (Watson et al., 2013). In an influential article, Dane Ward (2006) stressed "the absolute necessity of integrating the library into the curriculum" and added that "the one-shot library session has never been an adequate approach to teaching and learning the research process," particularly when one considers the combination of reasoning and "inward understandings" that take place in the learning process (p. 401). In response to the increasing awareness of the limitations of one-shot library instruction, librarians and professors began to devise models of instruction. Xiao and Traboulay (2007) created a professor—librarian partnership in Blackboard that was used in a master's thesis seminar. Composition professors and teaching librarians created new teaching models that helped students integrate the research and writing processes. The literature emphasizes how composition and rhetoric theory inform the teaching of information literacy (Elmborg, 2006; Jacobs & Jacobs, 2009; Norgaard, 2004; Purdue, 2003). Mullins (2014) has written compellingly about the process of integrating information literacy into course design. Her IDEA model—interview, design, embed, assess—is a highly systematic elaboration of course design that shares a lot in spirit with the project presented in this paper. We independently came up with a strikingly similar process for designing and revising our course materials which is described here. IDEA is based on the well-known, highly successful ADDIE (analysis, design, development, implementation, and evaluation) method of instructional design developed by Robert Branch (Branch, 2009).

In 2007, Jackson wrote, "coordinated efforts to partner with faculty to use the LMS for teaching and learning information literacy and critical thinking skills remain less common and need to be developed" (p. 455). She made the case that educational technologies and campus partnerships could be effectively used to integrate library resources into the curriculum, for example "Blackboard can be used to scaffold instruction and infuse information literacy activities throughout subject specific courses" (p. 459). Scaffolding is an educational design concept involving the facilitation of learning via gradual steps and levels of support as created by Jerome S. Bruner and based on the ideas of psychologist Lev Vygotsky. David Wood, Jerome S. Bruner, and Gail Ross originally defined scaffolding as the guidance of a teacher or facilitator who helps a student complete a complex task by following these steps:

- Consider curriculum goals and the students' needs to design appropriate tasks.
- Engage students by establishing a shared goal.
- Accurately diagnose students' needs and levels of understanding.

- Provide tailored assistance through cueing, questioning, prompting, modeling, telling, or discussing.
- Give feedback to help students monitor their progress.
- Control for frustration or risk by beginning with something students can do with little or no assistance.
- Build in assignments that can be completed successfully, moving gradually to more complex tasks.
- Scaffolding (the support of the facilitator) should be removed gradually as students begin to demonstrate mastery.
- Know when it's appropriate to stop. Support is no longer provided when students can perform independently (Bruner, 1983; Larkin, 2002; Wood, Bruner, & Ross, 1976).

Less research has been published about the use of scaffolded lessons in the teaching of information literacy skills to undergraduate students. Otis's (2013) thought-provoking book chapter in *Successful Strategies for Teaching Undergraduate Research* describes specific activities that train students to evaluate source materials for value, content, and credibility before they embark on a research project. As she notes, teaching critical thinking takes time and attention from both the professor and the librarian and involves "more than simply dropping the activities into existing information literacy models or research instruction sessions" (p. 134). Similarly, creating research guides that support specific online course content, tutorial videos that target specific skills, and activities that demystify the research process require a great deal of planning and communication between the course professor and librarian. Not all librarians have the time to create or update these components due to their existing workload, so a section on sustainable services will be included later in this chapter.

10.3 ENGLISH 201 COURSES AT THE BOROUGH OF MANHATTAN COMMUNITY COLLEGE

The Borough of Manhattan Community College (BMCC), part of the City University of New York, has a population of more than 25,000 students from 163 countries and speaking 111 languages. Approximately half the students are first-generation college students. Approximately 72% of students identify as black or Latino. More than 90% are eligible for New York's tuition assistance program. In the 2013–14 academic year, 19,890 students received federal Pell Grant awards. Of first-time incoming freshman, 29% of students placed into developmental courses for reading and 44% for writing (BMCC Facts and Statistics, Spring 2015; BMCC Factbook 2014–15). In such a large and diverse student body, it is impossible to generalize about the students' background in research and writing. When asked, many students arriving at the in-person library instruction classes at BMCC admit they have never visited the library before and don't use the library website.

The challenges of the online learning structure, which requires more student writing—including blog posts and discussion board participation—and less time spent on campus, prompted English Professor Kelly Secovnie to request that librarian Lane Glisson work as her instructional partner in two online English 201 Introduction to Literature courses starting in the fall of 2012. English 201 is the second English course that all students at the college are required to take, and it is part of the general sequence that begins with English 101 Composition. Students are required to write four substantial analytical essays that address literary texts in a variety of genres and must learn to integrate research into their writing about literature and literary criticism.

10.4 PARTNERING TO DESIGN ONLINE RESEARCH INSTRUCTION

Having attended in-person library instruction courses, the professor recognized their value for her English students. The first meeting—to establish an effective library instruction plan—was an important first step in the process. The professor outlined her course and gave the librarian the syllabus and course schedule. The librarian's first task was to become familiar with the course, its assignments, and due dates. It was important for the librarian to listen carefully to the professor's goals for the course because they informed the design of reference instruction. Interpersonal communication was key in establishing a good collaborative relationship. The ability to reach consensus between the professor and the librarian ensured a happy, long-term partnership. At the initial meeting, they established the dates when links to library resources would be introduced; videos, online lessons, and discussion boards would go live; and essay assignments would take place. A similar process could work smoothly with any department's syllabi and does not pertain only to English courses.

Individual professors will have different expectations in terms of the extent of collaboration the course requires. The first meeting is a good time to establish those parameters. Strategizing how and when the librarian will be introduced to students is important. The first week of online courses is often dedicated to getting students comfortable with the Blackboard interface and the course syllabus and structure. Posting photographs and introductions of both the librarian and the professor helps make students feel welcome. Students must understand the different roles of the professor and librarian. Initial introductions clarify which questions should be fielded to the course professor, such as assignment and grading questions, and which questions should go to the librarian, such as research strategies, bibliographic suggestions, and citation styles.

10.5 IMPLEMENTING A SCAFFOLDED MODEL OF RESEARCH INSTRUCTION

Given the challenges that community college students face, a fresh approach that would engage and empower them and prepare them for college-level writing and research was necessary. While the articles and books cited have demonstrated the benefits and techniques of library instruction delivered in online courses, less has been written about the process of scaffolding library instruction into a course syllabus. To scaffold research across the course efficiently, the librarian was given teaching assistant status in the Blackboard LMS, allowing her to post announcements, send group emails, and create content in the course's weekly folders: research blogs, instructional videos, targeted lessons, and links to the BMCC library website.

Students need to feel that the librarian is an active presence in the course. This was achieved by including the librarian in the first discussion board in which students introduced themselves and interacted with the professor and their peers in a more informal conversation. It was beneficial to introduce the librarian early in the course and emphasizing her distinct role. This solved the problem of students posting questions to the librarian that were more properly assigned to the course professor. The library components were graded assignments to ensure that students completed them. Students place higher value on the parts of the course that impact their grades. This ensured that students would actively participate in research-related assignments. Even research discussion board threads were given extra credit points to encourage participation.

10.6 START SMALL AND BUILD OVER TIME

When integrating research concepts into the syllabus, it is easiest to start small with a few components and build more over several semesters. For example, the first semester of the project, the librarian created an 8-minute library day video using screen-cast software that showed students how to find basic research sources in the library website's databases and catalog; it also included a lesson about academic honesty. The professor designed a test to assess how well students understood and could apply the concepts discussed in the video. In subsequent semesters, the librarian added blog posts on related reading materials, instructional videos, an online SoftChalk lesson and two research discussion boards. Because BMCC's students are often enrolled in both face-to-face and online classes, they are also encouraged to visit the library building and contact the librarian for assistance, whether in person, by telephone, or via email.

Working as a team, the professor and the librarian augmented the syllabus to include research components that were specifically designed for the assigned readings and essays. They discovered the benefits of introducing library instruction over the course of the semester instead of trying to fit it into a single library instruction day. Simple searches were assigned first. For instance, students were asked to read a review of a play they had read and respond to that review in a short, low-stakes writing assignment. The review was found in a literature resource center database. Students were taken step-by-step through the process of logging into the library website, searching within the database for the article, and accessing the citation to include in their written response.

Over time, these exercises led to the more sophisticated tasks that supported the final project. Using an interdisciplinary perspective on research and writing, the professor and the librarian introduced students to the range of materials that can be used to write about literature, including literary criticism, long-form journalism, and historical scholarship. They not only wanted students to be exposed to these materials but also wanted them to critically engage in their understanding of these works. They found that when skills were introduced gradually, students reported less panic about doing research for their final research papers because they had been practicing smaller research tasks throughout the course.

10.7 A DIVERSE RANGE OF ACTIVITIES

Salyers, Carter, Cairns, and Durrer (2014) used the scaffolded e-learning model ICARE (introduction, connect, apply, reflect, and extend) to create learning modules that are designed in timed chunks and delivered in phases, creating opportunities for students to interact with the learning community and engage with the course material through diverse activities. The professor and librarian, taking a similar approach, varied the student's activities from week to week to promote student engagement.

A complete list of course modules and library interventions is included in Appendix A at the end of the chapter.

Highlights from the activities included the following:

- In Module 1, the librarian and professor posted the library day video, the test, and information about finding help at the library. This module included the essay 1 assignment, which asked students to compare two short stories, write about their common themes, and provide evidence directly from the text.
- In Module 2, since the students were reading an absurdist play, the librarian's notes suggested other plays in the theater of the absurd genre and additional resources for more in-depth study.
- An online lesson made with SoftChalk software taught students to use newspaper and encyclopedia databases for their research assignment in essay 2, which required students to compare the play *Antigone* to a contemporary conflict.
- The librarian led two research discussion boards during the semester—in Module 2 to support essay 2 and in Module 4 to support essay 4 and the final portfolio. She suggested a list of questions that students might ask to avoid drawing only questions about Modern Language Association (MLA) documentation (see Appendix A for the suggested list).
- Module 3 had no required research components. This was an important rest period for students, allowing them a bit of breathing room in the course. Essay 3 assigned students the task of responding creatively to the thesis of a work of creative nonfiction. They wrote their own creative nonfiction pieces that mimicked the rhetorical strategies of a piece they had read for class. While they were told that they could use their personal experience as a basis for their essays, several students used library research, including the Gale Virtual Reference Library database (required for essay 2), to enhance their creative nonfiction pieces. This showed that they had assimilated the database lesson.

- Module 4's main research assignment was a 3-minute screen-cast video created by the librarian that explained how to find and use good articles of literary criticism for students' essay assignment related to Chinua Achebe's novel *Things Fall Apart*. The professor followed up with an assignment that required students to find one article and one ebook from the online databases that could be used for their essay and write a works consulted page to show facility with citation. This is an example of the scaffolding principle in which the teacher–facilitator models the skill and the students then apply the skill independently.
- Module 4 also contained point-of-need bibliographic advice about where to find good books and articles about Chinua Achebe, essays written by Achebe, literary criticism, and articles on African Igbo culture and gender issues.

10.8 REVISION

At the beginning of each new semester, the professor and librarian met to discuss what did and did not work during the previous semester, adding elements to improve student experience and further scaffold the research process. In the second semester of the program, e.g., the original 8-minute library day video and test were moved from week 4 to week 2 of the course to get the students thinking about research earlier. One semester the modules had to be pushed later to accommodate a schedule problem. From that experience, the professor and librarian learned that it is best to introduce the librarian as soon as possible so that students become accustomed to consulting her.

10.9 ACCESSIBILITY

A personal touch is key to developing rapport with online students. In the same way that the professor kept office hours when students could talk about their writing assignments, the librarian encouraged students to email questions to her, set up a consultation in person or by telephone, or visit the library reference desk. Students noticed her accessibility and reached out for individual help as they prepared their final portfolios. The fall 2014 students who were introduced to their librarian in the first week and received the fully designed course were more active in contacting her than in earlier semesters of the project. During the time that essay 4 and the final portfolio were assigned, she helped several students per day with research, documentation, and how to incorporate research into their paragraphs, which is often a challenge for novice students. At the library, she continues to encounter students who introduce themselves with the words, "I know you. You helped me in Professor Secovnie's online English course." Bringing the library to online students is an excellent way to bring the students back to the library for research assignments throughout their academic career. From the librarian's point of view, the opportunity to collaborate with an experienced course professor and our diverse and lively students was highly rewarding. Likewise, the course professor appreciated the chance to work so closely with a research specialist who enriched the course content and showed careful attention to helping students in their times of need.

10.10 ASSESSMENT

The professor and librarian engaged in a continuous process of assessing students' assimilation of research skills using a variety of methods, including student survey information. Initially, however, their objective was to get something up and running using a face-to-face model of library instruction and applying it to an online course. The primary concern in terms of assessment that first semester was to determine whether students comprehended the materials presented and whether they were able to apply them to their writing, particularly their research papers. The first assessment test was designed to follow the 8-minute library day video, and students could take the test as many times as they needed.

10.11 LIBRARY DAY TEST

In spring 2013, the first semester the test was administered, one course section (a cohort of approximately 25 students) showed 12 completed attempts with an average score of 7.4/10. The other section showed 13 completed attempts with an average score of 6.8/10. Given that the library day video was the primary intervention in the class during that early semester of the project, they had hoped for better results. Since the course sections begin with 25 students, the completion rate was poor, with the grades being rather less than expected given that students could try the test as many times as they liked. Throughout the next two semesters, they added the scaffolded sections based on the idea that the students needed more interventions sooner in the semester so that students did not see research assignments as unrelated to the rest of the course.

The spring 2014 sections showed improvements, with one section of 17 participants having an average grade of 6.9/10 and the other boasting 15 participants with an average score of 7.7/10. Although the rates of participation were increasing, the scores were not significantly different than the earlier semester's results. Seeing better student participation, they sought to build on the scaffolded model of instruction and made several course changes such as integrating the librarian more effectively into the early course structure with a personal introduction video, smaller blog posts that broke up instruction into more digestible pieces, and connections between much shorter videos that were targeted to specific research assignments. The student test results demonstrated small, incremental improvement in completion rates and scores as a result. In fall 2014, for instance, one section of the class showed 20 completed attempts with an average score of 7.9/10, and the other showed 21 completed attempts with an average score of 7.5/10.

Overall, the completion rate for the library day test rose from roughly 50% of students completing the assessment in spring 2013 to 82% of students completing it in 2014. In that same period, the average score (averaged between both sections) rose from 7.1/10 in spring 2013 to 7.7/10 in the fall 2014 semester. While the second statistic is not likely significant, it is encouraging that students demonstrated better knowledge of the video, coupled with more participants completing the assessment. This could theoretically have led to a decrease in scores as the number of students of varying abilities increased, yet there was a slight increase in scores.

10.11.1 Discussion of Library Day Test Results

The results over time point to the problem for student engagement when faculty attempt to simply replicate the in-person one-shot library class in an online environment. In contrast, the preceding results demonstrate the power of scaffolded instruction, integrating library instruction into the class structure, and adding a personal touch to the relationship between the librarian and the members of the online class. When students feel they have a real connection to a person behind the videos, they seem more likely to engage. As more students completed the test, one might have expected to see the total average scores fall given that the students most likely to take the test in earlier semesters would have been the strongest students in the class overall. In fact, average scores rose slightly, even with more students taking the test. Also, students need to take English 101 Composition as a prerequisite, which includes a standardized final essay exam, taken by all students and worth 30% of their course grade. This helps standardize the students who are entering the English 201 course and allows instructors to surmise that the improved instruction, rather than student variance, accounts for the positive changes seen in their participation and achievement on the library day test.

10.11.2 Postcourse Survey

The professor and librarian wanted to find out more about students' research behavior after taking the course. In fall 2014, in addition to the library day test, they offered extra credit to students who took a postcourse survey. The survey was completed by 15/18 students in one section and by 13/14 students in the second section. The average response rate was 83% and 93%, respectively. In future semesters, they plan to evaluate data from both pre- and postcourse survey results, which will give a more precise picture of students' attitudes toward research before starting the course and after having worked with our program. At the start, though, they found the results useful in determining what students were "taking away" from the course. The survey, designed to address concerns about students' skills and attitudes, is included in full in Appendix B at the end of the chapter.

10.11.3 Types of Sources Accessed

The first question of the postcourse survey asked students to check all of the sources that they used to conduct research. (See Table 10.1 for the results.)

The most significant result shown in Table 10.1, particularly for BMCC's student population, is that 46% sought help from a librarian. According to Head and Eisenberg's Project Information Literacy Report, "Lessons Learned: How College Students Seek information in the Digital Age" (2009), when seeking course-related research, of 15 resources used, librarians ranked 14th on the list. "Eight in ten respondents—80%—reported that they did not use librarians for help with a course-related research assignment" (p. 23). Instead, their study found

Table 10.1 Types of sources accessed

Answers	Percent answered
Wikipedia or similar	23.0
Books or ebooks from the library catalog	92.3
Google or other search engines	61.5
Articles from library databases such as JSTOR and Academic Search Complete, among others	100
Google Scholar	15.4
Newspaper articles (print or online)	61.5
Blogs	7.7
Social media such as Twitter or Facebook	0
Encyclopedias (print or online)	61.5
Friends and family	15.4
Help from a librarian	46.2

that course instructors' readings played an important informational source followed by Google and then by scholarly research databases. A study of undergraduate education majors found that for a research paper or presentation, only 27.5% used library resources while 72% found their information on the Internet (Martin, 2008). That study and others observed that students rely on Google and Wikipedia because they value fast and convenient access over credibility and quality (Colón-Aguirre & Fleming-May, 2012; Connaway, Dickey, & Radford, 2011). The results of this project, in contrast, suggest that students had absorbed the lessons that were integrated into the research instruction, regarding the appropriate sources to use in academic research. While 27.5% of students mentioned in Martin's study reported using library resources, authors of this chapter saw rates of 92% using library books or ebooks, and 100% using library databases.

10.11.4 Evaluating Credibility in Source Material

In academic culture, it is understood that credible sources are vital for conducting research. Students, on the other hand, often do not appreciate the finer distinctions between sources. Given this concern, one of the survey questions asked, "How do you decide if a source is reliable to be used in academic research? Please write your answer in a few sentences." Students' answers varied, but most leaned toward citing library sources as more credible—some without giving much thought as to why—while others showed keen insights into understanding the issue of credibility. As one student wrote:

> Generally, if a source is from a library reference site, or is in a literary journal—it tends to be fairly reliable. These people are writing educated pieces that show proof of research, and also these works have been edited and checked by other people in their field. Sites such as wikipedia [sic], while are sometimes very thorough and accurate, are not held to the same quality standards as articles, books, journals, that are published. At wikipedia [sic], anyone can type up an article or add to a piece that is already there.

A second student delved more deeply into ways to judge the quality of information:

> In order to decide which if [sic] a source is reliable, I tried looking to see if the source has a.) A date in when the source was made, b.) If the site can be edited because if so, then I shouldn't take the source seriously and c.) If I can find the same piece of information repeatedly, but also see which one has more information in comparison to the other one.

When discussing reliability of sources, a third student mentioned the course librarian by name, stating:

> The best way to decide if a source is reliable to be used in academic research is to check with the research librarian (Prof. Glisson, in this case). Otherwise, sources that come from certified experts that have credibility. Most of the time, as long as sources have credibility, they can be considered reliable. However, some websites are very apt at copying credible sources, so we must be careful.

This student, noting the value of information professionals and "experts" such as scholars or journalists, also displays insight into the prevalence of "copying" on the free web, alluding to the fact that it is hard to cite from a website that may copy the writing of others and perhaps even alter it, or at the very least, not give credit for other writers' original work.

10.11.5 The Research Process

A sample student's answer to a question about his or her research process reveals a fairly typical response of students who completed the survey. The student writes:

> I first look at the question that I'm answering. Then I do a basic search on the BMCC online library. During my search I read the brief summaries of each article or book that comes up and do a process of elimination that way. If I find something I like or that I think could help me then I use the citation tool and copy and paste it on a word document for future refrence [sic]. Since I prefer to read on paper than online I print the article or part of the book to read on my

free time. Once I've skimmed through all the documents I decide on which reference material I'll be using and get rid of the rest. I then proceed to thoroughly read through the material and highlighting any interesting quotes along the way or sometimes I just write notes on a notebook about the readings to refer back to later.

This survey response shows the depth of engagement of many of these students in the research process, evidenced by the fact that they are really connecting with the instruction provided and organizing their own research in response. The student just cited shows that he or she understands the research process and initial broad gathering of materials, progressing to a more targeted approach as the student's ideas evolve. It was encouraging to see such thoughtful insights into the research process from freshmen. The students came away from the course with skills that are transferrable to courses in any department and can be used in their future work, which is an important consideration in a general education course.

10.11.6 Challenges for Student Researchers

The survey asked students, "What do you find most challenging in conducting research?" Students could choose more than one answer, and the results are tabulated in Table 10.2.

Students overwhelmingly found "knowing how to integrate research sources with my own ideas" to be the most difficult aspect of conducting research. This should be expected; freshman-level students are just beginning to get their feet wet in conducting library research. At the same time, students were more confident in areas that the professors had reinforced, such as deciding if sources were acceptable for academic research

Table 10.2 Challenges for student researchers

Answers	Percent answered
Finding sources to use	30.8
Deciding on how much research I have to do	46.2
Deciding if the sources are acceptable for academic research	15.4
Knowing how to integrate research sources with my own ideas	76.9
Deciding how to evaluate the usefulness of a source for my particular topic	15.4
Knowing how to use appropriate search terms to help me find information on my topic	23.1

(only 15% reported finding this to be their greatest challenge) and finding sources to use (30% struggled with this task). As these tasks are ones that had been scaffolded into research paper assignments, it was encouraging to find that students were able to devote more of their time to the trickier critical thinking skills that would be needed to effectively integrate ideas found in research with their own. The need for this type of support was reinforced at the end of the semester when several students emailed or phoned the librarian to ask her how to use research sources effectively in their writing.

10.12 WORKLOAD AND SUSTAINABILITY

Skeptics of this method of delivering information literacy training generally ask, "Is it sustainable?" When the course model was new, there was a lot of work to do to create the videos and other components that were delivered to students. Once it got rolling, the main administrative task was to update videos and the Blackboard folders at the beginning of the semester if databases or the website had changed. The professor and librarian shared these tasks, which could usually be accomplished fairly quickly. However, once the components are made, the course modules do not simply run themselves because the human element is a big part of successful online teaching. The course relies on the ability of both professors to make connections with the students and to be available in office hours when they have questions. The scaffolded lessons are there to support student researchers as they move toward independence, but human connection is a big factor in student engagement. The life of the course relies on an interdisciplinary model that is based on partnership between the professor, the librarian, and their students.

At the time this project began, only a few librarians at the City University of New York (CUNY) had experimented with placing library specialists in Blackboard. At many colleges in the system, there was no uniform library presence in the Blackboard LMS. Since 2015, however, the librarian from this project has worked with other CUNY colleagues on a university-wide working group chaired by Nora Almeida of New York City College of Technology, CUNY, to create a new librarian role in Blackboard that gives permissions in courses to all librarians who partner with professors. Last fall, the working group conducted its first pilot program of embedded librarians from many CUNY colleges to partner with professors in the Blackboard LMS. Recently, presenting at the ITC eLearning 2016 conference in Scottsdale, Arizona, the authors met a librarian who had served simultaneously in 17 Blackboard courses in one semester. She explained that some courses required less complex involvement than others, therefore balancing her workload. Some professors might request simple interactions that provide only links and LibGuides. Others require more dynamic scaffolding of research activities like the one described in this chapter.

10.13 REFLECTIONS ON THE METHOD AND THE PARTNERSHIP

Over three years, by adjusting the pace of the course and scaffolding of research instruction, the authors observed improved engagement with course content particularly in regard to research. Students' comments frequently mentioned the importance of credibility in the research they found. They began to understand that good research is the underpinning of stronger critical thinking and more engaging writing. Although the students were novice researchers, the project allowed them to gain confidence in their writing process and exhibit far less nervousness about academic research, particularly regarding their final project and portfolio. The professor noted less panicked end-of-semester emails with questions about how to do basic research. Instead, at her one-on-one conferences with students at the end of the semester, she was hearing more questions about how to effectively use the research students found to respond to the literature in their readings.

The portfolio system, which gives students the opportunity to revise their essays and other forms of writing completed during the online course, is an advantageous method to use with scaffolded research assignments. Putting together a final portfolio requires students to reflect on their writing and their development as analytical writers in an academic setting. The students' final portfolios are the culmination of the semester's work that they select and revise to demonstrate their research and writing skills. It enables students to consider the interpretive process that goes into research and writing. Research requires reflection in order to gather literature materials that can inspire students to think more deeply about their topics. As Jacobs and Jacobs (2009) have observed, "research is as much a process as writing. Like effective writing, effective research does not happen in just one sitting but involves iterative processes such as revision, reworking, rethinking, and above all, reflection" (p. 72). A portfolio system, paired with many opportunities to practice different forms of research in small lessons across the course of the semester, helps students become more engaged in the research and writing process.

With the popularity of instantaneous Google search habits, it is necessary to train students to reflect on their ideas and look more deeply into the themes that literature and history offer us. A scaffolded approach to research instruction breaks down the process for students who have not been previously trained to think in this manner, making the plethora of information that is available in online libraries more navigable. While this chapter describes how these methods are used in an online course using the Blackboard LMS, the principles have informed the librarian's reading of what goes into effective in-person library instruction. She has begun to use the same scaffolding methods in her collaborations with professors of on-campus classes as well in English and in other departments. The professor has used lessons and materials developed for the online class to also scaffold research assignments into other upper-level literature classes that she teaches.

In Norgaard's (2004) article "Writing Information Literacy in the Classroom," he convincingly argues for partnerships between professors and librarians that create a philosophy of pedagogy that develops the intellectual process in students. As he observes, "librarians can prove to be essential allies for writing teachers, insofar as they might reinforce several key underlying assumptions: that knowledge is dialogic, not just informational; that knowledge springs from inquiry motivated by genuine problems and cognitive dissonance; and that knowledge grows through a process of substantive revision" (p. 222). Failing to create these partnerships, he writes, carries a significant loss to both the disciplines of writing and of research as defined by our intellectual tradition. Having worked with community college students for several years, we clearly see that they appreciate research assignments that are grounded in the world they inhabit and problems that are relevant to their life experience.

As more courses go online, making library research real in the digital classroom is crucial to the quality of online teaching. The benefits of working in partnership extend to our students as well. The students see that the professor values the role of libraries in providing credible information for their assignments. In engaging with the librarian, the students begin to understand that an active librarian can act as a co-collaborator in developing the intellectual dialog that becomes their essay. Modeling partnership for our students helps them understand they are not alone in their projects and that they have professors and librarians they can turn to for advice, inspiration, and support. This motivates students to take that extra step to seek the help they need to succeed. Working as equal partners to help online students develop their knowledge of sources and their writing, professors and librarians can develop stronger courses and students who are more savvy users of information and who understand the connections between good research and good writing.

REFERENCES

Allen, E. A., & Seaman, J. (2016). Online report card: Tracking online education in the United States: Babson Survey Research Group and Quahog Research Group, LLC. Retrieved from <http://onlinelearningconsortium.org/read/online-report-card-tracking-online-education-united-states-2015/>.

Bozeman, D., & Owens, R. (2008). Providing services to online students: Embedded librarians and access to resources. *Mississippi Libraries*, 72(3), 57−59.

Branch, R. M. (2009). *Instructional design: The ADDIE approach*. New York, NY: Springer.

Bruner, J. S. (1983). Child's talk: Learning how to use language. Oxford: Oxford University Press.

Colón-Aguirre, M., & Fleming-May, R. A. (2012). "You just type what you are looking for": Undergraduates' use of library resources vis. Wikipedia. *The Journal of Academic Librarianship*, 38(6), 391−399. Available from <http://dx.doi.org/10.1016/j.acalib.2012.09.013>.

Connaway, L. S., Dickey, T. J., & Radford, M. L. (2011). "If it's too inconvenient I'm not going after it": Convenience as a critical factor in information-seeking behaviors. *Library & Information Science Research*, 33, 179−190. Available from <http://dx.doi.org/10.1016/j.lisr.2012.12.002>.

Daly, E. (2010). Embedding library resources into learning management systems: A way to reach Duke undergrads at their points of need. *College & Research Libraries News*, 71(4), 208−212.

Deyrup, M., & Bloom, B. (2013). *Successful strategies for teaching undergraduate research*. Lanham, MD: The Scarecrow Press, Inc.

Elmborg, J. (2006). Critical information literacy: Implications for instructional practice. *The Journal of Academic Librarianship*, 32(2), 192−199. Available from <http://dx.doi.org/10.1016/jacalib.2005.12.004>.

Filgo, E. H. (2011). #Hashtag librarian: Embedding myself into a class via Twitter and blogs. *Computers in Libraries*, 31(6), 78−80.

Getty, N. K., Burd, B., & Burns, S. K. (2000). Using courseware to deliver library instruction via the Web: Four examples. *Reference Services Review*, 28(4), 349−359.

Getty, N. K., Burd, B., Burns, S. K., & Piele, L. (2000). Using courseware to deliver library instruction via the Web: Four examples. *Reference Services Review*, 28(4), 349−359. Available from <http://dx.doi.org/10.1108/00907320010359678>.

Hatfield, A. J., & Brahmi, F. A. (2004). Angel: Post-implementation evaluation at the Indiana University School of Medicine. *Medical Reference Services Quarterly*, 23(3), 11. Available from <http://dx.doi.org/10.1300/J115v23n03_01>.

Haycock, L., & Howe, A. (2011). Collaborating with library course pages and Facebook: Exploring new opportunities. *Collaborative Librarianship*, 3(3), 157−162.

Head, A. J., & Eisenberg, M. B. (2009). *Lessons learned: How college students seek information in the digital age.* Project Information Literacy Progress Report. Retrieved from SSRN: <http://ssrm/cp,/abstract = 2281478>   <http://dx.doi.org/10.2139/ssrn.2281478>.

Held, T. (2010). Blending in: Collaborating with an instructor in an online course. *Journal of Library and Information Sciences in Distance Learning, 4*(4), 153–165.

Hemming, W., & Montet, M. (2010). The "Just for me virtual library": Enhancing an embedded e-Brarian. *Journal of Library Administration, 50*(5/6), 657–669.

Herring, S., Burkhardt, R., & Wolf, J. (2009). The "Just for me virtual library": Enhancing an embedded e-Brarian. *Journal of Library Administration, 50*(5/6), 657–669.

Ismail, L. (2011). Getting personal: Reaching out to adult learners through a course management system. *Reference librarian, 52*(3), 244–262.

Jackson, P. (2007). Integrating information literacy into Blackboard: Building campus partnerships for successful student learning. *Journal of Academic Librarianship, 33*(4), 454–461. Available from <http://dx.doi.org/10.1016/j.acalib.2007.03.010 >.

Jacobs, H. L. M., & Jacobs, D. (2009). Transforming the one-shot library session into pedagogical collaboration: Information literacy and the English composition class. *Reference and User Services Quarterly, 49*(1), 72–82. Available from <http://dx.doi.org/10.5860/rusq.49n1.72 >.

Kadavy, C., & Churppa-Cornell, K. (2011). A personal touch: Embedding library faculty into online English 102. *Teaching English in the Two Year College, 39*(1), 63–77.

Knight, v, & Loftis, C. (2012). Moving from introverted to extroverted: Embedded librarian services: An example of a proactive model. *Journal of Library and Information Sciences in Distance Learning, 4*(4), 153–165.

Kraemer, E. W., Lombardo, S. V., & Lepkowski, F. J. (2007). The librarian, the machine, or a little of both: A comparative study of three information literacy pedagogies at Oakland University. *College & Research Libraries, 68*(4), 330–342. Available from <http://dx.doi.org/10.5860/crl.68.4.330 >.

Larkin, M. (2002). Using scaffolded instruction to optimize learning. *Eric Digests.* n.p. Retrieved from <http://files.eric.ed.gov/fulltext/ED474301.pdf>.

Leonard, E., & McCaffrey, E. (2014). *Virtually embedded: The librarian in an online environment.* Chicago, IL: Association of College and Research Libraries.

Lillard, L. L., & Dinwiddie, M. (2004). If you build it, they will come, but then what: A look at issues related to using online course software to provide specialized reference services. *Internet Reference Services Quarterly, 9*(3/4), 135–145. Available from <http://dx.doi.org/10.1300/J136v09n03_10 >.

Love, M., & Norwood, S. (2007). Finding our way as "embedded librarians". *College & Undergraduate Libraries, 14*(4), 87–93. Available from <http://dx.doi.org/10.1080/10691310802128369 >.

Martin, J. (2008). The information seeking behavior of undergraduate education majors: Does library instruction play a role? *Evidence Based Library and Information Practice, 3*(4), 4–17. Retrieved from <http://ejournals.library.ualberta.ca/index.php/EBLIP/article/view/1838/3696>.

Montgomery, S. E. (2010). Online webinars! Interactive learning where our users are: The future of embedded librarians. *Public Services Quarterly, 63*(2), 306–311.

Mullins, K. (2014). Good IDEA: Instructional design model for integrating information literacy. *The Journal of Academic Librarianship, 40,* 220–226.

Norgaard, R. (2004). Writing information literacy in the classroom: Pedagogical enactments and implications. *Reference and User Services Quarterly, 43*(3), 220–226. Available from <http://dx.doi.org/10.1016/j.acalib.2014.04012 >.

Otis, S. (2013). Sources before search: A scaffolded approach to teaching research. In M. Deyrup, & B. Bloom (Eds.), *Successful strategies for teaching undergraduate research* (pp. 121—136). Lanham: Scarecrow Press, Inc.

Purdue, J. (2003). Stories, not information: Transforming information literacy. *Libraries and the Academy*, *3*(4), 653—662. Available from <http://dx.doi.org/10.1353/pla.2003.0095>.

Salyers, V., Carter, D., Cairns, S., & Durrer, L. (2014). The use of scaffolding and interactive learning strategies in online courses for working nurses: Implications for adult and online education. *Canadian Journal of University Continuing Education*, *40*(1), 1—19.

Schumaker, D. (2012). *Embedded librarian: Innovative strategies for taking knowledge where it's needed*. Medford, NJ: Information Today, Inc.

Stewart, V. D. (2007). Embedded in the blackboard jungle: The embedded librarian program at Pulaski Technical College. *Arkansas Libraries*, *64*(3), 29—32.

The Office of Institutional Research and Assessment, Borough of Manhattan Community College. (Spring 2015a). *BMCC factbook* 2014—15. Retrieved from <http://www.bmcc.cuny.edu/iresearch/factbook/index.jsp>.

The Office of Institutional Research and Assessment, Borough of Manhattan Community College. (Spring 2015b). *BMCC facts and statistics*. Retrieved from <http://www.bmcc.cuny.edu/iresearch/fact_stat.jsp>.

Tumbleson, B. E., & Burke, J. J. (2014). *Embedding librarianship in learning management systems: A how-to-do-it manual for librarians*. Chicago, IL: ALA Editions.

Ward, Dane (2006). Revisioning information literacy for lifelong meaning. *The Journal of Academic Librarianship*, *32*(4), 396—402. Available from <http://dx.doi.org/10.1016/j.acalib.2006.03.006>.

Watson, S. E., Rex, C., Markgraf, J., Kishel, H., Jennings, E., & Hinnant, K. (2013). Revising the "one-shot" through lesson study: Collaborating with writing faculty to rebuild a library instruction session. *College and Research Libraries*, *74*(4), 381—398. Available from <http://dx.doi.org/10.5860/crl12-255>.

Wood, D., Bruner, J. S., & Ross, G. (1976). The role of tutoring in problem solving. *Journal of Child Psychology and Psychiatry and Allied Disciplines*, *17*(2), 89—100.

Xiao, J., & Traboulay, D. (2007). Integrating information literacy into the graduate liberal arts curriculum: A faculty-librarian collaborative course model. *Public Services Quarterly*, *3*(3/4), 173—192. Available from <http://dx.doi.org/10.1080/15228950802110577>.

York, A. C., & Vance, J. (2009). Taking library instruction into the online classroom: Best practices for embedded librarians. *Journal of Library Administration*, *49*(1—2), 197—209.

APPENDIX A LIBRARIAN'S ACTIVITIES IN THE COURSE MODULES

English 201: Introduction to Literature (Online Course in the Blackboard LMS

Module 1 (Essay 1):
Week 1:

- Students become familiar with the course structure and Blackboard and begin their first course reading Chimamanda Ngozi Adichie's *Half a Yellow Sun*. In their first discussion board, students, the professor, and the librarian introduce themselves and respond to each other's posts to create a sense of community.

Throughout the course, there is a contact tab on the left sidebar with the professor's and the librarian's contact information and greeting messages.

- **Ideally, precourse information literacy survey is administered during week 1 or week 2 before the students experience any library instruction modules.**

Week 2:

- **Announcement** from the course professor introducing the librarian and describing her role in the course.
- **Librarian's notes** explain ways that students can reach her: email address, office number, and room number and also ways to get help at the library and on its website:
  - the link to the QuestionPoint chat reference service on the library website,
  - the location of the reference desk, and
  - the link to request individual consultations either in person or via telephone.
- **Library Day**: In the assignments folder, the students are assigned to watch the librarian's 8-minute instructional video *Hunting for Treasure at BMCC Library*, accessed via a link to a webpage. The video emulates the modeled searches they would receive in an in-person library instruction class with screen casts and pirate-themed illustrations. After viewing the video, students are given a test to assess what they have learned.
- **Two other short videos** are provided via links in week 2 for their use throughout the course: *Logging in to the BMCC Databases* and *Citing Research Sources Using MLA Style*. They are reposted during specific weeks during the course at point of need.

Module 2 (Essay 2):

Week 4:

- **Librarian's notes and video message:** A short (2-minute) video talk by the librarian to establish her as a familiar face and reinforce the services available on the library website or in person.

Week 5:

- **Librarian's notes:**
  - Resources for the theater of the absurd: other plays and authors available that students can access at the library.

- Links to the library catalog, ebooks, databases for articles, and databases for encyclopedias and reference books.
- Logging in from home and a link to my login video.

Week 6:

- **SoftChalk lesson**: The students are assigned a four-webpage research activity produced using SoftChalk software on how to use newspaper and encyclopedia databases that were required for their essay 2 assignment.
- Working with simple articles from newspaper and encyclopedia databases prepares them for later, more complex research assignments. They were given a short quiz at the end of the SoftChalk activity to assess how thoroughly they read the lesson.
- **Extra credit discussion board**: Students post questions in a special research discussion board set up for one week only and led by the librarian. A list of possible topics they might want to ask gives students some idea of what kind of questions were possible:
    - focusing your topic to get better research and using key words and subject words to find what you need,
    - specific articles or books that will help with your particular research assignment,
    - navigating the databases that your professor has assigned,
    - logging in from off-campus, and
    - citing references using MLA style.

    Usefulness: All students could access the threads of questions and answers whether or not they posted questions. Extra credit encouraged students to log in.

Module 3 (Essay 3):

Weeks 7–9: Research Rest.

Since Module 2 was so research and library heavy, we thought it best to give students a break to focus on other aspects of the course involving writing and interpretation. Although no research was required for essay 3, students sometimes include research in their essays, independently utilizing databases introduced in Module 2.

Module 4 (Essay 4):

Week 10–13:

- **Librarian's notes:** In the assignments folder, the librarian posted a list of good bibliographic sources (in print and online) that are available at the college library related to Chinua Achebe's novel *Things Fall Apart*, including books and articles about Achebe, essays by Achebe, literary criticism, and materials about African tribal culture. Links to the article databases and the catalog were included. The librarian's email address was included.

- **Advanced research video:** A short 3-minute video prepared by the librarian models how to look for articles of literary criticism, gender issues, cultural and historical information for the essay 4 assignment that focused on Chinua Achebe's novel *Things Fall Apart.*
- **Library research activity for essay 4:** Designed by the course professor, the activity instructs students to sign onto the library's website and follow the video's instructions to find an ebook on Achebe's novel *Things Fall Apart* that is not featured in the video that would be useful in their research and a useful article in the databases recommended in the video. The students are instructed to create MLA citations for the ebook and article in the format of a works cited page. The activity is graded on the quality of the research found and the accurate formatting of the works cited page.

Week 13:

- **Extra credit research discussion board**: A second research discussion board led by the librarian is offered in week 13 to support essay 4 and the final portfolio project in which students have the opportunity to revise their writing and reflect on their research and writing process. Again, a list of possible topics they might want to ask is provided. The discussion board is an opportunity to connect with the librarian and also can lead to individual consultations by phone, email, or in person.

End-of-course outreach:

- The librarian sends a general email to the students in the course via Blackboard offering help with any essay 4 research questions. The professor meets with students in one-on-one conferences to check their progress and offer suggestions for revising essay 4 and the final portfolio, allowing her to spot any research problems students are having.

End-of-course assessment activities:

- **Postcourse survey:** The course professor requests that students contribute to the research on this learning model by agreeing to fill out the postcourse survey. This allows the professor and librarian to track improvements to students' research attitudes and knowledge after having completed the course.
- **Student portfolio review:** From the reflections in the student portfolios, the quality of the citations, and the ability of students to

integrate their research into their writing, the professor and librarian can assess how well each student has assimilated the information literacy skills that the course is modeling. However, publishing students' writing requires the students' consent, and thus far consent has been easier to obtain for the pre- and postcourse survey than for portfolio reflections. Therefore, attitudes expressed in this chapter have been drawn from the surveys, from which we have obtained consent.

- **Debriefing:** At the beginning of every semester, the course professor and the librarian discuss what could have gone better the previous semester and make any changes to the design of the course that will improve students' acquisition of research skills such as designing new activities or adjusting the scheduling of the librarian's interventions.

APPENDIX B SURVEY QUESTIONS

- **Question 1: Multiple Answer**

  How do you find information for your research papers? Check all that apply.

  Possible answers:
  - Wikipedia or similar
  - Books or ebooks from the library catalog
  - Google or other search engines
  - Articles from library databases such as JSTOR and Academic Search Complete
  - Google Scholar
  - Newspaper articles (print or online)
  - Blogs
  - Social media such as Twitter or Facebook
  - Encyclopedias (print or online)
  - Friends and family
  - Help from a librarian

- **Question 2: Multiple Answer**

  Of the choices above, which three sources do you think provide the highest-quality information for research papers? Check your top three choices.

  Answers:
  - Wikipedia or similar
  - Books or ebooks from the library catalog
  - Google or other search engines

- Articles from library databases such as JSTOR and Academic Search Complete
- Google Scholar
- Newspaper articles (print or online)
- Blogs
- Social media such as Twitter or Facebook
- Encyclopedias (print or online)
- Friends and family
- Help from a librarian
- **Question 3: Short Answer**

    How do you decide if a source is reliable to be used in academic research? Please write your answer in a few sentences.
- **Question 4: Multiple Answer**

    What do you find most challenging in conducting research?

    Answers:
- Finding sources to use
- Deciding on how much research I have to do
- Deciding if the sources are acceptable for academic research
- Knowing how to integrate research sources with my own ideas
- Deciding how to evaluate the usefulness of a source for my particular topic
- Knowing how to use appropriate search terms to help me find information on my topic
- **Question 5: Multiple Choice**

    How often in the last two years have you used a library to research your papers or projects?
- Never
- 1–2 times
- 2–4 times
- More than 4 times
- **Question 6: Multiple Answer**

    If you have used a library for research, what materials did you use?

    Answers:
- Books or ebooks
- Articles from databases such as JSTOR and Academic Search Complete
- Newspaper articles
- Encyclopedia articles
- Other (please describe in question 7)

- **Question 7: Short Answer**

  If you chose "Other" in question 6, please describe which other materials you used from the library here.

- **Question 8: Multiple Answer**

  How important is it to give credit to the authors of the research sources that you gather when writing an essay?

  Answers:
  - Very important
  - Somewhat important
  - Worth doing only if the information is easy to find
  - Not really necessary
  - A waste of time

- **Question 9: Short Answer**

  What steps do you take to prepare to write your research paper? Please write down all the steps that you typically utilize.

CHAPTER 11

Forging Connections in Digital Spaces: Teaching Information Literacy Skills Through Engaging Online Activities

11.1 INTRODUCTION AND LITERATURE REVIEW

According to the US Department of Education, more than one-quarter of college students are taking at least one distance education course (2014). As online courses and distance education become more established as central modes for delivering higher education, librarians and other instructors must familiarize themselves with the unique needs, opportunities, and challenges of facilitating information literacy in an online context. These include the creative use and integration of new technologies and the creation of a sense of community to facilitate interaction and support learning. This chapter describes effective strategies and techniques for developing online information literacy instruction (ILI) that engages students and builds interaction and considers and responds to the new information literacy framework from the Association of College and Research Libraries (ACRL, 2015). Assignments and learning activities using these methods have been developed for courses taught at Metropolitan State University. These courses include the online versions of "Searching for Information," taught by the authors and others, as well as a book-publishing course and a management research methods course. While the assignments have been developed for these information literacy contexts, the strategies can be adapted to "virtual one-shot" sessions as well as ILI embedded in other courses aiming to forge connections with distance students.

Although online learning presents opportunities to use new technologies to engage students and facilitate interaction, this potential is not necessarily fully realized. Text-based individual learning in an online course may be perceived by students as unhelpful and disengaging, and

Distributed Learning.
© 2017 T. Maddison and M. Kumaran.
Published by Elsevier Ltd. All rights reserved.

social interactions and experiences based in the real world are more likely to be perceived as enhancing online learning (Boling, Hough, Krinsky, Saleem, & Stevens, 2012). Yet merely assigning group work and creating discussion boards may not be enough (Keengwe & Schnellert, 2012). Instead, interactivity must be built into the design of a course (Du & Durrington, 2013; Fink, 2013). To realize the potential for student engagement in online contexts, assignments must be thoughtfully constructed to facilitate interaction, and they can be designed to make effective instructive use of technologies in the process.

The literature supports numerous approaches to addressing these needs. One easily overlooked method to improve collaboration and engagement is to teach students how to interact online by clearly stating and managing expectations (Small, Arnone, Stripling, Hill, & Bennett, 2012; Stodel, Thompson, & MacDonald, 2006), and by modeling discussion behaviors (Du & Durrington, 2013). A second approach is to create learning experiences based in students' real lives. Research shows, for example, that it can be helpful if group collaboration occurs around real-world experiences, authentic tasks, and problem solving (Boling et al., 2012; Du & Durrington, 2013; Palloff & Pratt, 2007). In addition, making space for students to be themselves and explore personal interests together also enhances engagement and collaboration (Cuthbertson & Falcone, 2014; Lester & Perini, 2010). A third way to deepen engagement is to construct assignments that emphasize service learning and community engagement, which can facilitate civic education as well as student engagement (Guthrie & McCracken, 2010). The learning activities described in this chapter use several of these approaches making real-world personal connections for students, which is a key theme.

The use of technologies can support and enhance student connections with each other and with the course content. Widely adopted technologies may be selected to be especially engaging, such as social media sites that many students are already extensively using for personal reasons (Lester & Perini, 2010; Mitchell & Smith, 2009; Small et al., 2012), as well as popular mobile technologies (Cuthbertson & Falcone, 2014). Employing multiple approaches and using a diverse array of technologies may in itself increase engagement in online courses (Dixson, 2010; Stodel et al., 2006). The assignments and learning activities discussed in this chapter use multiple technologies in layered ways to create engaging learning experiences for students.

In addition to designing assignments that address many of the issues raised, the authors have also shaped and refined them in light of the development of the ACRL's (2015) Framework for Information Literacy for Higher Education. Only recently finalized, this framework lays out six nonlinear threshold concepts that can guide information literacy teaching and instructional design practices:

- Authority is constructed and contextual.
- Information creation is a process.
- Information has value.
- Research is inquiry.
- Scholarship is conversation.
- Searching is strategic exploration.

These concepts, combined with "knowledge practices" and "dispositions," make up the frames. The resulting framework provides new perspectives on developing, executing, and evaluating information literacy activities, as well as opportunities for new kinds of research into information literacy, as Gibson and Jacobson (2014) have described. The assignments and activities described in this chapter help students understand that authority is constructed and contextual and information creation is a process, with some support toward the understanding that information has value and the recognition that scholarship is conversation. Viewing ILI through these frames illustrates how the assignments contribute to student development and lifelong learning.

11.2 ASSIGNMENTS AND LEARNING ACTIVITIES

11.2.1 The Visiting Expert

One technique that can make online instruction more engaging is to incorporate a guest lecturer or visiting expert into the course. A visiting expert can introduce a new type of authority to the course in addition to that of the instructor. The visiting expert also adds a real-world element to the course, potentially making it more relevant—and thus more engaging—for students. The activity is set up so that students drive the discussion and determine the questions asked of the expert, thus empowering students and inviting multiple perspectives, diversity of opinions, and expanded social interaction.

Librarians can implement this technique in two ways. A librarian can serve as a guest lecturer or visiting expert for a course taught by another instructor, possibly within one of his or her liaison disciplines. Alternatively, a librarian can use this technique by including a visiting expert in courses he or she teaches. Concrete examples of these two types of implementations by librarians at Metropolitan State University are described in the following.

11.2.1.1 Example: Book-Publishing Course

In the first implementation example, a librarian set up a series of visiting experts to include in her online course, "The Craft and Commerce of Book Publishing." For example, during one week of the course she invited an author and a book editor to participate in the course. The author had a wealth of relevant experience that she was able to bring to the course, including writing, editing, publishing, blogging, and managing issues of intellectual property.

Students posed questions to the visiting expert in a Reddit "Ask Me Anything" style via a class discussion forum, which allowed students to drive the discussions. Reddit, the news and social media site, has popularized the online interview format in which forum users create the questions that interviewees then answer. This approach provides the opportunity for questions and responses—and the resulting conversations—to reflect where the learners are in terms of their understanding of the course material. The online format also makes it possible for busy experts to participate in the virtual classroom, whereas visiting a physical classroom often presents scheduling challenges. The follow-up discussion and student evaluation of the experience suggested that the technique was effective in bringing an immediate and practical level of understanding and awareness to aspects of publishing that could seem opaque or remote. While experience was emphasized for students, feedback from visiting experts corroborates the validity of this approach. Visiting author Margret Aldrich noted, "I liked that the online format gave everyone an opportunity to contribute to the discussion and gave them time to be thoughtful and meaningful with their comments" (personal communication, April 23, 2015).

Another visiting expert, publicist Amelia Foster stated, "This has been great. The questions about [book] distribution were really smart. These students are onto something" (personal communication, October 9, 2015). As the publicist in charge of promoting Margret's book, Amelia was intentionally brought into the course the same semester to provide additional depth and real-life examples to the discussions. This pairing effectively revealed the varied perspectives related to creation, publication, and promotion of the same book.

11.2.1.2 Example: Management Research Methods Course

In the second implementation example, two liaison librarians had been successfully delivering ILI sessions for an in-person management research methods course for several years. Based on this established partnership, an instructor from the university's College of Management worked with the librarians to incorporate them as guest experts in two modules of an online course titled "Practical Research Methods for Managers." The two modules addressed searching for and evaluating secondary research sources, as well as how to use research logs to document the process. The modules were included in both fully online and hybrid (part in-person delivery, part online delivery) modalities.

Students were tasked with responding to various discussion prompts related to the material covered in the two modules. While the prompts were provided by instructors, the ensuing discussion was student driven because it focused on research being performed by the students on topics of their choice. The librarians served as visiting experts for these discussions and provided feedback and conversation related to student progress such as evaluating search strategies formulated by the students. Students were also encouraged to ask questions of the visiting experts. Based on student and instructor feedback, this arrangement was quite successful and was continued over subsequent semesters. In addition to the level of course integrated instruction achieved with this model, other unexpected benefits were also realized. In this example, the instructor and his colleagues valued the librarians' contributions enough to invite them to participate in curriculum revisions for the course. Course instructor Larry Selin said this about the partnership:

> Not only did the collaboration add librarian expertise to the course, but it added a fresh 'voice' to the instruction and... reduced barriers that might exist between students and librarians. You [librarians] became more accessible to more students.
>
> **Personal communication, April 15, 2015**

The visiting expert technique most directly addresses two frames from the Framework for Information Literacy for Higher Education (ACRL, 2015). As students hear from and engage with visiting experts, they begin to understand how authority is constructed and contextual. The frame of scholarship as conversation reveals the value of students engaging in conversations, driven by their own problems and needs, with experts and outside sources of information.

11.2.2 Editing Wikipedia Assignment

This assignment, which was created for the "Searching for Information" course, required students to edit one or more Wikipedia pages. Students were given the choice to work individually or collaboratively. They were also given free rein to choose an entry to edit and the level of editing to be performed. Some edits were as simple as correcting grammatical errors, while others involved adding information or citations to an entry. Several students even chose to create new entries when they identified gaps in coverage and they had the relevant knowledge to support an entry on the topic. For example, one student updated an entry for a band of which he was a member, while another created an entry for her sister, a recent Miss Universe Ghana winner. Students were encouraged to incorporate information from more traditional library sources and research databases.

After editing or creating a Wikipedia entry, students were asked to post about their experience in a class discussion forum. They provided links to the entries they edited and then went on to describe their experience and share their thoughts about Wikipedia after participating in the editing process. The assignment required students to be creators of information while also allowing them to make choices and incorporate their own interests into their academic experience. Some students saw their edits quickly rejected or deleted by other editors or members of the Wikipedia community. Others noted that information creation is often an iterative process, as demonstrated by the page histories for Wikipedia entries and their own experiences editing or creating entries. One student commented in the discussion boards that,

I would have never guessed that a single page on Wikipedia could have three years of conversation that got it to where it is today. Being able to dispute a piece of information and have an impact on the outcome, I feel, would be such a rewarding experience.

Student communication, February 4, 2014.

This activity can be used as part of a series of scaffolded information literacy activities in a full credit-bearing online course, or it can be modified for use in a one-time virtual class visit or workshop. In addition to allowing students to pursue their personal interests, the activity requires students to interact with a well-known resource in greater depth and perform authentic tasks related to it. The assignment also makes use of multiple technologies—editing a wiki and conversing with others on a discussion board—thereby helping to diversify the array of technologies used in a course or workshop and potentially enhancing students' experience of online learning.

This assignment addresses two frames from the ACRL's (2015) Framework for Information Literacy. The frame that describes information creation as a process is addressed by this assignment as students participate in the cycle of adding and editing content in Wikipedia and seeing their additions changed or rejected by other editors. Consistent with the frame of authority is constructed and contextual, the work of editing Wikipedia allows students to begin to identify and evaluate authority within that context while also fostering the development of each student's own voice of authority. In addition, it adds nuance to the discussion about Wikipedia as a source of authoritative information. As the ACRL Framework (2015) notes, "novice learners come to respect the expertise that authority represents while remaining skeptical of the systems that have elevated that authority and the information created by it . . .," and we should "acknowledge biases that privilege some sources of authority over others, especially in terms of others' worldviews, gender, sexual orientation, and cultural orientations . . ." (Authority Is Constructed and Contextual section, para. 2). One student in the course chose to edit the Wikipedia section on females involved in break-dancing or "b-boying" after finding it to be lacking. She noted:

> *I think the issue of gender imbalance on Wikipedia makes me more likely to edit pages just because I don't think it'd be fair to the readers (audience) to get just the male perspective on a topic. But, I think if more female editors whether it be professors, doctors, or even college students like myself willingly participate in editing these wiki pages, it would create an even more diverse sense of community.*
>
> **Student communication, September 20, 2015.**

Wikipedia provides an ideal place for students to discuss issues related to authority and bias in an immediate and dynamic way.

11.2.3 Book Reviews and Book Talks

In this two-part assignment, students in the "Searching for Information" course selected a book to read and review in two different online venues.

They were required to choose and read a book from an extensive list of fiction and nonfiction books, all with themes that in some way relate to literacy, libraries, or information studies. Examples include *Fahrenheit 451*, *World War Z: An Oral History of the Zombie War*, and *Everything is Miscellaneous: The Power of the New Digital Disorder*. They were then asked to review their books on Goodreads and create a book talk of the book on YouTube. Creating content using public, online social media accounts is another method of giving real-world experience and adding an element of possible peer review and reinforcement. Allowing students to choose personally relevant books keeps them engaged.

11.2.3.1 Part One: Book Review Using Goodreads

For the book review, students created an account on the public book sharing and review social media site Goodreads.com. They were encouraged to first create a list of books that they had previously read and a list of books that they hoped to read, following guidelines presented in Steve Leveen's (2005) *The Little Guide to Your Well-Read Life*. They were invited to "follow" the course instructor's account on Goodreads, which gave them an added personal connection. For the graded assignment, they were required to review the book that they chose to read, making sure to touch on themes related to the course content and to share a link to their review with their classmates. While there were guidelines as to how they wrote the review, they were mostly encouraged to seek and use their authentic voice and to consider that they have a real audience for their writing that is broader than simply their classmates and professors. They were reminded that the general public, as well as their classmates, could read and comment on their reviews, so they should consider these audiences. The required length of the reviews (400—700 words) was relatively short, but the expectation was that the review be written in an engaging way and be modeled on best practice guidelines covered in the course. Clarity and concision were emphasized, and students were told that these are good writing skills to foster in an era of Twitter, blogs, and email.

11.2.3.2 Part Two: Book Talk Using YouTube

After the book review, students created a book talk about the same title. They were given a handout on how to create a book talk and why book talking is a valuable skill. The students recorded themselves giving the book talk and uploaded it to YouTube and then shared links to the videos with each other. While most of the students were familiar with YouTube, and most had filmed simple digital videos on their cameras or smartphones, few had written a script, edited a video, and uploaded it for public consumption. The instructor took care to assist students with finding the appropriate technology if needed. The library has cameras available for student use, and thus far all of our online students have been able to procure a camera or phone that can record video.

Outside of librarianship, the phrase "book talk" is not well known, but explaining a reading, making connections between the text and real life, and doing so in an engaging way is a valuable skill. The instructor gave examples of how the ability to "book talk" can translate into work, college, or personal situations. While students may not have many opportunities to discuss a book, they could find themselves in professional situations for which they must read reports or analyze data and then verbally summarize the content to convince others of its value.

The book review and book talk address several key themes related to successful, engaged online learning. Students appreciate feeling empowered to make their own choices and to incorporate their own interests. Most are able to find a book that in some way relates directly to their lives or is at least a compelling and interesting story. Many students express relief at getting to choose their own book instead of having it assigned to them. One student, in her discussion post reflecting on the course, noted that,

> I think the section that gave me the most enjoyment was the book talk and book review. The topics and book choices we were given were all very interesting and it actually got me to sit down and take some time for myself to actually read and enjoy what I was reading so I could learn from it. I'm sure most of you can relate when there is something to read that doesn't interest you that it is harder to understand it so it makes it more difficult to want to read it.
> **Student communication, April 29, 2013.**

There is also a sense that the assignment is an authentic task that has real-world application. The style of writing students use for their reviews is not unlike a well-written email meant to convince a coworker of a point. Also, the students are highly likely to be creators and consumers of similar reviews on online sites such as Amazon and Yelp, as well as YouTube videos in personal or professional contexts.

This assignment, while built around the intellectual work of reading and reviewing a book, also manages to integrate the use of several kinds of technology. As noted previously, employing multiple technologies may in itself be engaging for students (Dixson, 2010; Stodel et al., 2006).

In doing their book reviews and book talks, students use a social media site focused on books, video-recording technologies, and course discussion boards. Social interaction is increased through the discussion boards and the ability to comment on one another's work on Goodreads.

This learning activity, from book selection to review and talk, touches on two of the principles in the ACRL (2015) Framework for Information Literacy. Students experience the frame of information creation as a process when they select a book, reflect on its mode (fiction or nonfiction) and format (electronic or print), read it, write and publish an online review, script and record a book talk, and upload it to a popular public video site. Students experience the frame of scholarship as conversation when they engage in peer critique and peer feedback because classmates are able to comment on each other's reviews both on Goodreads and within the online discussion forum in the learning management system. They also reflect on how different mediums of conversation—formal writing, online reviews, and popular video sites—reach different audiences.

11.2.4 Social Media Plan Group Project

As a final assignment within the course "Searching for Information," students were required to work in small groups to successfully create and deliver a social media plan to a real-world "client." They first collaboratively developed a group charter to help guide their interactions with each other as their work unfolded. The charter created guidelines about participation and identified sanctions for members not participating in a manner consistent with the agreed-on guidelines. Having students create the charter together helps set a baseline for expectations, heads off common concerns students have about group work, and helps ensure they will realize the benefits of collaborative learning.

After the charter was finalized, the next part of the assignment focused on information gathering. Each student group selected an organization with which to work throughout the assignment. The organization could be a nonprofit or a small business that may or may not already have a social media presence but had no previously adopted social media plan. One or more students from the group then contacted the organization to assess its social media needs. Students were expected to conduct an investigation of the organization and research the industry or nonprofit sector to which the organization belonged. The students used skills they learned earlier in the course such as interviewing people and using library databases to conduct research on organizations and industries. They also needed to review the literature on creating social media plans and wherever possible find examples of social media plans for similar organizations online.

The final part of the assignment involved creating an actual social media plan that would be delivered to their client organization. There was no set document type or length for the social media plan; instead, students were told that it should be delivered in a format that would be most useful and desired by the specific organization. In most cases, the plan would include recommended goals and expected outcomes, suggested social media platforms, posting schedules, and content guidelines. Document design is an important factor at this step, with students applying what they have learned about how to effectively present information. They are encouraged to create a professional document and include images and graphics as appropriate.

Besides serving as an opportunity to integrate some of the information skills the students have learned over a semester, the social media plan project employs several approaches to enhance student engagement and connection. In many cases, students choose to work with an organization in their community or one with which they have a personal connection. There is a strong real-world or scholarship-in-action component, akin to community engagement, as some of the organizations will actually use and implement the social media plans. Students gain experience working as part of a team and in communicating with a client. Their use of the technology is different, too. Students do not just use or create content for social media in this assignment; they interact with an outside stakeholder to facilitate development through social media, thus engaging the information literacy frame of information creation as a process at a new level. They are required to problem solve in ways that have clear practical applications in a workplace environment. The external client lends more weight to the importance of good research and presentation and helps students understand ACRL's (2015) frame that information has value. Students also feel more accountable to their client organization than they might to their professor alone.

11.3 DISCUSSION

Online learning environments can make intentional and integrated use of technology and student interactions to forge strong connections and enhance student engagement in learning. The assignments and learning activities described in this chapter implement several approaches to assist in these endeavors while also addressing new frames for information literacy. A key theme of all of these activities is offering students a learning experience that is based in the real world and connected to their immediate interests and needs. In his well-known essay "The Rhetorical Stance," Wayne Booth argued against artificial assignments and stressed the importance of creating work for students that gave them a "genuine audience" (p. 143). While he wrote this in 1963, instructors often still create assignments that feel removed from real-world application. This can be addressed by encouraging peer feedback, bringing in visiting experts, having students create content for popular social media sites and wikis and having students work collaboratively to meet the needs of community organizations.

Using multiple technologies in layered ways also creates engaging learning experiences for students. The activities described in this chapter have students creating content and interacting in Goodreads, YouTube, discussion boards, Wikipedia, and other social media sites and using whatever video-recording equipment is convenient. Students stretch their skills with the technologies and find new ways to engage one another through them. This is consistent with the idea of metaliteracy, defined by Mackey and Jacobson (2014) as expanding "the scope of traditional information skills (determine, access, locate, understand, produce, and use information) to include the collaborative production and sharing of information in participatory digital environments (collaborate, participate, produce, and share)" (p. 1). Thus, using these technologies is not simply about introducing novelty, it is an active practice in developing multiple literacies.

Another important feature of these learning activities is that they reflect the true integration of interaction, collaboration, and information technologies into the courses. With these activities and assignments, librarians strived to employ technologies to achieve collaboration among students, the instructors, and outside experts and clients. As stated by Guthrie and McCracken (2010):

> When collaboration is the center around which instructional design and delivery is based, it is possible to create a cohesive approach to curricular implementation, maintain the integrity of the learning environment, and deliver contextually relevant experiences that are both individually and collectively meaningful... (p. 154)

Direct engagement and collaboration within a dynamic information environment affords opportunities for meaningful and engaged instruction and creates stronger learning connections among students and instructors.

Assessment of these instructional approaches and activities is, of course, an ongoing process. The information literacy assignments and practices described in this chapter have only been in use for a few years at Metropolitan State University, and no formal long-term assessment project has yet been undertaken to measure their impact. However, course instructors undertake various informal measures of assessing the students during and after each course. The learning management system used by the librarians includes options for surveys and anonymous discussion forums, where they are able to periodically check in with students about their experiences in the course. In addition, the assignments build on each other and are graded using rubrics as a way of assessing student learning. At the end of the "Searching for Information" course, there is a "final thoughts" discussion board where students are prompted to describe which parts of the course they felt were most valuable to them. Student response, both anonymous and shared, has been overwhelmingly positive.

The most important lesson I learned from this class is that there are so many different ways of obtaining good information. ... Up until this point my research was alarmingly predictable and one-tracked. Through our examination of the many different aspects of information and many different mediums of information. ... I now see more value than ever in enjoying the journey part of research as opposed to being singularly focused on the destination.
Student communication, April 29, 2013.

Although instructors and students alike have found many interesting and beneficial outcomes to these instructional activities and approaches, there are some challenges. One challenge is protecting the privacy of students in social media environments. Not all students wish to have their activity on social media sites be visible to the class or the public or have their coursework open to public review. Students sharing a private link or using pseudonyms in social media have generally addressed this concern. Another challenge has been the highly integrated use of technology while working with student populations that may have a lack of access to and experience with some of the technologies. We have attempted to address this by making technology available through library circulation, encouraging students to use and build on the technology they already know, and encouraging them to embrace the steep learning curve as a way to help build their digital competence. A final challenge has been one familiar to many librarians and faculty: keeping up with new technology. We are always looking for fresh content and new digital tools to explore and work with, which helps ensure that students are familiar with and engaged with the material. For example, this semester we have begun to explore ways to integrate live-streaming apps such as Periscope into the course content and have added newly published books to the reading selection list.

11.4 CONCLUSION

Online course environments provide an array of opportunities to engage students in cultivating information literacy skills and building competency in the ACRL Information Literacy threshold areas. Learning management system features such as discussion boards can be used in ways that facilitate active learning and allow students to drive the course content. At the same time it can be highly beneficial to direct students out of the learning management system toward a variety of social media sites to complete tasks that teach students how to more effectively collaborate online and write and create content meant for public consumption. This creates a fully immersive experience that requires students to both analyze and create content, which can broaden student awareness and provide transferable real-world skills related to technology and social media. While some librarians may be apprehensive about teaching online if they are accustomed to in-person instruction, the online environment lends itself particularly well to the transfer of certain skills and techniques and the forming of collaborative spaces. As instructors become more proficient in successful methods of online instruction, they will continue to find ways to teach information literacy that capitalize on the strengths of digital spaces.

REFERENCES

Association of College and Research Libraries. (2015). *Framework for information literacy for higher education*. Retrieved from <http://www.ala.org/acrl/standards/ilframework>.

Boling, E. C., Hough, M., Krinsky, H., Saleem, H., & Stevens, M. (2012). Cutting the distance in distance education: Perspectives on what promotes positive online learning experiences. *Internet and Higher Education*, *15*, 118–126. <http://dx.doi.org/10.1016/j.iheduc.2011.11.006>.

Booth, W. (1963). The rhetorical stance. *College Composition and Communication*, *14*(3), 139–145. Retrieved from <http://www.ncte.org/cccc/ccc>.

Cuthbertson, W., & Falcone, A. (2014). Elevating engagement and community in online courses. *Journal of Library & Information Services in Distance Learning*, *8*(3–4), 216–224. Available from <http://dx.doi.org/10.1080/1533290X.2014.945839>.

Dixson, M. D. (2010). Creating effective student engagement in online courses: What do students find engaging? *Journal of the Scholarship of Teaching and Learning*, *10*(2), 1–13. Retrieved from <http://www.iupui.edu/~josotl>.

Du, J., & Durrington, V. A. (2013). Learning tasks, peer interaction, and cognition process: An online collaborative design model. *International Journal of Information and Communication Technology Education, 9*(1), 38−50. Available from <http://dx.doi.org/10.4018/jicte.2013010104>.

Fink, L. D. (2013). *Creating significant learning experiences: An integrated approach to designing college courses* (2nd ed.Hoboken, NJ: Jossey-Bass.

Gibson, C., & Jacobson, T. E. (2014). Informing and extending the draft ACRL Information Literacy Framework for higher education: An overview and avenues for research. *College & Research Libraries, 75*(3), 250−254. Available from <http://dx.doi.org/10.5860/0750250>.

Guthrie, K., & McCracken, H. (2010). Making a difference online: Facilitating service-learning through distance education. *Internet and Higher Education, 13*, 153−157. Retrieved from <http://www.journals.elsevier.com/the-internet-and-higher-education/>.

Keengwe, J., & Schnellert, G. (2012). Fostering interaction to enhance learning in online learning environments. *International Journal of Information and Communication Technology Education, 8*(3), 28−35. Available from <http://dx.doi.org/10.4018/jicte.2012070104>.

Lester, J., & Perini, M. (2010). Potential of social networking sites for distance education student engagement. *New Directions for Community Colleges, 150*, 67−77. Available from <http://dx.doi.org/10.1002/cc.406>.

Leveen, S. (2005). *The little guide to your well-read life: How to get more books in your life and more life from your books.* Delray Beach, FL: Levenger Press.

Mackey, T. P., & Jacobson, T. (2014). *Metaliteracy: Reinventing information literacy to empower learners.* Chicago, IL: Neal-Schuman, an imprint of the American Library Association.

Mitchell, E., & Smith, S. S. (2009). Bridging information literacy into the social sphere: A case study using social software to teach information literacy at WFU. *Journal of Web Librarianship, 3*(3), 183−197. Available from <http://dx.doi.org/10.1080/19322900903113381>.

Palloff, R. M., & Pratt, K. (2007). *Building online learning communities: Effective strategies for the virtual classroom* (2nd ed.San Francisco, CA: Jossey-Bass.

Small, R. V., Arnone, M. P., Stripling, B. K., Hill, R. F., & Bennett, B. (2012). The three C's of distance education: Competence, creativity and community. *School Libraries Worldwide, 18*(2), 61−72. Retrieved from <http://www.iasl-online.org/publications/slw/index.html>.

Stodel, E. J., Thompson, T. L., & MacDonald, C. J. (2006). Learners' perspectives on what is missing from online learning: Interpretations through the community of inquiry framework. *International Review of Research in Open and Distance Learning, 7*(3). Retrieved from <http://www.irrodl.org>.

U.S. Department of Education, Institution of Education Sciences, National Center for Education Statistics. (2014). *Enrollment in distance education courses, by state: Fall 2012 (NCES 2014-023).* Retrieved from <http://nces.ed.gov/pubsearch/pubsinfo.asp?pubid = 2014023>.

CHAPTER 12

Innovation Through Collaboration: Using an Open-Source Learning Management System to Enhance Library Instruction and Student Learning

12.1 INTRODUCTION

Purchase College, State University of New York (SUNY) has been fortunate to have a meaningful and effective collaboration between the Library and the Teaching, Learning, and Technology Center (TLTC). The TLTC administers, manages, and has expert knowledge of Moodle, the college's learning management system (LMS). Parallel and complementary missions and skills, congenial working relationships that allow for open and constant dialog, a shared ethos of flexibility and intrepidness, and geographic proximity (the TLTC is physically and administratively embedded in the library) have made this collaboration not only possible but also deeply interwoven into the culture of the two departments. The addition of a digital media developer and the use of Moodle, an open-source LMS with no licensing fee, have also played a critical role in facilitating the library and TLTC's alliance and shared effort to support the entire campus community. Hosting the college's instance of Moodle on campus, as opposed to contracting with a third-party hosting solution, allows for increased flexibility and ongoing refinements because self-hosting affords the TLTC access to Moodle's core code.[1] The innovations and customizations discussed in this

---

[1] Core code, sometimes referred to as *source code*, is defined by the Linux Information Project as "the version of software as it is originally written" (2004). Since Moodle is an open-source project, its core code can be modified to suit to the needs of its users. For an explanation of what is included in Moodle's core code, please see Moodle, 2014b, *Moodle Architecture*.

Distributed Learning. © 2017 T. Maddison and M. Kumaran.
Published by Elsevier Ltd. All rights reserved. **221**

chapter would not be nearly as straightforward and timely without this level of access to the code base. An environment conducive to partnership and the exchange of ideas has fostered numerous projects that have enhanced student learning, deepened relationships with teaching faculty, and improved workflows. Gibbons (2005) states that "without a positive working relationship between the library and ITS [information technology services], the task of library and CMS integration is nearly insurmountable" (p. 19). The projects that are discussed include customization of user roles, embedding tutorials and learning modules, digitization of the student project process, and using rubrics for authentic assessment, all within Moodle.

12.2 CUSTOMIZED USER ROLES IN MOODLE

The library is integrated into the college's LMS in one fundamental way: by the creation of three customized user roles—course librarian, service desk staff, and reserves librarian. TLTC created a suite of customized roles for librarians and library staff within Moodle that allow greater access to courses and course content and provide additional functionality within virtual learning spaces. As Gibbons points out, "academic libraries must go where the students and faculty are. More to the point, libraries need to be where the learning is happening, even if that is the virtual environment of a CMS," (2005, p. 12). The close working relationship between the library and the TLTC allowed the development of these three custom roles to be done in consort with bilateral input from both parties.

12.2.1 Course Librarian Role

Prior to 2008, Purchase College used Blackboard as its LMS, and the library had little involvement with its use across campus. The environment was such that Blackboard was considered the domain of teaching faculty and the purview of the instructional technology center; the library did not recognize any connections between its mission and the use of the LMS. This attitude probably stemmed in part from the limited nature of proprietary LMSs at that time. The version of Blackboard that Purchase utilized offered only the rigid and dichotomous roles of instructor or student, and did not lend itself to the unique needs of librarians.

Shortly after choosing and adopting Moodle as the college's virtual learning environment in 2009, the TLTC initiated conversations with the library about how to bring librarians into the Moodle environment.

From those initial discussions, the need for a custom course librarian role in Moodle was discovered. In the past, librarians asked the TLTC, administrators of the college's LMS, for access to certain courses to aid in teaching information literacy sessions. With the understanding that librarians should not be given full teacher privileges in Moodle courses, the course librarian role was designed. The course librarian role creates a clear distinction between the librarian and the instructor so that the course instructor maintains full autonomy to use the LMS to augment and complement face-to-face interactions. The instructor must add the librarian to the course at the course librarian role, which requires the librarian to interact with the faculty and reinforces a culture of self-service on the part of instructors because they do not need to contact Moodle administrators to control who has access to their course.

The course librarian role allows librarians access to courses with enhanced permissions including the ability to post to course discussion forums, contribute resources and activities, add blocks, and communicate directly with students (both individually and as a class) before, during, and after instruction sessions. The librarian can access all of the submitted assignments, but the role does not allow the librarian to see student grades outside of assignments created or added to the course by the librarian.

With the level of access afforded to course librarians, they can more meaningfully and seamlessly integrate information literacy into the course, alongside all other course materials and interactions. Librarians can create "assignment-specific library resources embedded within course pages to supplement or replace traditional face-to-face instruction, as a communication tool for providing online reference assistance to students at their point of need, and for making information literacy a collaborative effort between faculty and librarians in a virtual learning environment" (Pickens-French & McDonald, 2013, pp. 53–54).

Specific examples of how Purchase College librarians have used the course librarian role in Moodle to enhance bibliographic instruction include the following:

- posting a discussion forum entry that summarizes the main points of an instruction session,
- adding a block comprising tailored links to resources for a particular topic (e.g., specific databases, LibGuides, journals),
- creating a Moodle page resource to brainstorm search terms as an in-class activity,

- developing an interactive database activity in which students work collaboratively to collect resources on a shared topic, and
- conducting authentic assessment to evaluate Association of College and Research Libraries (ACRL) information literacy skills based on completed assignments.

The specialized course librarian role expands how librarians can interact with courses without sacrificing class time, which is a common obstacle for academic librarians. In addition, the integration of the librarian into the digital learning environment allows for information literacy to be presented outside of the designated "library session." Activities, follow-up, assessments, and so on can be completed before, during, or after the students meet with the librarian for traditional one-shot instruction sessions. This integration solidifies the librarian's presence within the context of the course as a whole, even if the students do not receive face-to-face instruction with the librarian.

12.2.2 Service Desk Staff Role

As Moodle was embraced by instructors at Purchase College, the number of questions both the library and the TLTC received from faculty and students increased in kind. Library staff at the circulation and reference desks received questions from students that they were unable to answer because they did not have access to specific Moodle courses. In addition, the TLTC, which only has two full-time employees, looked to the library for basic, level-one Moodle support. To address the needs of the library, as well as the TLTC, the service desk staff role was created. When users are assigned at this role, usually at the "category" level, they are able to find and access every course, regardless of the course's availability (i.e., whether or not the instructor has made the course visible to student view).[2] Similarly, service desk staff can view every resource and activity, regardless of its availability to students but cannot interact with the course in any way (e.g., see submitted assignments or create course content).

[2] Category level is distinct from system level access since Moodle is used for purposes other than to deliver course content. For example, committees, departments, students groups, and others use Moodle to share resources and communicate. Moodle spaces used for these purposes exist in their own categories (e.g., programs, departments, and committees). Assigning the service desk role at the category level ensures that users with that access have read-only access to all spaces within specific categories; in this case, only official course sections (e.g., fall 2015).

Each semester, the TLTC manually assigns librarians and other service and help desk staff to the service desk staff role at the applicable category level in Moodle (current and previous semesters). Since this role is only assigned to specific course categories, its capabilities become inherent to the user's account when he or she accesses courses within those categories and does not require any intervention on the part of the instructor. Users assigned at this role do not appear in an instructor's course roster within Moodle, and they are not disruptive to the administration of the courses and thus prevent student and instructor confusion.

The service desk staff role alleviates the barrier of access, an oft-cited hurdle that can prevent library involvement in the LMS. Meredith Farkas argues that "while some systems offer user roles with limited access, in many cases, access is an all or nothing proposition" (2008, p. 60). The service desk staff role exists outside of the "all or nothing" dichotomy by allowing library staff assigned to that role view-only access but limited interactivity with the course content.

Librarians experience Moodle courses through the lens of the service desk role and benefit most during reference interactions due to how integral Moodle is to the academic experience at Purchase College. Librarians routinely field questions at the service desks, via email, and by text that can only be answered if the librarian has access to particular courses in Moodle. The service desk staff role circumvents the need for librarians to request to be added to individual course spaces by providing them access to all courses within particular course categories (e.g., Fall 2015) without the need to be explicitly "enrolled" in the course. This elevated access to all courses allows librarians and library staff the ability to effectively troubleshoot issues as they arise and enhances the level of service provided to students and faculty. Service desk staff are able to access all course resources and activities, even if the instructor has them hidden from student view, which is an invaluable asset when troubleshooting issues at a busy service desk.

Librarians also routinely use the service desk role to assist with collection development. Librarians can consult syllabi posted in Moodle courses relevant to their subject areas, as well as monitor assigned readings, identify important figures and topics in specific disciplines, and determine students' research needs and requirements. Access to course content at this level grants unparalleled insight that is independent of faculty participation and essential to informing collection and resource needs. For example, a comprehensive syllabi analysis project was completed for both new media and Latin American art using the information provided in relevant Moodle courses, which directly informed monograph fund allocations and collection-development decisions.

Implementation of the service desk role among library staff greatly elevated the quality of service provided to patrons. The advent of the service desk role is one of, if not the most, important developments that has impacted the day-to-day service desks interactions at all levels, from answering simple questions about books on reserve to conducting a comprehensive syllabi review for a particular major. After receiving positive feedback from library service desk staff, the TLTC extended the role to include members of the campus technology services helpdesk to broaden "Level 1" support.

12.2.3 Reserves Librarian Role

To streamline and digitize the physical library reserves process, the TLTC developed a custom library reserves block in Moodle where faculty can initiate requests and track progress and students can see the status of reserve items and their call numbers as they become available.[3] Centralizing the reserves process in Moodle required the creation of the reserves librarian role, which is granted at the system level to the two library staff members who process all physical library reserve requests, including books, DVDs, and CDs. All requests are collected and managed via an interactive dashboard in Moodle that is only accessible to those assigned to this role. From the dashboard, reserves librarians receive and process all physical library reserve requests and update the status so that instructors can monitor requests via the library reserves block in their Moodle course.

An instructor initiates a physical library reserve request using the library reserves block, which searches the local holdings in IDS (Information Delivery Services) Search, the union catalog for participating New York state public and academic libraries. Since it is not open source, Purchase College's version of Aleph (Integrated Library System - ILS) does not allow access to its data, so Moodle is not able to get holdings information directly from Aleph. Fortunately, IDS Search is open to developers and Moodle can communicate directly with Purchase College Library's local holdings as reflected in IDS Search. The integration of this fundamental library function into the LMS creates a seamless experience

---

[3] Please see the library's reserve guide at http://purchase.libguides.com/reserves/requests for a detailed explanation, with images, of how to place physical reserve requests within Moodle.

for students and faculty. The course instructor adds the block to his or her course and initiates reserve requests for books and media. The library reserves block is also designed to allow previous requests to be quickly reused from past semesters. Managing print reserves requests via the LMS is quite unique; as Black (2008) states, one of the most common ways libraries are integrated into the LMS is in the context of electronic reserves. The reserves librarian's role and the corresponding library reserves block in Moodle have revolutionized the physical library reserves process at Purchase College Library and has reinforced the library's overall effort to encourage faculty self-service.

Customized user roles in Moodle are demonstrative of both macro- and microlevel library courseware involvement, methods introduced by Shank and Dewald in their 2003 article "Establishing Our Presence in Courseware: Adding Library Services to the Virtual Classroom." They state that "Macro-Level Library Courseware Involvement … entails working with the developers and programmers of courseware to integrate a generic, global library presence into the software… Micro-Level Library Courseware Involvement… involves individual librarians teaming up with faculty as consultants to participate in developing a customized library instruction and resource component for the courseware-enhanced courses" (p. 38).

The reserves librarian role, along with the library reserves block, and the service desk role are the only systemwide, macrolevel presences of the library in the LMS due to the limitations imposed by the college's enterprise resource planning (ERP) system. The college's ERP controls course creation, instructor assignment, and student enrollment within Moodle. Accurate Moodle section creation and course enrollments were more important than the ability to use a predefined Moodle template for all course creation. Prior to the implementation of the new ERP, which replaces a homegrown student information system, the library and TLTC had collaborated on developing a custom Moodle course template with a library resources block that included links to certain databases and research guides on which all course sections were created. Currently, librarians participate almost exclusively in Moodle at the microlevel, which is facilitated by the course librarian role.

12.2.4 Embedded Tutorials and Learning Modules in Moodle

Leveraging the library's established bibliographic instruction program with the college's first-year composition course, College Writing, the instruction

coordinator worked with other instruction librarians to develop information literacy tutorials on skills commonly covered in college writing
one-shot library sessions, such as: finding books, finding articles, creating
search terms, using the library's discovery layer, and finding different types
of sources. The tutorials cover fundamental skills (e.g., where to click),
which allows librarians to spend face-to-face class time on higher-level
concepts and active learning activities employing the pedagogical flipped
classroom technique

> that inverts the traditional lecture-plus-homework formula. By moving the deliv-
> ery of foundational principles to digital media, such as video lectures or tutor-
> ials, class time is freed up for engaging activities that allow students to apply
> these basics to practical scenarios in the presence of their instructor.
>
> **(Arnold-Garza, 2014, p. 10)**

A key element to successfully "flipping the classroom" is the use of
video lectures, or, in this case, video tutorials, which "puts lectures under
the control of the students: they can watch, rewind, and fast-forward as
needed" (EDUCAUSE Learning Initiative, 2012). Purchase College
Library's tutorials are produced using Camtasia, enhanced with voiceover
narration, and then uploaded to the library's YouTube channel for free
hosting and easy distribution. Accompanying each tutorial is a transcript
that can be interpreted by screen readers to ensure accessibility.

Recognizing that the most effective way to convey supplemental
course content is within the context of the LMS, the library's instruction
coordinator reached out to the TLTC to explore methods to integrate
the tutorials into Moodle. Over the course of several conversations, the
TLTC and the instruction coordinator decided that the instruction coordinator will be enrolled into the Moodle course for each section of
college writing in the course librarian role so that she can embed all four
tutorials, along with their short companion quiz, into a week or topic
area that is then hidden from student view but still visible to the instructor. Typically this hidden week or topic is located at the bottom of the
Moodle course to distinguish it from the instructor's course content. The
library and the instruction coordinator have an established relationship
with the college writing coordinator, so the inclusion of preloaded course
content was a welcome addition to the college writing curriculum.

When instruction librarians are assigned to their sections of college
writing and then added to the Moodle space at the course librarian role
by the instructor, the tutorials are moved to the appropriate week and
topic or are linked to from the hidden week using Moodle's autolinking

filter, which "create[s] links to an activity whenever the name of the activity is written in texts within the same course in which the activity is located" (Moodle, 2014a). Generally, librarians embed the tutorials in the week or two before the scheduled library session to allow the librarian time to review the quiz results before meeting with the class. To prompt completion of the tutorials, librarians communicate directly with students via discussion forums and course announcements to remind them to watch the tutorials and complete the quizzes. The persistent presence and visibility of the librarian in the course's virtual learning environment helps to combat the perception that the library and the librarian are merely one-time visitors to the course; instead, librarians are active participants and directly connected with students.

Using Moodle's robust quizzing feature, librarians developed a bank of quiz questions based on each tutorial. To present the tutorials and the quiz as one resource, the YouTube-hosted video tutorials are embedded in the description area of each quiz. The quizzes and their results allow librarians to monitor which skills and concepts require more explanation and emphasis during face-to-face instruction sessions. Each college writing library session is informed by quiz results and tailored to the needs of that particular group of students. The quiz results also serve as an assessment device so that the library can adjust its instruction program for college writing appropriately. Quizzes in Moodle are self-grading, which provides immediate and qualitative feedback to the students on their performance. College writing instructors have incorporated the tutorials and quizzes into the curriculum, where they serve as a required graded assignment. The library's use of Moodle's quiz activity allows for easy integration of quiz results into the course grade book, which requires little effort by the instructor.

In the 2011 article "Learning to Leverage: Using Moodle to Enhance F2F Interaction in One-Shots," Rebecca Oling, the coordinator of library instruction, and Marie Sciangula first described the library's use of embedded tutorials within the LMS for college writing. As part of that process, both authors elicited feedback from college writing instructors, who "felt that the tutorials allowed the library session to be more focused, ... a function of the higher order skills the students were able to practice, like actually getting past the execution of searches to focus on limiting and evaluating the relevancy of results" (Oling & Sciangula, 2011, p. 10). Incorporating information literacy tutorials into the college writing curriculum, facilitated via Moodle, has had a positive and measurable impact on classroom teaching from both the librarians' and instructors' perspectives.

12.2.5 Digitization of the Student Project Process Using Moodle

One of the most innovative and significant collaborations between the library and the TLTC is the transformation of the year-long student projects process. At Purchase College, *student projects* is the blanket term used to refer to the "culminating student experience" and includes senior projects, capstone papers, and master's theses. Librarians, catalogers, developers, and TLTC staff partnered to create a workspace in Moodle that has taken student projects from a traditional paper-based process to a digital, dynamic, student-centered venture that is interwoven with reference, instruction, assessment, and access. Prior to 2011, student projects, a requirement for graduation, were submitted to the library either in print or on disc. This manner of collecting student projects, the culminating work of a student's academic experience at the college, was problematic for several reasons: metadata was inconsistent and arbitrarily designated since the college had not mandated any standard; approval signatures from project readers were difficult to obtain because they needed to occur in person; and the submission workflow was questionable in that a reader could sign off on a project but a student could submit a completely different project to the library since the student was responsible for the final physical submission. The library was seen not only as the final destination for all student projects but also as the authority in terms of due dates and other details, a misleading assumption that caused confusion for students and staff. The TLTC approached the director of the library to begin discussions about how Moodle could facilitate a more effective student project submission process.

The result was the development of a specialized shared Moodle space with enhanced functionalities, including research, support, and online submission of student projects. The TLTC partnered with key librarians to develop an enhanced Moodle course shell that serves as a collaborative workspace, as well as the mechanism by which student projects are submitted. Undergraduate students are given their own student project Moodle space at the beginning of their junior year when planning for the student project ideally occurs; graduate students are automatically provided with a student project space when they enroll in a graduate program. Permissions to the student role within the student projects course category were tweaked to allow students additional functionalities (e.g., students can upload files, add activities, and post resources to share with their readers) so that they would feel a greater sense of ownership over their project space.

The functionalities of the student project spaces in Moodle evolved in response to tangible problems with the traditional submission process. From within the student project space, the administration of the formal submission process was digitized: students now invite readers to the space electronically; drafts and other documents can easily be exchanged in the Moodle environment; metadata is collected via a standardized form and approved by the reader, ensuring validity of the metadata, which is essential for discovery and archiving; a uniform file format (pdf) was established to enable access; and perhaps the most important feature is that the final approval of the student project is completed within Moodle using an electronic signature that readers can use to sign off on a project from any location. To safeguard the legitimacy of the process, all signatures are invalidated if any metadata are changed or if a new final document is uploaded.

To best control the quality and completeness of the metadata entered for each student project, a series of conditional controls were built into the final submission process. For example, students are unable to submit their final project until project type (e.g., senior project, capstone paper, or master's thesis), major, and readers have been designated. In addition, project title and abstract both need to be entered and approved by the first reader before the final project can be submitted. These dependencies are essential to ensuring that metadata for each student project are controlled, consistent, and complete in order to facilitate discoverability and cataloging by the library.[4] The creation of the customized student project space and the detailed programming that administers its functions was made possible by the agility of Moodle and the PHP programming skills of the TLTC's digital media developer. With the library's input on procedural and cataloging considerations, the student project space was developed to meet the needs of the college in terms of practical online submission and those of the library as the repository of all student projects.

Once the student project spaces in Moodle were designed, tested, and refined, the library and the TLTC turned their attention to developing a mechanism by which end users search for and access student projects. Before the adoption of Moodle as the primary environment of the

---

[4] Please see the library's student projects guide at http://purchase.libguides.com/student-projects/setup for a detailed explanation, with images, of each step of the student projects' setup and submission process from both the reader's and student's perspective.

student project process, print student projects were held in the library's archives and were only accessible on request. Although print student projects were cataloged and discoverable in the library's local catalog, patrons would have to physically visit the library to access a student project or pay a small fee to have a scanned version sent via email. Because Moodle became the accepted and only way to submit student projects to the library in order to graduate, these projects were already in an electronic format; however, the library still needed a way to make them publicly accessible and searchable. The library and the TLTC experimented with a number of ways to achieve this ranging from hosting projects on a shared network drive that was then linked to a catalog entry to developing a simple search function in Moodle that only searches items in the student project category. As a result projects from different years were scattered across multiple digital platforms because none of the solutions could be combined to provide a centralized searchable repository. To accomplish this, the library and the TLTC looked to a third-party solution, CollectiveAccess, an open-source cataloging and digital asset manager. CollectiveAccess was chosen because it is open source, is compatible with Moodle, and has an integrated full-text search that is able to search the content of text documents even if they were not originally created as optical character recognition documents.

To disseminate student-generated scholarship, the library and the TLTC hired Whirl-i-Gig, the developers of CollectiveAccess based in New York City, to create an automated workflow and content-mapping schema so that projects submitted via Moodle are automatically published in the college's instance of CollectiveAccess three times each year in correspondence to graduation dates (i.e., spring, summer, and fall). CollectiveAccess comprises both a back-end cataloging interface (Providence) and a front-end website (Pawtucket) that communicate with one another and allow content searching and browsing through custom fields. In this case, the library defined its own cataloging standards to match the information gathered during the student project submission process. In CollectiveAccess, student projects are searchable and browsable both in the back and front ends by student name, student type, reader name, major, title, and keyword, using CollectiveAccess' standard full-text search. The library was instrumental in determining the searchable and browsable fields as they reflect the practical needs of patrons.

In addition to mapping Moodle data to CollectiveAccess to capture, catalog, and make new projects available, the library and the TLTC plan to digitize the college's archive and create a single comprehensive digital repository of student projects. At the time of publication, plans for large-scale digitization are being developed to make nearly 40 years of projects accessible, allowing for better and broader access to this collection of important student scholarship and creativity. Both the LibAnswer and LibGuide pertaining to these projects are the second most visited of all the library's content on both platforms. Since 2013, the LibGuide has been viewed 36,850 times. Since 2013, the LibAnswer ("Where can I find information on senior projects?") has been accessed 15,032 times. In addition, the library reference desk consistently fields questions in person and via email, chat, and text from current students, faculty, and staff, as well as from alumni about how to access student projects. The need for a comprehensive, user-friendly, and searchable repository of all student projects is clear.

The library and the TLTC hope that once all student projects are accessible via one repository in CollectiveAccess, additional discussions can be held about developing a comprehensive institutional repository of both student and faculty work that is accessible to the public and fosters academic discourse. Individually, neither the library nor the TLTC could accomplish changes on this scale; it was the combination of the library's and the TLTC's expertise, perspectives, and commitment that enabled the transformation of the student project process and will facilitate future discussions to advance the development of an institutional repository at Purchase College.

12.2.6 Using Moodle Rubrics for Authentic Assessment

As part of an ongoing effort to increase awareness and assessment of information literacy skills on campus, the library recognized a need for a longitudinal view of Purchase College students' research skills. After evaluating several major testing instruments, it became clear to the Library Assessment Working Group (a team of three librarians) that only meaningful data based on actual student research products and not test scores would allow full assessment of the level of information literacy skills demonstrated by students. The working group developed a rubric-based assessment of Purchase College's senior projects (to which the library has both print and online access). The rubric is based on the ACRL's *Information Literacy Competency Standards for Higher Education* and is also mapped to SUNY student learning outcomes.[5] The working group identified the most measurable criteria in the ACRL standards, matching the rubric categories and benchmarks to specific standards, indicators, and outcomes. The final criteria include the presence of a thesis statement; the authority of references; the variety of references; the consistency of attribution; the quality of citations (in-text and works cited); the ability to paraphrase, summarize, and quote effectively; integration of resources to support a thesis; the overall organization of content; and the limitations of research. The complete rubric is available in Appendix 1.

The rubric underwent several norming sessions to ensure it would be consistently interpreted and applied by anyone.[6] The norming stage was of crucial importance because the long-term goal of the working group is to encourage use of the rubric by instructors and senior projects readers across the college. The rubric was initially developed to be used by librarians to assess the information literacy skills of graduates and thus gauge the effectiveness of the library's instruction program. Looking beyond its exclusive use by librarians, the working group brainstormed ways to get instructors to use the rubric while grading research papers throughout the curriculum. Since the librarians only have access to senior projects these are the only ones readily available for assessment. To truly evaluate the information literacy skills of Purchase College students, the rubric needs to be applied to research papers produced at all academic levels and for a variety of academic disciplines. Wider utilization of the rubric is facilitated using Moodle's rubric function. Moodle users "can save their [advanced] grading forms as shared templates on the site. Such forms can be then picked and re-used by all teachers in their courses," (Moodle, 2015). Moodle uses the terms *rubric* and *advanced grading forms* interchangeably. To allow the rubric to be used by other instructors in Moodle, the "Save rubric and make it ready" option must be clicked. After saving the rubric as ready, the rubric can be published as a new public template as seen in Table 12.1. Without selecting this option, the rubric is available for use only within the context of the course in which it was created and is not accessible to any other Moodle user. Designating the rubric as ready allows any user

---

[5] For the entire text of the SUNY Student Learning Outcomes please see Office of the Provost, State University of New York, 2010.

[6] For a detailed description of the norming process, please see Gervasio, Detterbeck, and Oling (2015, pp. 724–725).

Table 12.1 Detail of senior projects information literacy skills assessment rubric as it appears to faculty in Moodle

Thesis statement: students will produce coherent texts within common college level forms. ("writer presents an easily identifiable and focused controlling purpose or thesis.") (SLO1, ACRL 1.1.b)	Exceeds: writer presents easily identifiable, focused, original thesis statement *4 points*	Meets: clear and concise thesis statement based on information need *3 points*	Meets some: defines a thesis Statement based on information need (it is present, but may not be clear and concise) *2 points*	Meets few: unclear or absent thesis statement or research question *1 points*
Authority of references: students will locate, evaluate, and synthesize information from a variety of sources ("all sources are of good quality, well-selected and appropriate to the topic") (SLO 9, ACRL 1.2.b)	All references are peer-reviewed professional journals or other discipline appropriate sources (See http://bit.ly/IXLcRU) *4 points*	References are primarily peer-reviewed professional journals or other discipline appropriate sources (See http://bit.ly/IXLcRU) *3 points*	Although most of the references are professionally legitimate, a few are questionable within that discipline (e.g., trade books, internet sources, popular magazines,...). *2 points*	Most of the references are from sources that are not peer-reviewed and have uncertain reliability *1 points*
Variety of references: students will locate evaluate and synthesize information from a variety of sources ("student has located adequate information from a variety of print and electronic sources.") (SLO 9, ACRL 1.2.b)	Uses a mix of sources, appropriate to the discipline, and balances use of these sources and source types throughout *4 points*	Uses a mix of sources, does not rely too heavily on one source or source type *3 points*	Too heavily relies on one source or source type (as appropriate to discipline) *2 points*	Uses only one type of reference or obtains all reference from one source (i.e., all articles from the same journal or same author) *1 points*

who has the ability to create an assignment (i.e., any user assigned with the Teacher role) to change the grading method of that assignment to rubric and then select a predefined template—in this case, the information literacy rubric.

The rubric was just finalized in early 2015, and the library is yet to actively promote its wider use. Going forward, the library plans to pilot the rubric with a group of instructors who already have an established relationship with the library so it can develop an effective strategy for marketing the rubric to a larger audience. Librarians are able to share the information literacy rubric with the entire college in the virtual learning environment where assignments are already being collected and graded only because of Moodle's flexibility and the willingness and ability of the TLTC to respond to this feature request.

12.3 LOOKING AHEAD

These projects, born of the meaningful and effective collaboration between the library and the TLTC, have had a significant impact on teaching and learning at Purchase College. However, the projects discussed in this chapter do not represent the totality of the library and TLTC's partnership. In addition to maintaining and improving on the initiatives mentioned here, the library and the TLTC have several other collaborations on the horizon. Most immediately, they are creating a pre-populated menu of information literacy assignments and learning activities developed by librarians that allows instructors to select and directly import the activity into their Moodle course. In addition, the library and the TLTC are working together to determine how to best collect student projects that include digital media such as films, audio recordings, and high-resolution images via the student project Moodle space. The library and the TLTC are also exploring ways to leverage Moodle to more effectively deliver the library's streaming content in support of both online and face-to-face instruction. Finally, the library wants to further incentivize student participation in the information literacy tutorials via badging administered in Moodle. The library and the TLTC are only able to innovate and create new initiatives because of a shared sense of responsibility to improving teaching and learning at the college and a mutual respect for creating a true of partnership of equals.

REFERENCES

Arnold-Garza, S. (2014). The flipped classroom: Assessing an innovative teaching model for effective and engaging library instruction. *College & Research Libraries News*, *75*(1), 10−13.

Black, E. L. (2008). Toolkit approach to integrating library resources into the learning management system. *The Journal of Academic Librarianship*, *34*(6). <http://doi.org/10.1016/j.acalib.2008.09.018>.

CollectiveAccess. (2016). Retrieved from: http://www.collectiveaccess.org/.

EDUCAUSE Learning Initiative. (2012). *7 things you should know about flipped classrooms.* <http://www.educause.edu/library/resources/7-things-you-should-know-about-flipped-classrooms> Accessed 05.11.15.

Farkas, M. (2008). Embedded library, embedded librarian: Strategies for providing reference services in online courseware. In S. K. Steiner, & M. L. Madden (Eds.), *The desk and beyond: Next generation reference services* (pp. 53−64). Chicago, IL: Association of College and Research Libraries.

Gervasio, D., Detterbeck, K., & Oling, R. (2015). *The slow assessment movement: Using homegrown rubrics and capstone projects for DIY information literacy assessment.* <http://www.ala.org/acrl/sites/ala.org.acrl/files/content/conferences/confsandpreconfs/2015/Gervasio_Detterbeck_Oling.pdf> Accessed 04.11.15.

Gibbons, S. (2005). Integration of libraries and course-management systems. *Library Technology Reports*, *41*(3), 12−20.

Moodle. (2014a). *Activity names auto-linking filter.* <https://docs.moodle.org/29/en/Activity_names_auto-linking_filter> Accessed 04.11.15.

Moodle. (2014b). *Moodle architecture.* <https://docs.moodle.org/dev/Moodle_architecture> Accessed 02.23.16.

Moodle. (2015). *Advanced grading methods.* <https://docs.moodle.org/29/en/Advanced_grading_methods> Accessed 04.11.15.

Office of the Provost, State University of New York. (2010). *Guidelines for the approval of State University general education requirement courses.* <http://system.suny.edu/media/suny/content-assets/documents/academic-affairs/general-education/GenedCourseGuidelines_20120530.pdf> Accessed 04.11.15.

Oling, R., & Sciangula, M. (2011). Learning to leverage: Using Moodle to enhance F2F interaction in one-shots. *LOEX Quarterly*, *38*(3), 8−10.

Pickens-French, K., & McDonald, K. (2013). Changing trenches, changing tactics: A library's frontline redesign in a new CMS. *Journal of Library & Information Services in Distance Learning*, *7*(1−2), 53−72. Available from: <http://dx.doi.org/10.1080/1533290X.2012.705613>.

Shank, J. D., & Dewald, N. H. (2003). Establishing our presence in courseware: Adding library services to the virtual classroom. *Information Technology and Libraries*, *22*(1), 38−43.

The Linux Information Project. (2004). *Source code definition.* <http://www.linfo.org/source_code.html> Accessed 02.23.16.

APPENDIX 1 SENIOR PROJECTS INFORMATION LITERACY SKILLS ASSESSMENT RUBRIC-PURCHASE COLLEGE LIBRARY

http://tinyurl.com/SPILSAR.abric

ACRL Standard	SUNY Learning Outcome	Performance	Exceeds Expectations (4)	Meets Expectations (3)	Meets Some Expectations (2)	Meets Few or No Expectations (1)
1.1.b	SLO 1: Students will produce coherent texts within common college level forms. ("Writer presents an easily identifiable and focused controlling purpose or thesis.")	Thesis Statement	Writer presents easily identifiable, focused, and original thesis statement based on information need	Clear and concise thesis statement based on information need	Defines a thesis statement based on information need (it is present, but may not be clear and concise)	Unclear or absent thesis statement or research question
1.2.b	SLO 9: Students will locate, evaluate, and synthesize information from a variety of sources. ("all sources are of good quality, well-selected and appropriate to the topic")	Authority of References	All references are peer-reviewed professional journals or other discipline appropriate sources	References are primarily peer-reviewed professional journals or other appropriate sources	Although most of the references are professionally legitimate, a few are questionable within that discipline (e.g., trade books, internet sources, popular magazines...)	Most of the references are from sources that are not peer-reviewed and have uncertain reliability
1.2.b	SLO 9: Students will locate, evaluate and synthesize information from a variety of sources. ("student has located adequate information from a variety of print and electronic sources.")	Variety of References	Uses a mix of sources, appropriate to the discipline, and balances use of these sources and source type throughout	Uses a mix of sources. Does not rely too heavily on one source or source type	Too heavily relies on one source or source type (as appropriate to discipline)	Uses only one type of reference or obtains all references from one source (i.e. all articles from the same journal or same author)
2.5.c	SLO 3: Students will research a topic, develop an argument, and organize supporting sources in an accepted style. ("Carefully documented sources in an accepted style")	Attribution	Attribution is clear in all cases. Source of ideas and information is always identified.	Attribution is clear in most cases. Source of ideas and information is mostly identified.	Occasional attribution, but many statements seem unsubstantiated. Source of ideas and information is sometimes identified.	Little to no attribution. Reader is confused about the source of ideas and information.
2.5.c	SLO 3: Students will research a topic, develop an argument, and organize supporting sources in an accepted style. ("Carefully documented sources in an accepted style")	Quality of Citations (in text & works cited)	All in-text and bibliographic citations are complete and consistent	Most in-text and bibliographic citations are complete and consistent	Some in-text and bibliographic citations are incomplete/inconsistent	Most in-text and bibliographic citations are incomplete/inconsistent
3.1.b	SLO 3: Connects paraphrases and quotations to the larger thesis. SLO 9 "resources are consistently well-integrated into the paper"	Paraphrasing, summarizing, quoting	Expertly (in both quality and quantity) summarizes, paraphrases, or quotes in order to integrate the work of others into their own (ex: are graphs, tables, and other images well integrated)	Adequately summarizes, paraphrases, or quotes in order to integrate the work of others into their own (ex: are graphs, tables, and other images well integrated)	Does not consistently summarize, paraphrase, or quote, and does not always select appropriate method for integrating the work of others into their own	Does not summarize, paraphrase or quote in order to integrate the work of others into their own
3.7.a	SLO 3: "conclusions demonstrate writer's conscious attempt to integrate his or her own thinking with an analysis of outside sources"	Addresses & relates to thesis statement and research question	All parts of paper relate to thesis statement in a meaningful way.	Most parts of paper relate to thesis statement in a meaningful way.	Some goals of thesis are not met. Some parts of paper don't relate back to thesis.	The goals (if any) outlined in the introduction are not met. Reader unclear what conclusions are drawn
4.7.a	SLO 6: Students will identify, analyze, and evaluate arguments as they occur in their own and others' work. SLO 7: Students will develop well-reasoned arguments	Limitations of Research	Substantively discusses limitations of research and outlines areas of further inquiry.	Discusses limitations of research and outlines areas of further inquiry.	May mention, but does not substantively address limitations of research or outline areas for further inquiry.	Does not address any limitations of research or outline any areas for further research.
4.1.a	SLO 3: Students will research a topic, develop an argument, and organize supporting sources. SLO 7: Students will develop well-reasoned arguments	Organization of Content	Organizes content in a manner that expertly supports the thesis. Always provides clear and meaningful transitions or sections.	Organizes content in a manner that adequately supports the thesis. Generally provides clear transitions or sections.	Organizes content in a manner that somewhat supports the thesis. Provides some transitions or sections but not consistently.	Does not organize content in a manner that supports the thesis. Lacks clear or meaningful transitions or sections.

Kimberly Detterbeck, Darcy Gervasio, and Rebecca Oling

CHAPTER 13

From Technical Troubleshooting to Critical Inquiry: Fostering Inquiry-Based Learning Across Disciplines Through a Tutorial for Online Instructors

13.1 INTRODUCTION

With the growth of distributed education initiatives at colleges and universities, the instructional roles of librarians continue to transform dramatically. While these changes often foster student-centered pedagogical approaches, developing teaching collaborations that move beyond the mechanics of information access remains a common challenge for librarians. Increasingly, the widescale adoption of learning management systems (LMS) enables teaching faculty and librarians to collaborate on a wide array of information literacy and outreach initiatives that benefit students and strategic institutional priorities alike. As LMS use continues to grow and evolve, libraries are reevaluating and restructuring librarian roles in campuswide collaborative endeavors, enhancing student learning and integrating information literacy across the curriculum. Educational institutions have adapted to paradigm shifts in teaching and learning pedagogies over many centuries. One of the more significant institutional change across universities in the late 20th and early 21st centuries has been the emergence of e-learning, partly to maintain viability and partly to recognize cultural and societal changes over time. There have been several exhaustive overviews of the transformation of distance education and the adoption of e-learning in higher education (Bates, 2005; Courtney & Wilhoite-Mathews, 2015; Garrison, 2011; Garrison & Vaughan, 2008; Oliver & Herrington, 2001). While the quick adoption of e-learning technologies has been criticized by some (Guri-Rosenblit, 2005; Njenga & Fourie, 2010) who suggest that the widespread

Distributed Learning.
© 2017 T. Maddison and M. Kumaran.
Published by Elsevier Ltd. All rights reserved.

enthusiastic embracing of technology can occasionally prove deleterious to the teaching and learning process, it has more often been viewed as a logical evolution of traditional classroom approaches to learning (Herrington, 2006; Meyen et al., 2002; Salmon, 2005). Aided by rapid advances in both digital technologies and electronic communication modes, distributed education has undergone noticeably deep and systemic change, forcing a need to rethink how course content is delivered both in combining digital delivery with the traditional classroom or eschewing the in-person lecture method entirely. The "deep shift in educational metaphor," as R. J. Amirault (2012) described it, "has caused most universities to rethink the methodological approaches they have always used to deliver instruction" (p. 159). Despite this sea change, one of the more significant barriers to distance education has been a general resistance or hesitation by teaching faculty to adopt new modes of course content delivery. Ruth Gannon-Cook (2010) points to several factors that inhibit teaching faculty involvement in distance education: "no perceived need to participate; not enough [distance education] training; no perceived benefit to the faculty; and, the fear of being replaced by technology" (p. 63). More specifically, online instructors are often not trained sufficiently in using e-learning technologies, and there tends to be a general lack of understanding of these technologies across all levels—students, teaching faculty, and so on. While libraries are not always afforded the ability to address such factors as participatory need (often because of departmental or institutional directive), they are uniquely positioned to foster open dialogue and collaboration to assuage inadequate training and other hesitations. This unique position can help remove barriers that impede adaptation to new teaching and learning environments. Indeed, there are seemingly limitless opportunities for librarians to engage with teaching faculty in online environments. As Pamela Jackson (2007) surmised, "the seamless integration of library resources, information literacy, and librarian/faculty collaboration in the online classroom is lacking" (p. 455).

To effect change and make the highest impact on student learning, librarians should be embedded in the environments that students use most. Even the best efforts by librarians collaborating with teaching faculty in this integration and implementation only reveal half of the ongoing narrative. There remain real impediments to success in teaching with technology among novice teaching faculty members who are underprepared or reticent to do so. Seeking to address such challenges, librarians in the Department of

Teaching and Learning, Indiana University Libraries, collaborated with a systemwide office of online education and an online instructional design and development team to develop a tutorial module for online course instructors. The module moves beyond traditional library orientation to focus on strategic use of library resources and on assignment design. The instructor tutorial, which is accessed through the Canvas LMS, is informed by a knowledge of student research behaviors, an emphasis on research as a recursive and rhetorically situated process, and pedagogical strategies that foreground student reflection and metacognitive thinking. The tutorial itself presents information on effective assignment design and poses reflective questions to instructors that relate to the design of their courses and research assignments. The remainder of this chapter will discuss the origins, design, and implementation of the tutorial project as well as implications for future initiatives to support instructors in integrating information literacy and library resources into their teaching and development of a related tutorial for students.

13.2 INSTITUTIONAL CONTEXT

Indiana University Bloomington (IUB), a large, four-year research institution, is the flagship campus of an eight-campus public university system that had an enrollment of almost 115,000 students in associate, baccalaureate, master's, and doctoral degree programs in the fall 2015 semester. The university has significantly generated a wide array of technological innovations that have furthered its efforts in online distributed education. A university-wide study to develop comprehensive strategies for online education course and curricular development began in 2010. It led to the creation of the Office of Online Education and Indiana University (IU) Online (http://online.iu.edu) in 2012. Currently, almost 24,000 students are enrolled in at least one online course, an increase of almost 40% in distance education program offerings since 2013 and a more than 70% increase in the number of distance education credit hours taken between 2011 and 2015 (Indiana University Online, 2015). While graduate distance education programs account for the largest student enrollment in online programs at IU generally, there is an increased emphasis in undergraduate online education evidenced by the consistent and significant growth in enrollment across baccalaureate degree programs. Since the unveiling of IU Online and an increased emphasis on new modes of online course content delivery, the IUB Libraries has revised and reenvisioned its services and support for distance education. Previous obstacles and challenges for librarians supporting distance education at IUB, such as the decentralization of online course offerings and insufficient methods for identifying populations of online learners, made prior assessment efforts particularly problematic. Shifts in strategic priorities and efforts to centralize distance education initiatives, as well as administrative commitment across the university, have supported the libraries' efforts to establish key partnerships that support online learning literacy for students. Perhaps the most encouraging development among those efforts has been a collaborative project embedded in Canvas, the university's LMS, which is aimed at new and early career distance education instructors in developing proficiency with teaching in online environments.

13.3 TECHNOLOGY

Canvas (https://www.canvaslms.com), an open-source LMS developed by Instructure is seeing wide adoption across higher education institutions since being launched in 2010. A cloud-based course management system, Canvas offers the standard functionality of an LMS, providing course content management and delivery. Canvas also excels in newer features such as modules for organizing content in a variety of structures, creating a more linear flow of information for student work. Learning outcomes, rubrics, migration tools, peer review, screen sharing, video chat, and e-portfolios are among the many enhancements to the learning environment that Canvas employs, along with SpeedGrader, which places a document, spreadsheet, video, or blog, e.g., into a browser-enhanced viewable format that allows the course instructor to grade and comment on a student's work through markup. Canvas is highly integrated with a variety of web applications such as Google Docs, Google Calendar, RSS, and a wide array of social media. Such features provide many options for teaching faculty to communicate with students and continuously assess learning outcomes and objectives.

13.4 COLLABORATION

Following the establishment of the IU Online initiative, the eLearning Design and Services (eDS) unit (formerly Online Instructional Design and Development) was created to provide support through all stages of online course creation and implementation. The instructional designers in the unit work with many campus units and departments to support each online course and program. In addition to university information technology, they work with assistive and adaptive technologies support units, campus centers, and beyond. The eDS team has collaborated with the IU libraries in a wide variety of ways. The initial stages of this collaboration included the libraries' Department of Teaching and Learning (T&L) on two specific projects: collaborative development of reusable learning objects (RLOs) within a local course management system (pre-Canvas implementation) and exploring how a library orientation for instructors RLO would work within the LMS. Initially, T&L librarians were called upon to develop a set of milestone questions that eDS could use when meeting with teaching faculty and online instructors regarding the integration of library resources into online course curricula. Addressing questions of copyright, exam proctoring, and student access to licensed library content, the librarians created adaptable library learning aids for one course (J524: Homeland Security), a test case for exploring the learning platform and investigating potential problems with licensed resources and library-specific content. This test case was also instrumental in developing a systemwide support network among librarians, connecting instruction design and technology teams with regional campus librarians who would be better positioned to address questions about e-resource access and licensing at the local level. For the course, T&L librarians developed a checklist (see Appendix A) for online instructor that is intended to address potential information access issues for students. Librarians charged online instructors with recognizing the inherent difference in students' access to research resources through the IU Online environment. Students enrolled in cross-campus distance education courses were limited by access to those resources associated with their home campus. Online instructors were also asked to consider the research resources needed for their courses (databases, discovery systems, journals, etc.), as well as the potential for discovering outside sources required to support deeper inquiry in a disciplinary context. Ideally, this effort should ensure that all students enrolled would be able to access the breadth of information required to meet expected course learning outcomes. Much of the early stages of collaboration centered on student access to licensed library content. This focus proved necessary for obvious reasons, notably the complications that arise from a multicampus, multilibrary university system attempting to centralize the process of online learning. T&L librarians were tasked with understanding how student access to information works within a centralized environment, given the complications of vendor licensing restrictions. Over time, a centralized proxy server (an intermediary system channeling requests from clients seeking resources from other servers) was enabled, allowing a student enrolled on one campus but taking a course from a different campus to access resources via that particular campus's proxy server. For example, an Indiana University Northwest student taking an online course for which IUB is the campus of instruction will be able to access IUB-licensed library resources through the IUB libraries' website, online catalog, or IUB proxy server links placed in the LMS. In this example, the method for placing such links in the LMS would be no different than for a traditional IUB class, and the IUB Libraries would be able to advise teaching faculty, instructional designers, and consultants on how to do this (such as through a set of instructions provided on the libraries' website: http://libraries.iub.edu/linking-library-resources). Alternatively, when courses are based out of a campus other than IUB, the instructions would be similar but would place the appropriate proxy server prefix for the campus of instruction in front of the URL. As might be expected, while the early stages of our collaboration proved essential and vital to student success, they were only related to one aspect of teaching and learning in online environments—access. T&L librarians actively sought ways to expand the conversation beyond a traditional library orientation for instructors, focusing more on the strategic use of library resources and on assignment design.

The collaboration between the librarians and the eDS instructional designers moved in an altogether new direction with the pilot of an online series of instructional modules for distance education teachers labeled IU Teaching Online Series. Hosted as its own course in the IU instance of Canvas, the series is designed to open a conversation about what can be accomplished while teaching in an online environment, specifically within the LMS. While it does not reflect all the technical and creative possibilities for online instruction, it does focus on learning activities essential to help students succeed: strategies for student engagement and practical pedagogical approaches. The T&L department that fosters an integrated approach to information literacy education viewed this pilot as not simply an extension of its previous collaboration with eDS but as a potential method of teaching faculty engagement by introducing strategies for effective and meaningful research assignment design. At IUB, librarians view information literacy as a process that involves using information critically in a variety of contexts and for discrete, specific purposes. While it includes library resources, it also extends well beyond to include

> *a complex range of interrelated abilities and dispositions, it may be broken into the following dimensions: inquiry (exploring and developing research questions; identifying and evaluating information needs); evaluation (assessing functions and uses of sources); knowledge creation (using information to construct and communicate knowledge); and, conversation (engaging in the exchange of ideas and knowledge) (Indiana University Bloomington Libraries, 2015).*

13.5 THE PILOT SERIES

The Teaching Online Series pilot has initially focused on two overarching objectives for instructors: (1) developing an online course and (2) teaching an online course. Six modules make up the development of an online course aspect of the series, while an additional two modules focus on a variety of teaching concerns such as student engagement and communication, grading, participation, and the potential for student conflict in an online environment. Specifically, the modules address online course design, assessment, active and interactive learning, structuring course content, usability and visual design, as well as accessibility. An additional four modules constitute the teaching component of the series. These modules focus on online presence, being actively involved in the online classroom, efficiency, and course management. When eDS approached the T&L librarians with a proposal to develop a library orientation module for the series, much consideration was given to maximize the opportunity. While discussions between the two units initially focused on defining how a library orientation for instructors would look, the conversation moved beyond the traditional approach (proxied access to electronic resources, how to distribute licensed library content, etc.) and toward effective research assignment design. All faculty members who teach an online course at the affiliated campuses, including library faculty, will be invited to complete the module, which concludes with a feedback survey about instructors' perceptions of and potential applications of concepts introduced in the module. Survey responses will inform revision of the module, future initiatives to support instructors in integrating information literacy and library resources into their teaching, and development of a related tutorial for students.

The T&L librarians began by establishing a set of learning outcomes for the library module. It was determined that after completing the learning module, an online instructor would be able to "identify the contextual factors that determine access to library resources and services; recognize common challenges students have with research assignments in a system-wide eLearning context; and identify aspects of assignment design and pedagogical approaches that support students in successfully completing research-based assignments" (Indiana University Libraries Bloomington Department of Teaching & Learning, 2015). The module began by covering the circumstances related to information resource access across a multicampus system:

> When teaching an online course as part of a multicampus institution, there are many contextual factors that determine student access to library resources and services. There are several considerations [one] must make before planning [a] syllabus and developing student learning outcomes and expectations of how students engage with information sources. Each … campus library has different licenses to resources and students' access to these resources is determined by their home campus (Indiana University Libraries Bloomington Department of Teaching & Learning, 2015).

As previously mentioned, a centrally managed proxy server allows online students from other campuses to access resources from their campus of instruction. Discussion of good practices for avoiding obstacles to student access to licensed resources was a central feature of this topic. The librarians also took the opportunity to address an important issue with designing effective research assignments: whether students have access to the appropriate resources needed to accomplish the task. Common access issues related to library resources, and ways to mitigate those problems were addressed in the module. As librarians wrestled with attempting to convey potential problems and pitfalls associated with paywalls and blocked access to e-resources, the opportunity to move beyond simple resource access issues and toward effective research assignment design seemed a logical and necessary step, one for which eDS was most receptive and encouraging. The module does indeed shift in focus away from simple resource access concerns to considerations when designing research assignments. The librarians established a set of learning outcomes related to this topic: "what does [the instructor] want the students to learn through the assignment; how does the assignment align with the course learning outcomes; and, do the assignment guidelines communicate to students how the assignment supports their learning?" (Indiana

University Libraries Bloomington Department of Teaching & Learning, 2015). The role of research and information sources became a unique focus of the module, requiring online instructors to consider "the purpose of the research task … the role of sources in accomplishing this research … the audience with whom students would likely share their work … and, what implications … this has for the kinds of sources students use and how" (Indiana University Libraries Bloomington Department of Teaching & Learning, 2015). Alternative research assignments, which frequently align better with course learning outcomes or with the larger purpose of the assignment than a traditional paper, were also discussed in the module. The module stresses to online instructors that it is often more effective to emphasize the

> rhetorical function sources will play ([such as], supporting an argument or providing background information on a subject) and to offer examples of types of sources or research tools that may be useful starting points—are certain types of information likely to be most useful for accomplishing the given task? Would one recommend certain starting points for exploring sources? (Indiana University Libraries Bloomington Department of Teaching & Learning, 2015).

Over the several months required to construct the library module (see Appendix B for an outline of the module), the concepts of disciplinary context, discursive language, and sequencing in effective research assignment design proved too important not to include. The module as designed asked instructors to consider the following:

> Might certain terminology be unfamiliar to students (e.g. citation, databases, scholarly/peer-reviewed sources, primary/secondary sources)? Might students need further guidance about certain disciplinary practices? (e.g. reading a scholarly article in your field, structuring a literature review, citing sources). If certain terms, concepts, or processes need further explanation, what additional information or instruction is needed? (Indiana University Libraries Bloomington Department of Teaching & Learning, 2015).

The module continues in this vein when covering the concept of research assignment sequencing. When thinking about the research process in which students will be asked to engage, the instructor must consider what stumbling blocks they might encounter and, if the assignment requires multiple stages or steps, "it may be useful to break the assignment into multiple parts (such as background research, class discussion, annotated bibliography, etc.)" (Indiana University Libraries Bloomington Department of Teaching & Learning, 2015). When instructors break down complex research assignments into sequences of more reasonable and manageable parts, it "models how to approach a research question and how to manage time effectively, empowers students to focus on and master key research and critical thinking skills, provides opportunities for feedback, and deters plagiarism" (Indiana University Libraries Bloomington Department of Teaching & Learning, 2015). Finally, the module addresses assessment criteria—"what would a well done assignment look like, e.g., how will it be evaluated, and how will that evaluation criteria be communicated to students" (see Appendix C for communicating assessment criteria to students through rubrics) (Indiana University Libraries Bloomington Department of Teaching & Learning, 2015). As with the previous topics of discussion in the library module, external links to supplemental content (typically hosted in the LibGuide platform: http://iub.libguides.com) were provided throughout for concepts such as rubrics and beyond. The module also points to the very real need for ongoing help by collaborating with librarians at all stages of the teaching and learning process. Instructors are advised that "librarians can help with determining relevant resources, developing instructional materials related to the research process, or addressing common challenges students face with research" (Indiana University Libraries Bloomington Department of Teaching & Learning, 2015). This help can manifest itself in a variety of ways, be it through library reference services (and interlibrary lending services indirectly), instructional consultations with librarians, hosted instructional materials (on the library website, on a library guide platform, etc.), library course and research guides, or, in many cases, via a librarian embedded in the LMS course site. When librarians were developing the module for the Teaching Online Series, it was also determined that it might be useful to add a feedback survey at the end like other modules in the series with similar components. While the feedback survey is still in development, several potential questions for instructors completing the module will be addressed: will information from the module influence future development of research assignments or use of library resources—and, if so, how? What about the module was useful for the teaching process? What remaining questions or concerns might the instructor have about designing research assignments or incorporating library resources into an online course?

As the completed module, along with the Teaching Online Series, was still in the pilot phase, it had nevertheless been shown to be instrumental to the ever-growing number of systemwide online instructors. There has been an increased interest among online instructors in collaborating with librarians, especially teaching librarians, at all of the various system campuses. In addition, there has been an elevated focus on supplemental learning objects such as online guides that support finding articles and other information generally, as well as engaging in the research process across disciplines more specifically. It is anticipated that this tutorial project will open new doors for meaningful teaching partnerships that foster student-centered pedagogies. As this suggests, pursuing partnerships with localized teaching and learning centers and with other institutional units opens opportunities for integrating information literacy across the curriculum and within and beyond distributed learning environments. Going forward, an interesting partnership has begun as a result of the work completed on the library module. It was proposed that the T&L librarians begin working with a writing faculty member to design sharable learning activities for the classroom such as concept mapping, peer-review forms for student writing, and a variety of group projects and group presentations. While these learning activities would initially focus on writing courses, they would be designed to have much broader appeal across disciplines. Based on the content that was created for the Teaching Online Series library module, there is much information that corresponds with the writing curriculum. The discussion of this project would be remiss in not including a future objective with the piloted Teaching Online Series project. The T&L librarians hope to transform the library module intended for online instructors into a student module with similar yet differing concerns.

13.6 CONCLUSION

Teaching in online environments presents many challenges and obstacles for traditional instruction. Online environments present many barriers to access and understanding for students. Embedding in the LMS, as was done in Canvas in this case, affords librarians the opportunity to effect change at a more granular level rather than being passive observers left to have many difficult conversations with online teaching faculty. While the potential for "difficult" conversations will never be eliminated, allowing librarians to be part of the transformation process that instructors undergo when switching modes of content delivery to new and unfamiliar environments will afford the opportunity to shape those conversations in more meaningful and constructive ways. Tumbleson and Burke (2013) assert that "when embedded librarians provide ready access to scholarly electronic collections, research databases, and Web 2.0 tools and tutorials, the research experience becomes less frustrating and more focused for students" (p. 6). More powerfully, though, when librarians engage with instructors in the process of effective teaching and assignment design and help shape and inform those practices, the potential for student success and lifelong learning increases exponentially.

REFERENCES

Amirault, R. J. (2012). Will e-learning permanently alter the fundamental educational model of the institution we call 'the university'?. In L. Visser, Y. Visser, R. Amirault, & M. Simonson (Eds.), *Trends and issues in distance education: International perspectives* (pp. 157–171). Charlotte, NC: Information Age Publishing, Inc.

Andrade, H. (2012). *Understanding rubrics.* <http://learnweb.harvard.edu/alps/thinking/docs/rubricar.htm>. Accessed 19.07.12.

Bates, A. W. Tony (2005). *Technology, e-learning and distance education* (2nd ed.). New York, NY: Routledge.

Courtney, M., & Wilhoite-Mathews, S. (2015). From distance education to online learning: Practical approaches to information literacy instruction and collaborative learning in online environments. *Journal of Library Administration, 55*(4), 261–277.

Gannon-Cook, R. (2010). *What motivates faculty to teach in distance education? A case study and meta-literature review.* Lanham, MD: University Press of America, Inc.

Garrison, D. R. (2011). *E-learning in the 21st century: A framework for research and practice.* New York, NY: Routledge.

Garrison, D. R., & Vaughan, N. D. (2008). *Blended learning in higher education: Framework, principles, and guidelines.* San Francisco, CA: Jossey-Bass.

Guri-Rosenblit, S. (2005). Eight paradoxes in the implementation process of eLearning in higher education. *Higher Education Policy, 18*(1), 5–29.

Herrington, J. (2006). Authentic e-learning in higher education: Design principles for authentic learning environments and tasks. In T. Reeves, & S. Yamashita (Eds.), *Proceedings of e-learn: World conference on e-learning in corporate, government, healthcare, and higher education 2006* (pp. 3164–3173). Chesapeake, VA: Association for the Advancement of Computing in Education (AACE).

Indiana University Bloomington Libraries. (2015). *Teaching & learning.* Retrieved from <https://libraries.indiana.edu/teach>.

Indiana University Bloomington Libraries Department of Teaching & Learning. (2015). *Library: Teaching online series* (unpublished tutorial). Retrieved from <https://canvas.iu.edu>.

Indiana University Online. (2015, Fall). *IU ONLINE at a glance.* Retrieved from <http://online.iu.edu/_assets/docs/iuo-glance.pdf>.

Jackson, P. A. (2007). Integrating information literacy into Blackboard: Building campus partnerships for successful student learning. *The Journal of Academic Librarianship, 33*(4), 454–461.

Maki, P. (2004). *Assessing for learning: Building a sustainable commitment across the institution.* Sterling, VA: Stylus.

Meyen, E. L., Aust, R., Gauch, J. M., Hinton, H. S., Isaacson, R. E., Smith, S. J., et al. (2002). E-learning: A programmatic research construct for the future. *Journal of Special Education Technology, 17*(3), 37–46.

Njenga, J. K., & Fourie, L. C. H. (2010). The myths about e-learning in higher education. *British Journal of Educational Technology, 41*(2), 199–212.

Oliver, R., & Herrington, J. (2001). *Teaching and learning online: A beginner's guide to e-learning and e-teaching in higher education.* Mt. Lawley: Centre for Research in Information Technology and Communications, Edith Cowan University.

Salmon, G. (2005). Flying not flapping: A strategic framework for e-learning and pedagogical innovation in higher education institutions. *Alt-J: Research in Learning Technology, 13*(3), 201–218.

Tumbleson, B. E., & Burke, J. J. (2013). *Embedding librarianship in learning management systems: A how-to-do-it manual for librarians.* Chicago, IL: American Library Association.

APPENDIX A ONLINE COURSE DEVELOPMENT: LIBRARY RESOURCES CHECKLIST

For allowing students' access to research resources through IU Libraries online:

- Recognize difference in students' access to licensed library resources.
  - Identify students' home campus (ask students or view course roster).
  - If it is different than that of the faculty, take note of usernames.
  - Send usernames to comments@indiana.edu with course name/number and faculty's home campus.
  - Home campus = campus on which the course originates.
- Access to research resources:
  - Consider databases—discovery systems students will need for research-based assignments.
  - Consider journal titles, resources, databases students will need to support their learning through discovering "outside sources."
  - Ensure required databases are available to all students.
  - Create links to campus libraries or subject- or resource-relevant pages on library websites.

  For Uploading Course Content:
- Determine type of resource needed:
  - Faculty-owned content to replace textbook.
  - Print materials.
  - Online resources (web or library).
- Access to course readings (articles):
  - Search for known items or new items in OneSearch if your home campus has a subscription. If it does not, search in Academic Search. Otherwise, contact a subject librarian for a subject-relevant database.

- Identify Permalink.
- Link within Oncourse or Canvas using Permalink.
- If not available, reconsider or create PDF access via Courseload.
- Access to course readings (books):
  - Is book only available in print?
    - If yes, scan chapter and obtain copyright through Courseload.
  - Is book available electronically?
    - Contact subject librarian at your home library to identify availability of eBook and number of users.
    - Request access for multiple simultaneous users (1, 3, or unlimited are the options).
  Research Support for Student Learners:
- Access to research support from librarians.
  - Link to Ask a Librarian or Reference Services (or similar service) from your home campus library.
  - Contact librarian from your home campus regarding the possibility of a research guide or course guide (i.e., LibGuide) for your course.

APPENDIX B TEACHING ONLINE SERIES: LIBRARY MODULE TABLE OF CONTENTS

- Learning Outcomes
- Accessing Library Resources Across Campuses
  - Issue: Paywalls and Blocked Content
- Access-Related Concerns When Designing Research Assignments
  - Issue: Students from Different IU Campuses Who Do Not Have Access to Certain Online Resources.
  - Issue: Resources That Limit the Number of Simultaneous Users
- More Considerations When Designing Research Assignments
  - Examples of Alternative Research Assignments
- Planning Research Assignments
  - Identifying Learning Outcomes
  - Articulating Purpose and Context of the Research Assignment
  - Types of Sources
  - Stumbling Blocks
  - Assessment Criteria
- Collaborating with Campus Librarians

APPENDIX C COMMUNICATING ASSESSMENT CRITERIA TO STUDENTS: RUBRICS

Rubrics are one way to communicate assessment criteria to students. The Department of Teaching and Learning, Indiana University Libraries, provides online course instructors with the following guide to creating and using rubrics which was adapted by the Center for Innovative Teaching and Learning at Indiana University (http://citl.indiana.edu/resources_files/teaching-resources1/rubrics.php).

Rubric Creation: Four Simple Steps

A rubric articulates the expectations for an assignment by listing the criteria, or what counts, and describing levels of quality from excellent to poor (Andrade, 2012). Rubrics can make visible to students what they might be missing in an assignment and can help them from going wrong. The creation of a specific rubric is one way that experts can draw out their own unconscious competence.

To Create a Rubric:

1. Select the student assignment you would like to create a rubric for—be it a project, paper, lab session, or homework. This step is most effectively done with examples of student work on such an assignment, including high-, medium-, and low-level responses.

2. List on paper five to seven major mistakes students might make on this assignment. If you do not have examples of student work, what do you anticipate that students will get wrong on this assignment?

3. Turn the list of mistakes into a list of positive attributes by using nouns so that the descriptor describes what you want, not a deficit. In this argument and evidence rubric for a poster in history, the argument is not "Weak and vague" but "Succinct and clear." Use the power of the nonexpert to check on the clarity of your attributes. Ask someone from outside your field if they could do what is described in each attribute. For example, "Problem Solving" may not explain clearly that the instructor means "Correlate the results of physical actions with electronic outputs and develop a system to bring a signal from one place to another." Or, if the attribute is "Lots of Detail," it means "Select a critical incident from your observation notes and describe it in detail."

4. Calibrate each attribute (Exceeds/Meets/Does Not Meet Expectations). The high, medium, and low student examples might be useful at this

point for determining the different levels of work. If there are no degrees between success and failure, it is difficult to help students move toward mastery.

Using the Rubric:

The instructor can use the rubric to score student work—but it may be even more useful if the students are assigned rubrics to use on their own work or on the work of their peers individually or in teams.

Students at Alverno College, e.g., are asked to assess themselves on every assignment before they turn it in; they often score their own work using rubrics.

However a rubric is used, we have found the most powerful effect of a student using it comes from metacognitive reflection at the end of the process: "Now that you've evaluated X (e.g., another team's poster, your own homework), what do you see as strengths and weaknesses in your own X? What will you and your team do to improve your work on your X?" Such reflection can help students who are used to merely recalling or regurgitating information shift to a more conceptual mode of learning (Maki, 2004).

Create Rubrics in Canvas:

Canvas allows instructors to create rubrics to grade assignments, discussions, and quizzes. These rubrics can include learning outcomes, among other criteria. Visit Canvas' Instructor Guide to learn how to use rubrics within Canvas.

CHAPTER 14

Embedding the Library in the LMS: Is It a Good Investment for Your Organization's Information Literacy Program?

14.1 INTRODUCTION

In the early 2000s, the University of Arizona libraries (UAL) launched a multiyear initiative to embed targeted content into every course that was using the campus's main learning management system (LMS) called Desire2Learn (D2L). The goal of this initiative known first as the library "widget" and eventually as the Library Tools Tab (LTT) was to ensure that all students enrolled in courses that used D2L were offered library resources and services that would support their learning. The authors discuss the history and development of the embedded system, explore the types of library materials most used within the system, compare how usage of the embedded system differs from usage of the library's stand-alone course guides, and analyze the value that developing and embedding an integrated system can have on librarians' abilities to structure and scaffold instructional programs on campus.

14.2 LITERATURE REVIEW

Libraries have long used course and subject guides to connect students to the resources, services, and information literacy instruction necessary to their academic success. Initially developed in print and then moving online in the 1990s, these guides represent an attempt to assist users in navigating the world of information to locate subject or discipline-specific scholarship (Vileno, 2007). However, despite their present-day ubiquity (Ghaphery & White, 2012), research has shown that subject guides—as standalone resources without connection to a course through context or content—are of limited use to students (Courtois, Higgins, &

Kapur, 2005; Leighton & May, 2013; Little 2010; Reeb & Gibbons, 2004; Staley 2007). Reasons for their ineffectiveness include the fact that subject guides are not reflective of the mental models of undergraduates (Reeb & Gibbons, 2004), are likely to induce cognitive overload (Little, 2010), and are often difficult to find or use effectively without instructional intervention (Barr, 2010; Collard & Tempelman-Kluit, 2007; Staley 2007). Reeb and Gibbons (2004) argue that rather than subject- or discipline-based guides, "library resources organized or delivered at a course level are more in line with how undergraduate students approach library research" (p. 123). Course guides fare better in the literature than subject guides in terms of usefulness, although some argue that subject guides introduced as part of an in-person instruction session can be effective (Leighton & May, 2013; Staley, 2007).

If context is key for research guides, and if the guide content should be tied to a course rather than subject or discipline, a fitting home for these resources is in the online campus LMS environment where students are already working. As early as 2002, Cohen posed the question, "course-management software: where's the library?" Since then, academic libraries have explored integration of library resources and instruction with their campus LMS as a way to deliver targeted library content at the course level. Shank and Dewald (2003) outlined two different approaches to LMS integration: macrolevel (generic, global links to library content) and microlevel (customized and course-targeted links to library content), and the literature since then illustrates the evolution of modes of library embeddedness in the LMS environment.

Numerous case studies discuss the challenges, processes, and value in placing library resources and services directly in the LMS (Adebonojo, 2011; Barr, 2010; Black, 2008; Black & Blankenship, 2010; Bowen, 2012; Chesnut, Henderson, Schlipp, & Zai, 2009; Daly, 2010; Dygert & Moeller, 2007; Foley, 2012; Hristova, 2013; Jeffryes, Peterson, Crowe, Fine, & Carrillo, 2011; MacIsaac, 2011; Pickens-French & McDonald, 2013; Solis & Hampton, 2009; Washburn, 2008). These efforts represent a range of micro- and macrolevel integration: some libraries developed widget functionality to connect to the LMS (Casden, Duckett, Sierra, & Ryan, 2009; Hristova, 2013), others created and embedded modules within the LMS (Adebonojo, 2011; Collard & Tempelman-Kluit 2007; Foley, 2012; Washburn, 2008), and still others explored embedding librarians directly in an LMS course, giving them an active role in course activities (Becker, 2010; York & Vance, 2009).

Another subset of the literature describes the benefit of integrating library instructional content into an LMS for information literacy programs (Adebonojo, 2011; Dewan & Steeleworthy, 2013; Fabbro, Daugherty, & Russo, 2013; Henrich & Attebury, 2012; Hess, 2014; Jackson, 2007; Karplus, 2006; Kelley, 2012; Xiao, 2010; Zhang, Goodman, & Xie, 2015). Aside from embedding point-of-need library resources into the LMS—course-specific books and articles, databases, catalog links, ask-a-librarian chat services—the potential for scaffolding information literacy instruction in a scalable way is a significant reason to pursue LMS integration. As Karplus (2006) writes, "integrating information literacy instruction into the [LMS] for a given course is probably the most effective means of reaching the student in an electronic venue that they visit every day" (p. 7). Similarly, Jackson (2007) points out that an LMS "can be used to scaffold instruction and infuse information literacy activities throughout subject-specific courses" (p. 459).

Other studies focus on the assessment of various implementations of LMS integration either by surveying faculty, student, or librarian perceptions of and experience with library content in the LMS or through usage data of the embedded library resources (Bowen, 2012; Jackson, 2007; Leeder & Lonn, 2013; McLure & Munro, 2010; Murphy & Black, 2013; Xiao, 2010). Overall, the literature supports Leeder and Lonn's (2013) claim that "in an environment of decreasing use of libraries by students, decreasing rates of IL instruction, and increasing use of online information resources by students, linking the LMS to libraries becomes even more critically important" (p. 642).

Despite these strides, course and subject guides that live on library websites separate from an LMS are still the norm. Ghaphery and White (2012) found that among 99 Association of Research Libraries (ARL), all 99 had subject guides and 75 had course guides listed on their websites (p. 22). UAL is no exception. To the authors' knowledge, there has not been a comparative study of the usage of standalone course guides versus course guides embedded within the LMS. This chapter is an attempt to do so, with the hope of uncovering implications for student learning.

14.3 HISTORY AND DEVELOPMENT OF THE LIBRARY TOOLS TAB AT UAL

The ubiquity of user-friendly search engines, as well as students' access to and comfort with technology resulted in students bypassing library websites while conducting course-related research. To address this issue, UAL actively identified ways to seamlessly integrate library resources into the student research process. Since the mid-2000s, UAL has investigated and piloted several different approaches to reconnect students to the library. One of the more successful approaches is UAL's incorporation of library resources—including databases, ebooks, and library instructional materials—into the campus LMS.

UAL's first foray into LMS integration came after long negotiations with the campus administrators of D2L. Although the benefits of providing easy access to library resources to students within the LMS seemed obvious, the University of Arizona (UA) information technology (IT) unit that was responsible for managing the campus LMS was not as eager to integrate library resources as might be expected. Discussions revealed that the library was not alone in wanting to reach students through the LMS. UAL found itself in a position to distinguish the research and instructional resources that it intended to make available to students through the LMS from other types of information and resources that the rest of the campus wished to provide. After a number of discussions and negotiations, UAL was success-ful in making a case and secured a small space, initially known as the library "widget," on the student LMS home page (Fig. 14.1). This was the start of UAL being embedded in the campus LMS.

Figure 14.1 Original library widget on student LMS home page circa 2008.

While this was a good start, the library materials that could be made available were frustratingly generic and limited in quantity. In addition, once students moved away from the landing page and into course content, discussions, or other course features in the LMS, the widget was difficult to navigate back to and largely forgotten. In spite of the short-comings, the initial embedded widget allowed UAL and the campus IT unit to establish a collaborative relationship from which future successful iterations of the embedded library system would be developed.

During the years that the original widget was in place, the campus began to approach the management of instruction and instructional tech-nology in a more programmatic and centralized manner and established the Office of Instruction and Assessment (OIA). The creation of this unit, along with the successful implementation of the original widget, opened the door for UAL to make significant improvements to both the func-tionality and placement within the LMS. In fall 2011, the second genera-tion of the widget, now dubbed the "superwidget" (SW), was integrated into the campus LMS through a link on the top menu bar navigation within each course site. The placement of the link within the unchanging top navigation was important because it meant that students could easily access the SW from anywhere within a course site (Fig. 14.2).

The improved SW consisted of a page that provided students access to library instructional materials as well as links to other library services and resources. The SW was one of the first steps taken in using the LMS to offer students library resources and services that would support their learning in both a scalable and targeted way.

During the 2011–12 academic year, the team that developed and imple-mented the SW conducted a needs assessment and usability testing to iden-tify issues with the interface, functionality, and content. Using this data, the team planned and designed the next iteration of SW, which was renamed the Library Tools Tab (LTT) and rolled out in fall 2012 (Fig. 14.3).

It included an expanded suite of content and services that could be added to the pages, as well as a faculty and librarian drag-and-drop administrative interface through which changes could be made to course groupings, single courses, or even individual course sections (Fig. 14.4).

Figure 14.2 The superwidget link in the campus LMS, circa 2011.

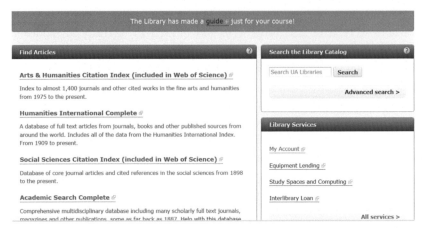

Figure 14.3 The current Library Tools Tab, 2012.

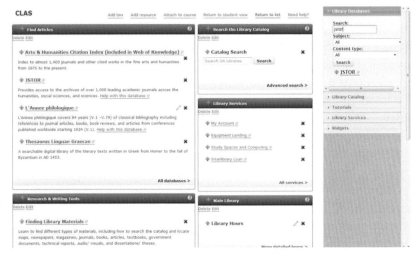

Figure 14.4 The faculty and administrative interface for the LTT.

These improvements have allowed librarians to collaborate with faculty more closely, customize, and scaffold course content so that students receive integrated support at the course section level. Specifically, the LTT allows UAL to provide a minimum common offering of resources and services to all courses using the campus LMS. By default, resources for each course prefix or department are provided (e.g., English—ENGL), and daily notices of new course prefixes added to the campus course catalog are received. Where librarians have greater knowledge of a particular program's curriculum, the LTT is customized with resources specific to an individual course (e.g., ENGL 102). Where there is time for active collaboration with instructors, librarians can customize the resources for an individual section of a course (e.g., ENGL 102, section 50). When students view the LTT, the resources appear automatically with no need on their part to select a particular course or section. This approach provides the foundation of the scalable system that allows UAL to deliver targeted support to departments when possible and to deliver baseline resources to nearly all LMS users.

This foundation is immaterial if the link is not present in the students' LMS course sites. Though there was some initial resistance from OIA, UAL successfully negotiated to have the link present in every course site, with only a few exceptions (e.g., departments with custom navigation), and with the ability for instructors to opt out by deleting the link. This opt-out nature of the LTT link in the LMS is the basis for LTT's success.

The LTT link is placed on every D2L navigation bar when the course is built. From there an instructor has the ability to edit the course navigation bar and remove the LTT link entirely. Due to privacy concerns held by OIA D2L administrators, the report does not indicate which instructors removed the link. Once a semester, UAL receives a general count of sites that have or do not have LTT. A stable average of around 77% of course sites leave it in place. This baseline allows librarians to monitor any abnormalities.

14.4 ANALYSIS: HOW IS THE VALUE OF THE EMBEDDED SYSTEM MEASURED?

In 2015, before embarking on the next phase of development work, the project team realized that it was imperative to get a sense of how the LTT was used. Google Analytics provided the means to identify which disciplines made the most use of LTT, highlight the most accessed library resources, and finally analyze data to compare LTT access with the access of course guides living on the UAL website. Through this analysis, members hoped to discover data illustrating the value of the embedded system. In this chapter, the authors use the terms *access* and *usage* interchangeably.

Beginning with frequency, daily use data was evaluated for the LTT's entire existence. These statistics reveal that—with the exception of 2 days in the summer of 2012—students accessed library content via the LMS on a daily basis since Aug. 2011. Various factors such as different technical versions of the LTT for main campus and health sciences users, faculty–librarian collaborations, and departmental regard for information literacy prevent extensive comparison between disciplines. However, the authors made a few general observations from the usage data. Taking advantage of the technical foundation that allows the capture of courses that use or do not use the LTT, a holistic review of LTT data shows that the disciplines that use the LTT the most are English, business, nursing, theater arts, and history. This was not a surprise as these departments have a well-developed relationship with the library. In conjunction with identifying which disciplines make the greatest use of the LTT, a closer look was taken to determine which resources receive the most use. Among the top used were databases, links to course guides, tutorials, the library catalog, and ebooks. The most commonly used databases included PubMed, Academic Search Complete, Business Source Complete, Education Abstract, IBIS, ERIC, FIS, Web of Science, JSTOR, and PsycINFO. It is not surprising to see those databases at the top, as the UA's online nursing program and English composition courses are among the heaviest users of the LTT. Other highly accessed features were the administrative interface used to customize content and links to library services, contact information for liaisons, and the main library website.

14.5 HOW DOES UAL COURSE GUIDE USE COMPARE WITH LTT USE?

In spring 2012, a UAL instruction librarian made an initial comparison of the use patterns between LTT pages and course guides hosted on the UAL website. While still too early to make any substantive analysis, she detected that the LTT pages consistently reflected increased use by students (DeFrain, 2012). Through comparing usage data on course guides hosted on the UAL website against LTT usage data for fall semesters 2011–14, the work described here tests the assumption that library access from the LMS encourages higher and more consistent use of library resources. The fall semesters were determined to be a better baseline due to the library's strong collaboration with the English department that yields high usage during spring semesters, as well as the noticeable absence of students from the university during spring break.

As seen in Table 14.1, during fall 2010 course guides were accessed slightly more than 15,000 times. With the launch of the LTT the next year, standalone course guide usage increased for the last time. From this

Table 14.1 Usage statistics for course guides and embedded LTT pages

	Fall 2010	Fall 2011	Fall 2012	Fall 2013	Fall 2014
Views of course guides hosted on library website	15,042	25,643	22,101	14,656	13,802
Views of LTT pages appearing in D2L	N/A	16,997 (LTT launch)	22,962	42,165	30,806

point, course guide usage decreased 13.8% (fall 2012), 33.7% (fall 2013), and 5.8% (fall 2014). While the course guide usage was decreasing, embedded LTT page usage more than compensated, with increases of 35.1% (fall 2012) and 83.7% (fall 2013) and a decrease of 26.9% (fall 2014). This data show that students access library resources through the embedded link within their LMS more often than through the course guide on the library website. It is not entirely clear why LTT usage increased so dramatically in fall 2013 only to decrease in fall 2014. One possibility for the increase is that UAL began embedding ebook versions of course textbooks in the LTT in 2013. The decrease also occurred following a unilateral decision made by OIA that ultimately moved the LTT link to a less noticeable position in fall 2014. This persisted for a time until the library's instructional services department head discussed the negative impact and convinced OIA to move it back. Once this happened, the LTT page views continued the upward trend seen in the previous semesters.

Perhaps equally telling is the data that show the difference in consistency of use between course guides linked on the UAL website and library resources embedded in the LMS. As illustrated in the following graphs (Figs. 14.5—14.8), LTT pages yield more consistent usage over the duration of the semester than course guides. The initial spike in usage of course guides presumably correlates with beginning-of-the-semester

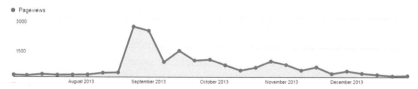

Figure 14.5 Course guide usage in fall 2013.

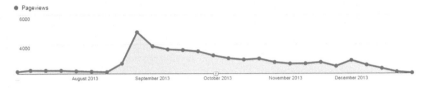

Figure 14.6 LTT usage in fall 2013.

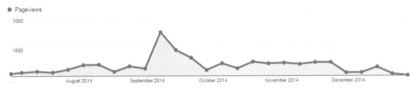

Figure 14.7 Course guide usage in fall 2014.

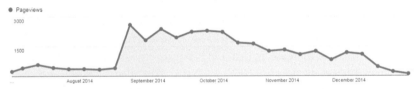

Figure 14.8 LTT usage in fall 2014.

instruction sessions or orientations when librarians demonstrate access to course guides in person. After the first couple of weeks, the usage of these guides drops precipitously—not a good indication that students are accessing library resources from course guides consistently throughout the semester.

LTT pages, on the other hand, show a similar spike in usage near the beginning of the semester, but without as significant a drop. Usage holds steadier, tapering off as the semester nears completion. From a broad view, this data indicate that the LTT encourages higher and more sustained usage of library resources. The behavioral implication behind this usage is that students are repeatedly jumping from the LMS to the library, which, at the very least, carries with it the potential to reinforce information literacy skills through increased exposure to library resources.

In terms of scale, at the height of course guide production, librarians published around 200 course guides, reaching approximately 259 courses sections. In the same semester (Spring 2014), the LTT reached an equivalent of 4983 sections, representing a 1824% increase in sections touched.

14.6 CONCLUSION

The analytics show that as a result of easy access students use library resources from the LTT more frequently and more consistently throughout the semester than they do from the guides they have to locate from the library's main webpage. The burden of having to navigate the UAL website or remember a long, confusing course guide URL is removed when access to library materials and instructional resources are placed within the course LMS. As Solis and Hampton point out, "one of the reasons students have chosen not to use library resources is that there are too many options, too many different places in which to search, and too many rules for searching ..." (2009, p. 89). The consistent increased overall use of LTT pages is good news for the UAL's ability to provide instructional support materials and resources to students. More frequent use of library resources and instructional tools allows students to gain the academic benefits that come from interacting with the library, which include strengthening critical thinking skills and increasing their likelihood of persisting in their academic studies (Soria, Fransen, & Nackerud, 2013; Thorpe, Lukes, Bever, & He 2016). Students also benefit from repeated exposure and practice. This increased exposure helps build muscle memory and skill and allows students to develop a firm foundation from which to initiate and complete research assignments.

Besides the greater reach, the analytics also reflect the variety of ways in which UAL is meeting the instructional needs of students and faculty. This model of LMS integration adds a new angle to the traditional time-intensive faculty–librarian collaborations and course guide design efforts. In this system, resources such as ebooks, course-specific tutorials, and access to library chat can be easily positioned within an LTT page, supporting the course work of hundreds or even thousands of students. The simplicity of deploying relevant resources to all students at the point of need in the LMS allows librarians to direct greater effort to those courses needing more complex library support. The LTT's ease of use and access has cleared the way for UAL to establish a baseline of information literacy support across a campus that teaches more than 40,000 students. Given the relatively small numbers of UA librarians compared to the size of the UA student population, it would be impossible to directly support students in all of the nearly 7000 courses represented in D2L. The LTT software and its underlying service model allow librarians to build on existing strong faculty relationships while also extending the instructional reach of liaison librarians by providing basic library resources and targeted materials to programs where a strong relationship has not yet been built.

The authors found that the LTT initiates, as well as supports collaboration with instructors in several ways. For example, as the service has grown, campus faculty regularly contact liaison librarians to request that links to course ebook be added to their LTT pages. These requests have opened the door to greater collaborations between liaison librarians and campus faculty in a number of departments. In addition, the existing, unmodified LTT pages can be pulled up for faculty during initial instruction consultations and used to structure planning for either in-class or integrated instructional support.

Overall, the data shows that the investment UAL has made in building and deploying an LMS-integrated instructional support system is benefiting faculty and students and allowing UAL to be successful in meeting its goal to embed content, tools, and services into the curriculum to increase student success and jump-start lifelong learning.

14.7 FUTURE DEVELOPMENT

The LTT has provided the means to deliver exceptional instructional services at the University of Arizona. However, in the world of software development, no program is ever truly complete. Its survival and likelihood for continued use demand constant tweaks, modifications, and feature enhancements. This chapter describes examples of LTT pages that were created with the functionality to add library-owned or subscription resources. UAL experienced significant organizational changes since the last release of the LTT. There was a lapse in LTT work until workforce capacity was reestablished and strategic directions gained clarity. During that time, librarians had firsthand opportunities to see gaps or flaws in the tool and gather user data to inform needs. For example, librarian experiences since the implementation of the LTT reveal that, with the exception of course ebooks and main disciplinary databases, UA faculty and instructors at UA rarely come to the table with suggestions of resources they want their students to access. This circumstance is actually fortuitous as it provides librarians carte blanche when it comes to influencing information literacy efforts and affirms their value as educators. Current requests from faculty represent a desire for the capability to add web links and free text to the LTT, which require newer enhancements and a heavy dependence on technical resources. Current work is underway to program new functionality focused on embedding additional instructional services. These additional services, combined with the lessons learned from previous iterations of the LTT, will allow librarians to better collaborate with faculty in embedding information literacy concepts and activities into departmental curricula.

Although no summative assessment has been completed, liaisons consistently receive feedback from individual faculty that student research papers improve when they are directed to use the LTT. Future work therefore entails intentionally gathering user feedback through focus groups and surveys—from faculty as well as students.

REFERENCES

Adebonojo, L. G. (2011). A way to reach all of your students: The course management system. *Journal of Library & Information Services in Distance Learning, 5*(3), 105–113.

Barr, D. (2010). Reaching students where they go: Embedding library resources in course content. *Science & Technology Libraries, 29*(4), 289–297.

Becker, B. W. (2010). Embedded librarianship: A point-of-need service. *Behavioral & Social Sciences Librarian, 29*(3), 237–240.

Black, E. L. (2008). Toolkit approach to integrating library resources into the learning management system. *The Journal of Academic Librarianship, 34*(6), 496–501.

Black, E. L., & Blankenship, B. (2010). Linking students to library resources through the learning management system. *Journal of Library Administration, 50*(5-6), 458–467.

Bowen, A. (2012). A LibGuides presence in a Blackboard environment. *Reference Services Review, 40*(3), 449–468.

Casden, J., Duckett, K., Sierra, T., & Ryan, J. (2009). Course views: A scalable approach to providing course-based access to library resources. *Code{4}Lib Journal*, (6). Available from <http://journal.code4lib.org/articles/1218>.

Chesnut, M., Henderson, S. M., Schlipp, J., & Zai, R., III (2009). Value-added library resources & services through Blackboard. *Kentucky Libraries, 73*(1), 6–12.

Cohen, D. (2002). Course-management software: Where's the library? *Educause Review, 37*(3), 12–13.

Collard, S., & Tempelman-Kluit, N. (2007). The other way in: Goal-based library content through CMS. *Internet Reference Services Quarterly, 11*(4), 55–68.

Courtois, M. P., Higgins, M. E., & Kapur, A. (2005). Was this guide helpful? Users' perceptions of subject guides. *Reference Services Review, 33*(2), 188–196.

Daly, E. (2010). Embedding library resources into learning management systems A way to reach Duke undergrads at their points of need. *College & Research Libraries News, 71*(4), 208–212.

DeFrain, E. (2012). *More than a pathfinder: Are we getting the most out of online course guides?* <http://hdl.handle.net/10150/221691>.

Dewan, P., & Steeleworthy, M. (2013). Incorporating online instruction in academic libraries: Getting ahead of the curve. *Journal of Library & Information Services in Distance Learning, 7*(3), 278–296.

Dygert, C., & Moeller, P. (2007). Linking the library and campus course management system. *The Serials Librarian, 52*(3-4), 305–309.

Fabbro, E., Daugherty, A., & Russo, M. (2013). *Seamless integration of library service points throughout the learning management system. Embedded librarianship: What every academic librarian should know* (pp. 59−72). Santa Barbara, CA: Libraries Unlimited.

Foley, M. (2012). Putting the library at students' fingertips. *Journal of Electronic Resources Librarianship, 24*(3), 167−176.

Ghaphery, J., & White, E. (2012). Library use of web-based research guides. *Information Technology and Libraries, 31*(1), 21−31.

Henrich, K. J., & Attebury, R. I. (2012). Using Blackboard to assess course-specific asynchronous library instruction. *Internet Reference Services Quarterly, 17*(3−4), 167−179.

Hess, A. N. (2014). Online and face-to-face library instruction: Assessing the impact on upper-level sociology undergraduates. *Behavioral & Social Sciences Librarian, 33*(3), 132−147.

Hristova, M. (2013). *Library Widget for Moodle. Code{4}Lib Journal, 19*. Available from <http://journal.code4lib.org/articles/7922>.

Jackson, P. A. (2007). Integrating information literacy into Blackboard: Building campus partnerships for successful student learning. *The Journal of Academic Librarianship, 33*(4), 454−461.

Jeffryes, J., Peterson, K., Crowe, S., Fine, E., & Carrillo, E. (2011). Integration innovation: Launching the library into a course management system. *Journal of Library Innovation, 2*(1), 20−34.

Karplus, S. S. (2006). Integrating academic library resources and learning management systems: The library Blackboard site. *Education Libraries, 29*(1), 5−11.

Kelley, J. (2012). Off the shelf and out of the box: Saving time, meeting outcomes and reaching students with information literacy modules. *Journal of Library & Information Services in Distance Learning, 6*(3−4), 335−349.

Leeder, C., & Lonn, S. (2013). Faculty usage of library tools in a learning management system. *College & Research Libraries, 75*(5), 641−663.

Leighton, H. V., & May, D. (2013). The library course page and instruction: Perceived helpfulness and use among students. *Internet Reference Services Quarterly, 18*(2), 127−138.

Little, J. J. (2010). Cognitive load theory and library research guides. *Internet Reference Services Quarterly, 15*(1), 53−63.

MacIsaac, P.L. (2011). *Maximizing library online presence while minimizing maintenance.* < http://hdl.handle.net/2149/3081>.

McLure, M., & Munro, K. (2010). Research for design: Exploring student and instructor attitudes toward accessing library resources and services from course management systems (CMS). *Communications in Information Literacy, 4*(1), 29−60.

Murphy, S. A., & Black, E. L. (2013). Embedding guides where students learn: Do design choices and librarian behavior make a difference? *The Journal of Academic Librarianship, 39*(6), 528−534.

Pickens-French, K., & McDonald, K. (2013). Changing trenches, changing tactics: A library's frontline redesign in a new CMS. *Journal of Library & Information Services in Distance Learning, 7*(1-2), 53−72.

Reeb, B., & Gibbons, S. (2004). Students, librarians, and subject guides: Improving a poor rate of return. *Portal: Libraries and the Academy, 4*(1), 123−130.

Shank, J. D., & Dewald, N. H. (2003). Establishing our presence in courseware: Adding library services to the virtual classroom. *Information Technology and Libraries, 22*(1), 38.

Solis, J., & Hampton, E. M. (2009). Promoting a comprehensive view of library resources in a course management system. *New Library World, 110*(1/2), 81−91.

Soria, K. M., Fransen, J., & Nackerud, S. (2013). Library use and undergraduate student outcomes: New evidence for students' retention and academic success. *Portal: Libraries and the Academy, 13*(2), 147−164.

Staley, S. M. (2007). Academic subject guides: A case study of use at San Jose State University. *College & Research Libraries*, *68*(2), 119–140.

Thorpe, A., Lukes, R., Bever, D. J., & He, Y. (2016). The impact of the academic library on student success: Connecting the dots. *Portal: Libraries and the Academy*, *16*(2), 373–392.

Vileno, L. (2007). From paper to electronic, the evolution of pathfinders: A review of the literature. *Reference Services Review*, *35*(3), 434–451.

Washburn, A. (2008). Finding the library in Blackboard: An assessment of library integration. *MERLOT Journal of Online Learning and Teaching*, *4*(3), 301–316.

Xiao, J. (2010). Integrating information literacy into Blackboard: Librarian-faculty collaboration for successful student learning. *Library Management*, *31*(8/9), 654–668.

York, A. C., & Vance, J. M. (2009). Taking library instruction into the online classroom: Best practices for embedded librarians. *Journal of Library Administration*, *49*(1–2), 197–209.

Zhang, Q., Goodman, M., & Xie, S. (2015). Integrating library instruction into the course management system for a first year engineering class: An evidence-based study measuring the effectiveness of blended learning on students' information literacy levels. *College & Research Libraries*, *76*(7), 934–958.

CHAPTER 15

A Decade of Distributed Library Learning: The NOSM Health Sciences Library Experience

15.1 THE NOSM CONTEXT

The Northern Ontario School of Medicine (NOSM) was born from a need to address the disparities in health care for the population of northern Ontario, with an emphasis on rural, remote, Francophone, and Aboriginal contexts. As stated in the NOSM website (www.nosm.ca), the school is "the first Canadian medical school hosted by two universities, over 1000 km apart. NOSM serves the faculty of medicine for Lakehead University in Thunder Bay, Ontario, and Laurentian University in Sudbury, Ontario." ("About NOSM", n.d.). Northern Ontario is distinctive due to its vast geography and dispersed population, with many communities facing a shortage of health-care professionals, lack of infrastructure, and complicated socioeconomic issues. NOSM's academic programs are structured to specifically attract, train, and retain physicians for practice in northern communities. As noted in *The Making of the Northern Ontario School of Medicine*, "NOSM is uniquely positioned as the first Canadian medical school with a special mandate to train physicians to practice in Northern Ontario and/or in rural communities in other parts of Canada" (Tesson, Hudson, Strasser, & Hunt, 2009, p. 21).

NOSM's undergraduate medical program follows a distributed community-engaged learning model, with five key themes interwoven throughout the curriculum:

- Northern and rural health
- Personal and professional aspects of medical practice
- Social and population health
- Foundations of medicine
- Clinical and communication skills in health care

Distributed Learning.
© 2017 T. Maddison and M. Kumaran.
Published by Elsevier Ltd. All rights reserved.

According to NOSM's *Community Report 2014*, "No other Canadian medical school provides training in more than 90 communities across a geographic expanse of 800,000 km^2." (Northern Ontario School of Medicine, 2015). Most of those 90 communities are considered rural or remote or both. Learners include medical students in the four-year undergraduate program; residents in the postgraduate family medicine and specialty programs; and students in health sciences and interprofessional education programs, including clinical nutrition, occupational and physical therapy, speech language pathology and audiology, and physician assistant programs. NOSM also has a large faculty with more than 1380 full-time, part-time, and stipendiary faculty from three divisions: clinical sciences, medical sciences, and human sciences. Like the learners, NOSM faculty are located across and beyond northern Ontario.

NOSM is recognized for its guiding principles of social accountability and community engagement. Medical students are at their home campus for the first two years of training, with ongoing community-integrated learning including several community placements each year. For example, first-year students have a month-long placement in an Aboriginal community while second-year students have two four-week placements in small rural and remote communities. The library has the most contact with the students in their first two years of medical school. Medical students in their third and fourth years of training are located in larger communities, first for their eight-month community-based longitudinal clinical clerkship and then for specialist clerkship rotations in larger hospital settings. Residents may be located in one of five larger centers or in rural communities throughout the north.

The sheer size of the region and its remote location comes with its own challenges for presenting distributed educational content (Federation of Canadian Municipalities, 2014). For example, many rural and remote communities experience gaps in Internet connectivity, which can impact users' ability to access streaming videos. Clinical faculty may have access to reliable Internet connections at clinics and hospitals but not from their homes. Given these larger infrastructure issues, the reliance on a variety of educational technologies is inherent in NOSM's academic programs.

15.2 THE HEALTH SCIENCES LIBRARY

The Northern Ontario School of Medicine celebrated its 10th anniversary in 2015. In 2005, NOSM's charter class of 56 undergraduate medical students walked through its halls. The health sciences library (then known as the Health Information Resource Centre) officially opened at its two main campus locations, the Lakehead University campus in Thunder Bay, Ontario, and the Laurentian University campus in Sudbury, Ontario. The Lakehead campus is home to the education services librarian, the research support librarian, a library technician, and an administrative professional. The Laurentian campus is home to the library's director, the public services librarian, the access services librarian, a library technician, and an administrative professional. The education services librarian coordinates library instruction for NOSM users, including undergraduate medical students, residents, staff, and faculty. The majority of the instruction is delivered by the education services librarian, with additional instructional support from each of the librarians.

The library functions to support the academic and research efforts of NOSM learners, faculty, and staff, regardless of physical location. The library's physical collection is compact, and the preference is always to acquire electronic formats to ensure equitable and ready access for all users. Likewise, while in-person library instruction is provided, it is essential that the library employ technology that allows for meaningful and engaging library instruction and research support that can be delivered across northern Ontario. Over the last decade, this technology has included videoconferencing, commercial products for creating asynchronous video, and web-conferencing and collaboration tools.

15.3 VIDEOCONFERENCING

Videoconferencing is the cornerstone technology of NOSM's distributed education model. Because northern Ontario encompasses such a large geographic area, travel is often expensive, time consuming, and impractical. Over the years, the technology has evolved, and NOSM currently uses Polycom and Cisco tools for videoconferencing. Meetings, academic sessions, symposiums, and many other events are made possible across the school and with community partners across the region through videoconferencing. It continues to be an integral technology for the delivery of library instruction within NOSM's academic and continuing education programs. The school's meeting rooms are equipped with large-group or small-group dedicated videoconferencing capabilities, and support is provided by NOSM's Technical and Information Management Support (TIMS) unit. Organizational support of videoconferencing means the responsibility for infrastructure, training, or technical support does not lie with the library, making it a cost-effective option for distributed library instruction. Multisite library instruction can be delivered by a single librarian; given our small staff, this is also beneficial. Alternately, sessions can be co-taught with librarians and instructors from other locations. For several years, the library co-taught a class for second-year medical students on evidence-based medicine and critical appraisal with an epidemiologist and a statistician. The librarian was located in Thunder Bay and had 28 students in her classroom. The epidemiologist and the statistician were located in Sudbury and had 36 students in their classroom. Videoconferencing allowed each person to see every other student and to assume presenter duties, including sharing PowerPoint slides, leading learning activities, and fielding questions. Because videoconferencing is a part of the NOSM culture and all the large group lecture-style teaching sessions are delivered through videoconferencing, medical students quickly adapt to the learning environment.

Whether it is a point-to-point or multisite videoconference, each participating site is connected to the presenter via a conference bridge. The bridge is a technology that enables multiple sites and videoconferencing systems to join a conference with a secure and private connection. For multisite videoconferences, NOSM partners with services such as Ontario Telemedicine Network (OTN) (https://otn.ca/en), formerly the Northern Ontario Remote Telehealth Network (NORTH Network). OTN and other similar services facilitate videoconferencing and telemedicine between organizations, health-care professionals, and patients. For education purposes, OTN allows instructional sessions to be broadcast not just between the two main NOSM campuses, but to OTN sites across the north. To join an event, participants are provided with a call-in number and meeting ID, which is verified by a central OTN operator. While each site may only have one or two participants, it is an opportunity for synchronous library instruction that might not otherwise be available (Cameron, Ray, & Sabesan, 2014). Participants may need to travel to a specific location (a hospital, community health center, or health unit) for the session, but they do not have to leave their home communities. It is significant to note that many health-care professionals in the region do not have ready access to health sciences librarians. To address this obvious need for information services, the library has hosted a series of "lunch and learn" sessions that were 45—60 minutes in length. Selected topics included:

- Information resources for diabetes
- Aboriginal health resources

- Searching for health statistics on the web
- Travel health information
- Patient education: Best resources for busy clinicians
- Evidence-based medicine: Asking clinical questions
- Consumer health resources
- Point-of-care tools
- The principles of PubMed

Some of these sessions were also recorded and archived, so even those who could not attend were able to benefit from the information.

15.4 VIDEOCONFERENCING AND PARTICIPANT ENGAGEMENT

There are many advantages to videoconferencing, and certainly it is fundamental to the delivery of educational sessions for all NOSM users. That said, there are drawbacks to this presentation format, perhaps more so for instructors than for learners (Westberry, McNaughton, Billot, & Gaeta, 2015). Some topics and teaching styles are less effective when delivered via videoconference. Workshops, for example, can be challenging when instructors and participants are not in the same room. Attendance for curriculum-integrated library sessions is mandatory, but it is much easier for absences to go unnoticed if there is no instructor in the room. Further impact on participation rates is evident in the limited capability to develop a rapport with the group connected remotely. Without eye contact and other body language cues, engaging one's audience is especially difficult (Birden & Page, 2005). Whenever possible, a remote campus should be encouraged to focus its cameras so that they include the entire group. Ask the participants to self-identify if it is difficult to see who is speaking. The temptation can be to teach to those in the room, which can put the remote learners at a disadvantage (Chipps, Brysiewicz, & Mars, 2012). Often remote learners are muted to eliminate feedback and excessive noise during the presentation, but this too can be problematic by limiting immediate participation and interactivity. As a presenter, it is difficult to know if the remote participants are comprehending, questioning, or even listening to the instructional content. Librarians should be proactive in seeking input from participants. Asking for feedback is helpful, as is using a combination of didactic and problem-based instruction (Birden & Page, 2005). Many library sessions are necessarily didactic, using a combination of lecture, PowerPoint slides, and live demonstration with varying degrees of effectiveness. Real-time database searching is likely to resonate more with learners than a PowerPoint presentation with screen captures.

15.5 TECHNOLOGICAL CHALLENGES OF VIDEOCONFERENCING

NOSM's current configuration for multipoint videoconferencing does place some restrictions on presenters' creativity and spontaneity. Seemingly small things such as having to stand at a podium to speak into a microphone can cause stress and discomfort for some presenters. Participants may also struggle with the technology, particularly if there is not enough technology support at their location (Gallagher & Newman, 2011). At NOSM, the equipment setup may differ between rooms, and it is necessary for presenters to familiarize themselves with the technology in advance. Early issues with audio, connectivity, video resolution, and hardware have taught instructors the importance of backup slides, screen captures, handouts, and presentation sharing (Hortos, Sefcik, Wilson, McDaniel, & Zemper, 2013).

Although it remains a standard technology at NOSM, videoconferencing is no longer the primary medium for library-sponsored instruction. It is convenient and practical to use newer online conferencing tools that allow librarians to host instructional sessions via webcasting. These online conferencing tools, such as Cisco WebEx, require fewer resources and offer librarians greater autonomy and flexibility.

15.5.1 Webcasting

In 2008–09, NOSM introduced Adobe Connect, a popular webcasting tool. It was launched schoolwide, and the library quickly adopted it as a preferred tool for distributed instruction. Adobe Connect significantly impacted library instruction in that librarians were able to deliver high-quality programming to learners across the north from the convenience of their own offices. Likewise, participants could join the session from wherever their computer was located via direct link to the library's virtual classroom, a secure space in Adobe Connect. As hosts of the session, librarians could share documents, presentations, and provide real-time demonstrations to individuals or groups of participants. The audio and chat features allowed for feedback, questions, and interactivity during the session. Sessions could also be recorded and archived for repeated viewing or be uploaded to the library website. Adobe Connect greatly enhanced the ability to provide personalized search instruction because the software allows the host to share their screen. Participants are able to see the librarian's live searches on their computer screen as if they are conducting the search themselves. Participants were required to install Adobe Flash Player, which could be a significant deterrent to using this tool.

Adobe Connect was an essential tool for the library for several years, until NOSM officially transitioned to Cisco WebEx (Cisco, 2016) in 2015. In addition to Skype-like web conferencing, Cisco WebEx supports content sharing, teleconferencing, videoconferencing, and event recording. The basic features are similar to those of Adobe Connect, but one advantage that is of particular interest to the library is that individuals can create personal meeting rooms. This means there is no overlap or interruption of sessions, as there had been in the shared Adobe Connect "virtual classroom." With personal meeting rooms, librarians can hold ad hoc sessions with individuals or small groups without interrupting another library session. Each librarian is responsible for hosting his or her own sessions in a secure online meeting space. The interface for WebEx is clean and simple for hosts and participants, and it is easy to share hosting privileges for co-teaching. Sharing documents, presentations, videos, and screens is seamless, although some user feedback indicates that alternating between applications can be tricky. Participants can annotate and make notes about the content while the screens are shared. WebEx also allows participants to create whiteboards for real-time collaboration. This tool is especially useful for engaging participants in brainstorming activities such as formulating search questions and developing search terms. The sessions can be recorded, and content created or edited during the session can be saved for future reference. Instructors can see participants with webcams and also look to the participant list to ensure everyone is connected. Seeing a participant list may alleviate instructors' concerns that attendants will prefer to watch recorded sessions on their own time than attend scheduled live sessions (Gushrowski & Romito, 2014).

As with previous products, the tool is not perfect. There is a small learning curve for participants, particularly with audio setup. Some of the features are not intuitive to use, and it is common for participants to not notice they are unmuted or that their webcams are on. From the host's perspective, it can be difficult to engage multiple participants if their audio is muted. It is important to check in with the participants to ensure they are attentive and to encourage interactivity.

15.5.2 Screen casting

Seeking another means to disseminate library instruction, NOSM librarians began to look at tools that would allow them to create asynchronous tutorials for topics related to information literacy and public services. Many products were tested, including single-user software like Adobe Captivate and free web-based programs such as Jing. At the time (2007—08), Adobe Captivate was chosen because it afforded the ability to integrate screen captures, audio, and text for both brief and longer format tutorials.

Adobe Captivate Version 3.0 (Adobe Systems, 2007) allowed us to create high-quality professional tutorials that could be accessed anywhere with an Internet connection. Library users could view library tutorials as needed, and librarians could embed videos into presentations, subject guides, and online learning environments. The tutorials could be stand-alone resources or supplemental teaching tools. Recorded tutorials addressed a segment of users, particularly faculty members, who could not or preferred not to attend scheduled library sessions. The tutorials were available via the library website so that busy clinicians could watch the tutorials as needed. The format of the tutorials was quite simple and practical with clear instructions for the use of a resource or service. Adobe Captivate incorporates both audio and text for accessibility, with high-resolution screen captures of databases, library tools, and other web-based resources.

Although Adobe Captivate is an excellent product, there was a learning curve for our librarians. Navigating its many robust features required varying degrees of personal training and practice (J. Ganci, 2011; Brown-Sica, Sobel, & Pan, 2009). For best results, it is recommended that users develop a storyboard for tutorials beforehand. Consequently, each tutorial required a significant time commitment by one or more librarians. It was not feasible to quickly create tutorials as needed unless they were relatively brief videos. Once completed, the tutorials also required frequent updating and editing to ensure the informational content remained accurate and current. Staying current with the software was also a challenge. It was cost prohibitive for NOSM to update to new versions, particularly as there were multiple users in different locations.

Despite these challenges, by 2009—10, the library cultivated a roster of synchronous and asynchronous instructional sessions that addressed the varied needs of our users. Our suite of sessions covered a variety of topics, including:
- using MEDLINE and PubMed
- using point-of-care tools

- bibliographic-management tools like RefWorks and personal accounts in databases
- using Medical Subject Headings (MeSH)
- using the library catalog
- accessing resources from off campus
- accessing electronic resources from our two partner universities

Topics were selected based on library user feedback collected at previous sessions, in our virtual chat and library email accounts, and from members of the Health Sciences Library Advisory Committee.

15.6 LIBGUIDES, PADLET, AND POWTOON

As part of ongoing commitment to quality improvement, the library continues to explore new methodologies for information literacy instruction for distributed learners. For several years, the library has used LibGuides (or subject guides) to present relevant resources and content on curricular or library information topics. Recently, the library has been experimenting with using LibGuides as an active teaching tool. By creating subject guides for specific courses or classes, librarians have been able to integrate them directly into undergraduate instruction sessions and into Moodle, NOSM's learning management system. Students are asked to open the appropriate LibGuide during sessions so they have immediate access to readings, links to websites, and other content of interest. This not only creates awarenesss for our subject guides but also provides students with a reference for the content covered within the session.

The library has also recently increased participant interactivity by embedding Padlet into course-specific LibGuides. In some of our undergraduate classes, librarians noted that the students on the distributed campus were not as actively engaged as the students in the room. Padlet (www.padlet.com) is a web-based collaboration tool that is accessible and easy to use. Because it is web-based, there is no software to download or install. There are both free and paid versions of Padlet, and the library has had success with the free version. Padlet can be used as a standalone tool or embedded into LibGuides and other learning environments as a collaborative whiteboard. For longer videoconferenced sessions, we have sought ways to engage learners and encourage participation. In one recent session, second-year medical students were asked to suggest search terms. In past sessions, students tended to greet this request with reluctance and hesitation. With Padlet, individuals can contribute to the whiteboard anonymously and in real time. Unless they self-identified as the contributor, no one knew who made which suggestion. There is less pressure to be "right," resulting in some very funny but relevant responses. Participants only need to be given a link to the whiteboard, or the subject guide in this case. Because it is related to the curriculum, the guide is "private" and students are given the URL in class. As an instructor, it is easy to create and personalize whiteboards. Within the whiteboards, files such as documents and images can be uploaded; comments can be added, edited, and grouped. The content can also be saved and exported to Excel or PDF. In this case, none of the learners had seen or used Padlet before so it was also a fun tool with which they could play and learn.

Another tool that can be embedded into LibGuides is Powtoon (www. powtoon.com). Powtoon is cloud-based software for creating animated presentations; like Padlet, there are both free and premium subscription versions available. For this tool, the library opted for a subscription version, which allows control of privacy settings, removal of the Powtoon water-mark and closing slide, and creation of videos up to 60 minutes in length. Powtoon is an easy tool to use but does require practice. There are fun tutorials on using Powtoon for education, although much of it is geared toward K−12 teachers. As an academic health sciences library, we determined that some of the graphics and animations were not suitable for NOSM users. For these we created internal guidelines to ensure that all videos comply with the library's preferred look and feel.

Recently it was discovered that a tool to create a series of asynchronous modules called "Evidence-Based Essentials" was needed, and PowerPoint (2013 edition) was chosen as the medium. There are many other tools available, but the obvious advantage of PowerPoint is its familiarity and accessibility by most users. Creating a basic presentation is simple to do in PowerPoint, and using the tools to incorporate screen recordings and narra-tion are straightforward with some practice. The presentations can be saved as a PowerPoint file or as a video file that can then be uploaded to the web. The file was converted to video and uploaded to the library's YouTube page and to Moodle. Creating four, hour-long tutorials can be frustrating and time consuming, so it helped to use a tool that did not require much additional training. The resulting modules look professional and polished, the audio is clear and consistent, and the screen recordings of searches in PubMed add visual interest and depth to the overall content. The same process was also used to create a tutorial comparing the features of two popular point-of-care tools. The tutorial was sent to clinical faculty members who were participating in a product trial for the library. The tutorial was uploaded to NOSM's YouTube channel, which allowed for selective user access. Its ubiquity, ease of use, and streaming capabilities have made it a preferred product for long-form tutorials.

15.7 CONCLUSION

Innovative technology is paramount for distributed education and library instruction. It is inspiring to find solutions that are affordable, adaptable, and relatively easy for librarians and learners alike. The challenge is to develop instructional sessions that are stimulating, interactive, and appealing to multiple learning styles. To provide effective library instruction within a distributed education model requires a diverse suite of resources. In the health sciences library's first decade, our delivery methods have evolved to include videoconferencing, Adobe Captivate, Powtoon, and PowerPoint for asynchronous tutorials, and Adobe Connect and WebEx for synchronous individual and small group instruction. Librarians at NOSM have also experimented with layering these tools and integrating LibGuides and Padlet.

Technology that is available schoolwide such as WebEx and videoconferencing are the primary delivery mechanisms for library instruction at NOSM. Creativity, innovation, and experimentation is found in the tools used to create the content for library instruction. Small libraries such as the NOSM health sciences library must be savvy in selecting technologies that work with existing systems, are multipurpose, and can be implemented without ongoing external support from IT, graphic designers, or curriculum instructional designers. New applications such as Powtoon and Padlet are an inexpensive way to add visual interest and creativity to instruction. These applications are web-based, free or low-cost, and require minimal setup. They can be used independently or in conjunction with existing technology. It is feasible to compare a variety of applications simultaneously, which cannot be done with expensive subscription-based software products. Just as librarians remain current with information literacy frameworks and educational pedagogy, it is necessary to remain open and curious about new technologies for creating and delivering educational content. Even the most perfect tool is not perfect forever; the needs and demands of users evolve, the technology becomes outdated, trends change. Evaluation and assessment of current practices and products is beneficial for validating a technology's continuing usefulness.

Librarians should stay current with technologies for distributed education beyond the library context and consider how such tools can be adapted for library users. Twitter, blogs, journals, conferences, and fellow librarians are great resources for keeping abreast of innovative and useful technologies, even those that are not practical for one's own library. The temptation may be to implement new products for the sake of innovation, but it is more important that the product add value to enhance the delivery of library instruction.

As NOSM looks to the next 10 years, it is impossible to envision the new methods for delivering online content or to predict which tools will be passing fads. The best tools are the ones that work for both librarians and learners and encourage interactivity, creativity, and adaptability; provide for effortless accessibility; and are fairly easy for librarians to learn and use. As new technologies evolve, develop, and adapt to user demand, the opportunities for distributed and on-site learning with NOSM students, staff, and faculty will be exciting and diverse.

REFERENCES

Adobe Systems Inc. (2007). *Adobe Captivate 3 [computer software]*. San Jose, CA: Adobe.

Birden, H., & Page, S. (2005). Teaching by videoconference: A commentary on best practice for rural education in health professions. *Rural and Remote Health, 5*(2), 356.

Brown-Sica, M., Sobel, K., & Pan, D. (2009). Learning for all: Teaching students, faculty, and staff with screencasting. *Public Services Quarterly, 5*(2), 81–97.

Cameron, M. P. L., Ray, R., & Sabesan, S. (2014). Physicians' perceptions of clinical supervision and educational support via videoconference: A systematic review. *Journal of Telemedicine and Telecare, 20*(5), 272–281.

Chipps, J., Brysiewicz, P., & Mars, M. (2012). A systematic review of the effectiveness of videoconference-based tele-education for medical and nursing education. *Worldviews on Evidence-Based Nursing, 9*(2), 78–87.

Cisco Systems (2016). *WebEx [computer software]*. Milpitas, CA: Cisco Systems.

Federation of Canadian Municipalities (2014). *Broadband access in rural Canada: The role of connectivity in building vibrant communities*. Ottawa, ON: Federation of Canadian Municipalities. Retrieved from <https://www.fcm.ca/Documents/reports/FCM/Broadband_Access_in_Rural_Canada_The_role_of_connectivity_in_building_vibrant_communities_EN.pdf>.

Gallagher, P., & Newman, M. (2011). Feeling connected: Technology and the support of clinical teachers in distant locations. *Education for Health, 24*(2), 468.

Ganci, J. (2011). Seven top authoring tools. *Learning Solutions Magazine*. Retrieved from <http://www.learningsolutionsmag.com/articles/768/seven-top-authoring-tools>.

Gushrowski, B. A., & Romito, L. M. (2014). Faculty perceptions of webcasting in health sciences education. *Journal of Teaching and Learning with Technology, 3*(1), 72–89.

Hortos, K., Sefcik, D., Wilson, S. G., McDaniel, J. T., & Zemper, E. (2013). Synchronous videoconferencing: Impact on achievement of medical students, teaching, and learning in medicine. *Teaching and Learning in Medicine, 25*(3), 211–215.

Northern Ontario School of Medicine. (n.d.). *About NOSM.* Sudbury, ON: Northern Ontario School of Medicine. Retrieved from <http://www.nosm.ca/about_us/>.

Northern Ontario School of Medicine (2015). *Community report 2014: NOSM celebrates ten years.* Sudbury, ON: Northern Ontario School of Medicine.

Tesson, G., Hudson, G., Strasser, R., & Hunt, D. (2009). *The making of the Northern Ontario School of Medicine: A case study in the history of medical education.* Montreal, QC: McGill-Queen's University Press.

Westberry, N., McNaughton, S., Billot, J., & Gaeta, H. (2015). Resituation or resistance? Higher education teachers' adaptations to technological change. *Technology, Pedagogy, and Education, 24*(1), 101–106.

CHAPTER 16

Parallel Lines: A Look at Some Common Issues in the Development, Repurposing, and Use of Online Information Literacy Training Resources at Glasgow Caledonian University

16.1 INTRODUCTION

Designers will encounter eight common issues when creating training resources for online information literacy instruction (ILI). This chapter examines each issue and how they were dealt with at Glasgow Caledonian University (GCU).

The main issues or developmental stages are:

1. Market scan
2. Planning—including storyboarding, design, and project management
3. Pedagogical issues and ILI frameworks
4. Development software (HTML, Flash, and gamification)
5. Hardware and methods of delivery
6. Testing and pilot phases
7. Teaching—determining the best way to use online ILI resources
8. Futureproofing

Glasgow Caledonian University has approximately 17,000 students in various locations worldwide. It is one of the post-1992 British universities and was formed by the merger of two established intuitions, Glasgow Polytechnic and Queen's College, Glasgow. It delivers courses to students on campuses in Glasgow, London, and New York; and it serves students at international partner institutions, as well as other users through work-based learning arrangements provided by employers and professional associations.

Distributed Learning. © 2017 T. Maddison and M. Kumaran.
Published by Elsevier Ltd. All rights reserved.

GCU has three online ILI training packages: SMILE (Study Methods and Information Literacy Exemplars), PILOT (Postgraduate Information Literacy Tutorial), and SMIRK (Small Mobile InfoLit Realworld Knowledge). An appendix of all acronyms used throughout this chapter is available at the end.

16.2 MARKET SCAN

SMILE and PILOT were adapted from existing resources produced as a result of completed projects by UK higher education institutions; as such, a literature review was already done.

The first GCU online training resource, the 24/7 database tutor, was created in 2002 to meet an urgent need for large-scale training on the Cumulative Index to Nursing and Allied Health Literature database. The tutorials had a split-screen format with a live database session on top and step-by-step instructions below (Kelt, 2003). At the same time, the INFORMS project developed a similar format of tutorials with the strapline "the guide at the side." These were developed from the previous INHALE project and produced a range of split-screen format tutorials (Stubbings & Franklin, 2004; Thomas & Gosling, 2009).

The online training resource ran over the next few years and was adapted in response to user feedback. Different databases and additional content were added as needed. For example, instructions on exporting the results to RefWorks and formatting the citations into the required referencing style were added. The tutorials were incorporated into the information skills section of the main library website.

The next development in GCU's provision of online ILI came with the appointment of a new director of library services, Debbi Boden-Angell, who brought with her the results of SMILE, a project funded by the UK Joint Information Systems Committee (JISC). This online information literacy and communication skills training package was developed by Imperial College London, Loughborough University, and the University of Worcester. It used audio, video, and webpages to provide an integrated training package. Because this was the result of a completed project, a literature search and market scan were not repeated; instead, the focus was on restoring functionality and tailoring it to the needs of GCU. This was initially done by taking advantage of the contacts and knowledge of the team of subject librarians and running a pilot project with selected users in the schools of health and business. In 2014, GCU produced a mobile version of SMILE called SMIRK on a short timescale, which was primarily a mobile version of an existing product (Kelt, 2014).

In 2011, a project aimed at postgraduates called PILOT began. This was originally developed by Imperial College, London. Librarians at Imperial College shared files with GCU and provided permission to adapt as needed.

16.3 PLANNING: STORYBOARDING, DESIGN, AND PROJECT MANAGEMENT

An essential part of planning ILI is to establish a clear statement of the target audience and the aims of the project. At GCU, the target audience were split into two large groups: (1) undergraduates who would use SMILE and (2) postgraduates who would use PILOT. Few academic institutions cater to students based on 9—5 work hours alone. GCU has a wide range of learners, including part-time and day release students, night class students, work-based learners, and distance learners at remote campuses and overseas; in fact, this is common to most higher education institutions (Secker, Boden, & Price, 2007). Having online ILI resources available helps to "deliver equity by equipping all students with basic information-literacy skills" (Crawford & Broertjes, 2010, p. 195).

The GCU projects decided to work according to guidelines provided by the Plain English Campaign. Plain English guidelines helped create short clear text that is suitable for webpages and helps users who do not have English as a first language, thus meeting the needs of our overseas students. It also uses non—gender-specific language, which is especially relevant to the larger number of female students and mature students who make up the distance learning population (Lindsay, 2004).

The level of users' technology skills should also be considered. The myth of the digital native has now been largely disproved, so the planning needs to consider what type (if any) of technology skills training is available at the institution (Thomas & Gosling, 2009). Online ILI resource developers should also consider that since distance learners will not be able to take advantage of any on-campus provision, games and interactions should be mentally challenging yet easy enough to use by a student with a low level of technical skills. The time commitment required to complete an ILI resource and the attention span of the student are further characteristics to be considered (Lindsay, Cummings, Johnson, & Scales, 2006).

A major question at the project planning stage is whether to develop the training resource from scratch or adapt an existing item. Many ILI packages are available under Creative Commons licenses, and developers should be aware that "while adapting another institution's tutorial is undoubtedly less labor-intensive than developing a new one, it is still (depending, of course, on the extent of the customization) a complex process" (Bradley & Romane, 2008, p. 74). At GCU, all resources apart from the 24/7 database tutor were developed from the outputs of earlier projects. This had both positive and negative aspects, some of which such as menu design and accessibility will be discussed later. Thomas and Gosling (2009) highlight several related issues to adapting from preexisting tutorials such as ease of reuse, updating resources, and sharing resources as open educational resources (OERs).

The importance of storyboarding was stressed in the research literature. This allows project teams to retain a clear focus and build a detailed project plan that best uses staff time and expertise. As Bradley and Romane (2008) state, this was especially important when using members of different teams within an institution, teams across institutions, and outside contractors such as graphic designers or web developers. At GCU, a project team for the development of SMILE was established consisting of three subject librarians who were provided basic Dreamweaver training. The initial plan was for team members to fit the development work in to their day-to-day tasks, but this was a challenge and the team shrank to one librarian. This meant that although formal project initiation documents were submitted for SMILE and PILOT, further project planning was not possible because the development work was carried out in tandem with other day-to-day duties with fewer team members than had been anticipated.

Other common issues to consider at this stage are:

- **Accessibility:**
  - Build this at the start rather than retroengineering it.
  - Use meaningful file names rather than a numbering system to enable screen readers to function efficiently when viewing hyperlinks.
  - Consult with the disability team in the institution for practical help and advice.
    - Accessibility not only relates to differently abled students but also applies to students who may be disadvantaged by poor Internet connections or available technology. These issues should also be factored into the design (Bradley & Romane, 2008; Thomas & Gosling, 2009).

- **Look and feel of the package:**
  - Graphics-heavy sites take more time to download and use up data when displayed on mobile devices (see Fig. 16.1).
  - Speed of loading may be an issue for international users who do not have reliable broadband connections.
  - Ensure that graphics add something to the package rather than just look nice. Images are subjective and can date quickly, making the site less attractive to users. At GCU, we have moved away from using too many graphics when developing the mobile versions of SMILE and PILOT, resulting in a cleaner look with more white space.
  - The font (sans serif and dark type are best for accessibility) and size of text also need careful consideration. Some designers shrink the size to reduce scrolling (Craig, 2007), but this is only possible when working to a fixed template. Mobile friendly templates use flexible font sizes.
  - All of the projects that GCU examined ran in larger institutions that had institutional branding guidelines. This is an important

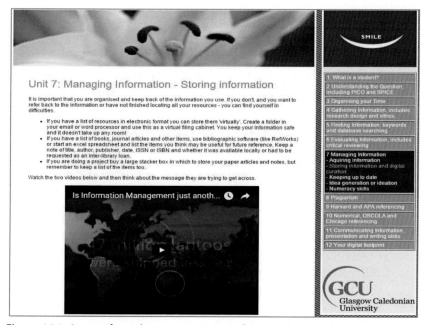

Figure 16.1 A page from the current version of SMILE showing the hard-to-update menu system and added GCU logo. This version did not run well on mobile devices.

issue when considering the design of the package. Most universities or colleges have marketing departments and institutional web teams who have strong opinions on the use of institutional logos and color schemes. It is best to consult with them at the start of the project (Thomas & Gosling, 2009). At GCU, this was not possible because we inherited fully formed resources with strong visual identities. This was not an issue when they were housed in our virtual learning environment (VLE) such as Blackboard or Moodle but did become a problem when they were made available on the open web. Adding the GCU institutional logo as part of the main menu design was the compromise. Later the problem was resolved when the final mobile versions using a mobile template were made.

- **Navigation through the package:**
  - Navigation is a major issue for all ILI package designers. Adjustments were made during the pilot phase or in response to user feedback. GCU found that it was a major issue in the earlier versions of SMILE when many users thought the forward and back icons at the foot of the page were just an attractive design feature.

  - Ensure that locations are "well signposted so that students know where they are at all times in the package" (Stubbings & Franklin, 2004, p. 4).
  - Provide a number of "escape routes" back to section start pages for students who become "lost" (Thomas & Gosling, 2009).
  - Design resources to be "platform agnostic" and to cope with the increasing number of browsers and operating systems in use (Thomas & Gosling, 2009).
  - Hospitality of platforms is relevant for those using mobile devices. It is an important issue when adding multimedia components for large numbers of mobile devices and institutional networks (Craig, 2007). Some platforms do not support Flash Player and may be reluctant to display PDF files.
  - Thorough testing on as wide a variety of devices as possible is recommended.

16.4 PEDAGOGICAL ISSUES, IL FRAMEWORKS

All of the IL resources were mapped against information literacy frameworks and other pedagogical models. The primary ILI frameworks were the Association of College and Research Libraries framework (ACRL, 2000) and Society of College, National, and University Libraries (SCONUL) seven pillars model (SCONUL, 2011), depending on the country of publication.

All of the GCU IL resources were built using the SCONUL seven pillars of ILI model (SCONUL, 2011) and were mapped against the requirements of several wider pieces of research. For SMILE, these were the GCU 21st-century graduate attributes, the Confederation of British Industry (CBI) graduate attributes (CBI, 2009), the National Information Literacy Framework Scotland (Crawford and Irving, 2013), and *The Future of Undergraduate Psychology in the United Kingdom* report (British Psychological Society, accessed 2012). The results showed gaps in SMILE's provision, so two new modules were created: "What Is a Student?" and "Your Digital Footprint." The former aimed to help students cope with the changes from school to higher education and introduced the concept of independent learning; the second dealt with web presence and the differences between personal and academic identity. This unit was introduced to teach Internet security.

PILOT was mapped against the National Information Literacy Framework Scotland (Crawford and Irving, accessed 2011) and the Vitae Researcher Development Framework (Vitae, 2011).

16.5 DEVELOPMENT SOFTWARE (HTML, FLASH, AND GAMIFICATION)

Once developers have a clear plan, a decision has to be made about which tools can be used to achieve it. The final format of the product will have a major influence on the choice of software. Many ILI resources use presentation software such as Xerte, Articulate, or iSpring, while others prefer to use a simpler form of HTML editor such as Dreamweaver. The choice at this stage may also affect the subsequent use of the resource as an OER. If the final resource is tied to a proprietary file format, it is harder for other users to adapt and develop.

At GCU, the choice of development software was shaped by the fact that both SMILE and PILOT were developed as HTML pages. Dreamweaver was chosen because it was intuitive and powerful. It also allowed for sharing various versions of the packages using Jorum, the UK higher education OER repository. Now that Jorum is no longer active, files will be shared using GCU's new repository edShare@GCU. Dropbox was used to share files with international colleagues.

Thomas and Gosling (2009) and Yang (2009) state that the growing move towards gamification and increased use of graphics over text demands that developers use more than just HTML development tools. Flash and Camtasia files or freely available add-in scripts are increasingly used to add more active demonstrations and instructional videos. Crofts and Hunter (2007) make the case that this type of presentation supports a wider range of learning styles. However, in some cases, the use of screen-capture software may result in a loss of interactivity, with the user becoming a passive viewer. GCU has moved away from the use of split-screen database tutorials to the use of both in-house and database suppliers' Flash and Camtasia videos. The newer versions of SMILE and PILOT link out to these videos. Developers should note that users may not be keen to view these resources if using their own mobile networks because they may be "data hungry." Students may prefer to wait till they can use their university network or eduroam.

Any future versions of GCU's online IL training will need to take greater notice of the "gamification agenda" (Yang, 2009) but should employ these techniques only if they are done in a professional manner (Fig. 16.2).

16.6 COMPATIBILITY WITH MOBILE DEVICES

At this point, the compatibility of future packages with a variety of mobile devices needs to be discussed. GCU's Google Analytics statistics revealed a rapid increase in the use of tablets and smartphones. In 2013, the university website was about to be redeveloped on a mobile-friendly template (see Appendix), and the library agreed to be one of the first microsites to move across to the new format. At that time, SMILE needed to be redeveloped and GCU wanted to use the same format as the main library website. However, the provision of the template took longer than anticipated, so a decision was made to create an interim version called SMIRK. This was achieved at low cost on a short timescale using the JQuery mobile tools

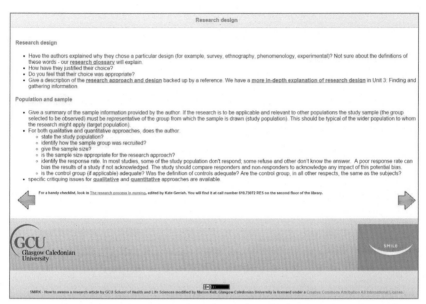

Figure 16.2 A sample page from SMIRK showing simplified "back and forward" navigation.

available in Dreamweaver. This demanded a thorough redesign, splitting SMILE into smaller component parts, rewriting the text, and changing the design to remove or drastically shrink the existing graphics (Kelt, 2014). However, it did allow the introduction of a more intuitive menu design with more hospitality for the addition of new content. It also proved to be a valuable learning experience in the development of a new package and in solving technical issues unique to the mobile environment. As such it was a worthwhile stepping stone toward the fully mobile version that will be available using Terminal 4 in 2016 (Fig. 16.3).

16.7 HARDWARE AND METHODS OF DELIVERY

A recurring topic in the literature is how ILI resources can be made available to users. Some studies (Crawford & Broertjes, 2010; Williams, 2010) opt for the use of the institutional VLE because it is easy to upload the resource and the system has a controlled log-in. Reliable user statistics may be available from the VLE (Stubbings & Franklin, 2004). Williams (2010) also points out the easy availability of other functions within a VLE such as discussion boards and blogs (Figs. 16.4 and 16.5).

Figure 16.3 Mobile version of PILOT showing home page and drop-down menu design.

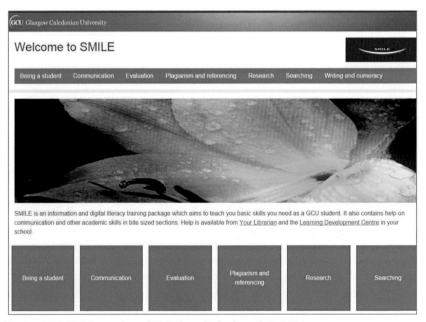

Figure 16.4 Mobile version of SMILE ready for launch in 2016.

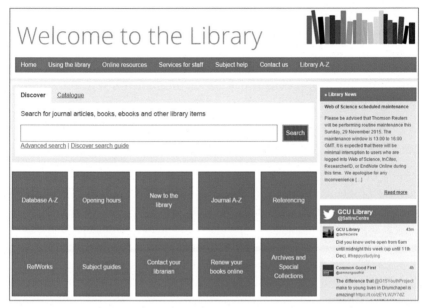

Figure 16.5 The library home page illustrating the GCU institutional colors, menu system, and navigation tiles.

GCU made the early versions of SMILE and PILOT available on the VLE. There was some disagreement on whether this was the best mode of delivery due to the lack of a quick and reliable method of obtaining user statistics from the VLE. Making the resources available on the open web would allow the use of Google Analytics to quickly and reliably obtain more detailed information on the user journey. It would also allow linking to user statistics on the Library website and main university site. This would let us plot how the users navigated through the resources available—e.g., did they navigate by using the menu system or by the use of search engines? The evidence and anecdotal feedback showed that GCU users did not find the menu system on either resource intuitive. The usage statistics rose once the content was searchable using the university website and Google. Another argument against the use of the VLE was that GCU's distance learners consistently had a problem logging in. If resources were made available with no passwords required, then this would remove a barrier to their use. After negotiation with the main university web team, SMILE and PILOT were made available on GCU web servers with the understanding that they would be redeveloped in the future using the main content management system Terminal 4 and templates to be consistent with the university's branding.

Stubbings and Franklin (2004) describe the problems in making IL resources available using external servers such as not having enough power to meet demands at peak times during the academic year. The resulting user frustration and failures during large organized training sessions are types of problems that can be magnified if several institutions are using the same server.

16.8 TESTING AND PILOT PHASES

Several of the articles in the literature review concentrate on methods of and the problems involved in assessing online ILI training. This is especially important in projects that have been formally funded. Among these aspects are the following:

- **User feedback** is an issue in higher education. It is important to keep note of it and to show how the ILI resource has been developed in response. The early versions of SMILE and PILOT only provided email links to gather feedback. Unsurprisingly, they were not heavily used. Later versions incorporated Google forms at the end of each section. These were easily tailored to match the content and automatically collated feedback into a spreadsheet. The first mobile version of PILOT relies on the standard university website feedback link, but it is planned to canvas opinions from the subject librarians once development of the new version of SMILE has been completed.
- **Formal usability testing** is an essential stage of development of any web-based resource, as it is often surprising how users navigate through a site. The GCU library website has been subjected to formal user testing giving feedback on how the tiles, breadcrumbs, and drop-down menus were used. This useful information was fed into the development of the mobile version of PILOT. Our draft departmental information literacy roadmap outlines plans to bring the future development of SMILE and PILOT under the control of the main library web group so maintenance and updating can become part of the annual cycle of reviews and regular management meetings.
- **Pilot exercises** are an extremely valuable source of information. At GCU, a pilot exercise for SMILE using selected staff in the schools of business and health was run. Feedback was sought from staff in the Effective Learning Service (ELS) and from student mentors. Detailed feedback from the last two groups was received. Student mentors'

feedback was highly positive. The information gained from the ELS showed the importance of carefully tailoring OERs to the target community. The wider pilot showed a need for a resource that could be used in small learning bites and thus allow students to study necessary skills before practical sessions or to remind them of essential skills for assignments such as referencing.

- **User surveys** are a heavily used mechanism to measure user reaction (Crawford & Broertjes, 2010; Stubbings & Franklin, 2004; Thomas & Gosling, 2009). Many online systems such as Survey Monkey allow users to provide opinions and rate the usability of a resource. In most cases in the literature, surveys were carried out before and after the use of ILI training resources. One study revealed that students were suffering from online survey fatigue, so a paper-based survey was administered at the beginning of lectures. This gave a higher response rate (Crawford & Broertjes, 2010). Although it can require considerable staff time to analyze their results, surveys give valuable written evidence of users' opinions and may also provide fuller qualitative data.

The feedback provided many areas for revision: for instance, where the text is unclear or areas that required more detail. This is where it is important to have the resource built in to the project plan to cover revision work. It is hard to anticipate how much time the team may need to devote to this, but some form of revision will most likely be required. At GCU, the main areas for revision were accessibility, navigation, and the provision of extra content defined by academics. The accessibility question was raised by the project team. The university disability team was asked to evaluate SMILE and PILOT and provide feedback. Unfortunately, SMILE was not built in a format accessible to screen-reading software. This led to major work such as renaming all files, providing alt text for images and rebuilding the menu system. However, a positive aspect of this was that extra navigation could be added to the menus and the footer of each page to address negative user feedback. PILOT did not need as much work because it was built using cascading style sheets, and changes could be made to the menus in the design templates, which then updated each section. The only way to deliver extra content requested by academics was to use OERs and tailor the resources to specific needs. This brings its own set of advantages and issues (Kelt 2012, 2013), but it was worth it to show the academics that their requests were met.

16.9 TEACHING: THE BEST WAY TO USE ONLINE RESOURCES

How do we build ILI resources into teaching? "In many cases, professors often do not have time in their syllabus to devote to a library session, or they believe that students became information literate at some prior point in their education and need no further library instruction" (Lindsay et al., 2006, p. 430). Persuading academics to free up time in their crowded timetables for ILI is extremely difficult, but it may be easier to persuade them to embed library sessions into existing modules as learning objects. Stubbings and Franklin (2004) describe how they embedded the INFORMS tutorials in several online modules provided in the institutional VLE. As these were database training modules, the students were either expected to use them in their own time or during scheduled computer lab time. The effectiveness was later assessed by a user survey.

Williams (2010) presents a useful comparison of three modes of instruction: online only, hybrid with a combination of online and face-to-face instruction, and face-to-face only. Her results show that all students showed a measurable improvement in skills in all three modes but that the hybrid approach gave the best results. This may present a way forward given the crowded timetables in many UK higher education institutions, though an online only option would have to be retained for distance learners.

The question of the structure of modules should be factored into the design of the ILI resource with a view to embedding ILI training into courses. It is worth bearing in mind that module structures can vary considerably between institutions and even schools within an institution. Once the target audience for the ILI resource has been defined, designers should consult the school(s) concerned to obtain further details on course structure and what the academics feel is appropriate.

User surveys to measure usability of the ILI resource were discussed earlier in the chapter, but there is also the question of assessing its actual value from a pedagogical point of view. "Is it more important to measure student learning or study how well the tool can be navigated and utilized?" (Lindsay et al., 2006, p. 431). Several authors (Crawford & Broertjes, 2010; Lindsay et al., 2006; Tronstad, Phillips, Garcia, & Harlow, 2009) recommend using a combination of assessment methods to not only measure student perceptions of their confidence and competence pre- and post-tutorial but also some type of assessment of their actual competence. These authors describe a study that combines survey data with actual assessments that have been built into the ILI resource. This type of examination gives a much clearer picture of the value of online ILI training as part of a module. The examination can be done using a formal assessment component that is either built in or administered separately. Many packages have informal quizzes that act in a formative way, helping students gauge whether they have fully understood the content of a section and suggesting areas for review. If sequential assessments are graded manually, then it should be made clear that the student should wait for results before proceeding to the next test. This will help prevent a small mistake in one test becoming a major misunderstanding as the student rushes ahead to complete all tests (Lindsay, 2004). It is up to the design team to decide whether they want to include a formal assessment or informal quizzes or leave the assessment component up to the academic module leader. At GCU, we found that it was extremely difficult to use the quiz function in the VLE without requiring the students to formally register on the IL course. However, a free add-on quiz function or informal quick questions could be added (Yang, 2009). There is also the question of provision of time and personnel for marking tests.

16.10 FUTUREPROOFING

Online ILI training is subject to the whims of fashion. An example of this would be the use of Second Life to provide ILI training. Many institutions have invested considerable resources in the development of and training on this platform. At GCU, we have found that it was not popular with students. This was partly due to the amount of computer processing power and time required to run the software (Aroche & Brown, 2009). There were also the issues of the steep learning curve required by both staff and students, "technological glitches," and the fact that "we cannot assume that every student of this generation will be either comfortable or savvy with online environments" (Rodrigues & Rehberg Sedo, 2008, para. 35). GCU experimented with ILI training in Second Life, but library staff found that it took far too long to provide a low level of student support compared with more "old-fashioned" but simpler methods. A related issue that should be considered is that it is often more effective to go where our users are. This can mean something as simple as using Twitter to provide a quick inquiry service. At GCU, we provide a Twitter-based inquiry service using @SaltireCentre. Sometimes students do not share our views on the best methods of providing support and training.

Only one of the papers in the review pointed out the desirability of making the ILI resource available to students after they have completed the individual module or course of study (Crawford & Broertjes, 2010). The continued availability of resources fits in to the larger current agenda of the transferability of ILI skills into the wider world of work.

It is wise to consider the ease of updating resources and, in some cases, whether it is necessary to develop a training resource at all. Over the years, database vendors have developed a wider range of high-quality training videos that can be built into library websites. At GCU, we have moved toward linking out to these rather than building our own. There may be an increased use of this in the future coupled with a higher use of adapted OERs.

16.11 AREAS FOR FURTHER RESEARCH

It would be an interesting project to trace the changing styles of online ILI training from frame-based tutorials to Camtasia or Flash on to presenting systems such as Xerte. Does the use of different formats have a sound pedagogical basis or are they just new trends? To conclude, the area of online ILI training development has been one of the most interesting phases of my career and all librarians need to explore it more fully.

REFERENCES

ACRL. (2000). *Information literacy competency standards for higher education.* Retrieved from <http://www.ala.org/acrl/standards/informationliteracycompetency>. Accessed 18.09.15.

Aroche, B. & Brown, H. (2009). *A second life for information literacy.* Retrieved from <http://conferences.alia.org.au/online2009/docs/PresentationA12.pdf>. Accessed 09.12.15.

Bradley, C., & Romane, L. (2008). Changing the tire instead of reinventing the wheel: Customizing an existing online information literacy tutorial. *College & Undergraduate Libraries, 14*(4), 73–86.

CBI. (2009). *CBI Future fit: preparing graduates for the world of work.* Retrieved from <http://www.cbi.org.uk/pdf/20090326-CBI-FutureFit-Preparing-graduates-for-the-world-of-work.pdf>.

Craig, E. (2007). Developing online information literacy courses for NHS Scotland. *Health Information and Libraries Journal, 24*(4), 292–297.

Crawford, J., & Irving, C. (2013). *National information literacy framework Scotland: The Right Information. Framework levels.* Retrieved from <http://www.therightinformation.org/temp-frameworklevels/>.

Crawford, N., & Broertjes, A. (2010). Evaluation of a university online information literacy unit. *Australian Library Journal, 59*(4), 187–196.

Crofts, K., & Hunter, C. (2007). *Using onscreen-action-capture tutorials to enhance student learning of MYOB software*. Retrieved from <https://www.academia.edu/297164/Using_Onscreen-Action-Capture_Tutorials_to_Enhance_Student_Learning_of_MYOB_Software>.

Kelt, M. (2003). Web based database tutorials at Glasgow Caledonian University. *Assignation-London*, *21*(1), 32–34.

Kelt, M. (2012). Developing SMILE using OERs and existing resources at Glasgow Caledonian University. *Journal of Information Literacy*, *6*(2), 135–137.

Kelt, M. (2013). Adapting PILOT for use at Glasgow Caledonian University using open educational resources and existing material. *ALISS Quarterly*, *8*(3), 26–33.

Kelt, M. (2014). SMIRK—Developing a new mobile resource from an existing project. *Library Hi Tech News*, *31*(7), 16–19.

Lindsay, E. (2004). Distance teaching: Comparing two online information literacy courses. *The Journal of Academic Librarianship*, *30*(6), 482–487.

Lindsay, E., Cummings, L., Johnson, C., & Scales, B. (2006). If you build it, will they learn? Assessing online information literacy tutorials. *College and Research Libraries*, *67*(5), 429–445.

PILOT. Retrieved from <http://www.gcu.ac.uk/library/pilot/>.

Rodrigues, D., & Rehberg Sedo, D. (2008). Experiencing information literacy in second life. *Partnership: The Canadian Journal of Library and Information Practice and Research*, *3*(1). Retrieved from < https://journal.lib.uoguelph.ca/index.php/perj/article/view/426/861 >.

SCONUL Working Group on Information Literacy. (2011). The SCONUL seven pillars of information literacy: Core model for higher education. <http://www.sconul.ac.uk/sites/default/files/documents/coremodel.pdf>.

Secker, J., Boden, D., & Price, G. (2007). *The information literacy cookbook: Ingredients, recipes and tips for success*. Oxford: Chandos Publishing.

SMILE. Retrieved from <http://www.gcu.ac.uk/library/SMILE/>.

SMIRK. Retrieved from <http://www.gcu.ac.uk/library/SMIRK/Start.html>.

Stubbings, R., & Franklin, G. (2004). A critical analysis of the INFORMS project at Loughborough University. *JeLit*, *1*(1), 31–41.

Thomas, J., & Gosling, C. (2009). An evaluation of the use of "Guides at the side" web-based learning activities to equip students in health sciences and nursing with information literacy skills. *New Review of Academic Librarianship*, *15*(2), 173–186.

Tronstad, B., Phillips, L., Garcia, J., & Harlow, M. (2009). Assessing the TIP online information literacy tutorial. *Reference Services Review*, *37*(1), 54–64.

Vitae. (2011). Vitae Researcher Development Framework. Retrieved from <http://www.vitae.ac.uk/researchers/428241/Researcher-Development-Framework.html>.

Williams, S. (2010). New tools for online information literacy instruction. *The Reference Librarian*, *51*(2), 148–162.

Yang, S. (2009). Information literacy online tutorials: An introduction to rational and technological tools in tutorial creation. *The Electronic Library*, *27*(4), 684–693.

APPENDIX

Alt Text

Alt text (alternative text) is a word or phrase that can be inserted as an attribute in a hypertext markup language (HTML) document to tell website viewers the nature or contents of an image. The alt text appears in a blank box that would normally contain the image. It is used by screen-reading software to explain the content of an image or icon to a visually impaired user.

Terminal 4 (T4)

Large institutions often use content management systems to manage the design and build of their institutional websites. At GCU, we use one called Terminal 4 (T4).

eduroam (education roaming)

The secured WiFi network service known as eduroam allows students, faculty, and staff to use their home institution's WiFi credentials to access WiFi network services when visiting other participating institutions without having to set up a guest account.

INFORMS

The INFORMS project ran from Oct. 2002 to Aug. 2003. It was a joint UK project carried out by the University of Loughborough and the University of Oxford. It built on the INHALE project (see below), which was previously funded by the JISC (see below). INFORMS aimed to build on the work of the INHALE project to provide resources to a wider range of institutions, allowing adaptation and development. The project also aimed to allow participants to "reshare" these adapted resources to the user community. Eventually, there were 400 units in the INFORMS database. There were efforts to put the project on a commercial footing, but these were not successful.

INHALE

The INHALE project ran from Sep. 2000 to Mar. 2003 and was funded by the JISC and aimed to:
• create a set of learning materials that embedded a variety of electronic services within a series of web-based units for use in VLEs, and
• demonstrate how the INHALE materials could equip students with transferable information skills to allow them to select, locate, use, and evaluate information for their studies.

The project produced a set of standalone web-based information skills units accessible to higher education and further education institutions in the United Kingdom. These were stored in a database of (learning) objects created from the initial standalone materials that were accessible, searchable, and extensible to other higher or further education institutions allowing individuals to re-create a tailored set of materials.

JISC (Joint Information Systems Committee)

The Joint Information Systems Committee is a UK-based organization. The website is at https://www.jisc.ac.uk/about. It champions the importance and potential of digital technologies for UK education and research and does three main things:
1. operates shared digital infrastructure and services,
2. negotiates sectorwide deals with IT vendors and commercial publishers, and
3. provides trusted advice and practical assistance for universities, colleges, and learning providers.

Jorum

Jorum (http://www.jorum.ac.uk/) was the UK's largest repository for discovering and sharing open educational resources (OERs) for higher and further education and the skills sector. It has more than 16,000 resources. Jorum forms part of the learning, teaching, and professional skills team in Mimas, part of the digital resources division of JISC (see above). The word *Jorum* is of biblical origin and means a collecting (or drinking) bowl.

Mobile-Friendly Template, Web Design Template, Template

When using a content management system to design and build websites for a large institution, the institution's main web team will provide web editors with a template to work to. This specifies design features, color schemes, institutional logos, and functionality for use with mobile devices. See Figs. 16.4 and 16.5 for examples.

Virtual learning environment is a learning management system such as Blackboard or Moodle.

SECTION IV

Case Studies

CHAPTER 17

Concept to Reality: Integrating Online Library Instruction Into a University English Curriculum

17.1 INTRODUCTION AND BACKGROUND

This is the chronicle of the vision, development, debut, and current status of a program of integrated online library instruction in a partnership between a university's library and English department. In addition to the generalities of the vision and its outcome to date, the study also examines other aspects of the journey from concept to reality, including technology selection, content selection, and evaluation techniques used in the development of the program; and the all-important human factors toward the outcome, including respect, cooperation, and shared vision and goals.

The University of Saskatchewan (U of S) is a medical- doctoral university in western Canada. It is the provincial university, a public research university, founded in 1907. Its University Library is a member of the Association of Research Libraries (ARL) and the Canadian Association of Research Libraries (CARL); it has a complement of approximately 145 full-time equivalent librarian faculty and staff serving more than 27,000 faculty, staff, and students (University of Saskatchewan Library Website, n. d.). The 2014 full-time enrollment student count was 18,987 (COPPUL Website, n.d.). The Department of English, within the university's College of Arts & Science, is the largest department within the college's Division of Humanities and Fine Arts, with students and programs ranging from first-year basic literature and composition through doctoral studies, and with the largest faculty complement (including tenure-track and contract instructors) in that division. It is also a department with critical ties to the college's Interdisciplinary Centre for Culture and Creativity (ICCC), which is the home of the master of fine arts in writing program, the digital culture and new media undergraduate program, and the women's and gender studies graduate and undergraduate

Distributed Learning. © 2017 T. Maddison and M. Kumaran.
Published by Elsevier Ltd. All rights reserved. 307

programs. The ICCC director, as well as the coordinators of the cited programs, are members of the English department faculty. The department also has ties to the classical medieval and Renaissance Studies program, the director of which and several of its fellows are members of the English department faculty, the Humanities Research Unit, as well as the Fine Arts Digital Research Center.

While the University Library had always known and considered the importance of liaison, collections, and instruction services to the Department of English and its students, faculty, researchers, and programs, due to librarian complement changes in the years immediately leading up to 2013, a series of part-time and contract librarians were responsible for those library services. In 2013, a senior humanities librarian (Donna Canevari de Paredes) who had not previously had direct responsibility for English literature was asked by the library's dean to take on the assignment of English liaison librarian. During the same time period, there were changes to the University Library's staff complement that allowed for the hiring of an instructional designer (David Francis), a position that was new and unique within the total employee complement of the University Library. That position was created in response to the University Library's strategic plan and specifically in line with the plan's first core strategy, that of "Learner and Teacher" and its dictum to "transform our services, collections, and facilities to contribute to the success of our learners and teachers and the library as a learning organization" (University of Saskatchewan Library Strategic Plan, n.d., Core Strategies Section).

Having observed how in-classroom instruction for English courses at all levels had been organized and delivered over a period of years by the sole librarian who had that assigned responsibility, the major concern for the English liaison librarian on taking on the English subject profile in 2013 was how to deal with the impossible task of serving the library literacy needs of the students in the many English courses at all levels that are delivered during a complete academic year not only at the main U of S campus in Saskatoon but also via distance education in other parts of Saskatchewan and online. The quantity made it unrealistic to promote or market to the English faculty in-person library instruction options for English literature courses at any level because it would have been impossible to deliver on the promise of instruction and service if all English course instructors had taken up the offer for an in-class library literacy instruction session. To that point, such library instruction was on a faculty request to the librarian basis.

At the U of S, there are currently five 100-level (first-year) English courses; four of those are one-term, three-credit courses; the fifth is a two-term, six-credit course. In a complete academic year, about 80 such course sections are taught both on campus (including those through St. Thomas More College, the Roman Catholic undergraduate college federated with the U of S) and at a distance through provincial off-campus locations and through web sections of these courses. To be clear, that is only speaking to the instructional needs of the 100-level courses. The English liaison librarian had and continues to have far-reaching library literacy teaching obligations in support of English; principal among those are teaching in support of the English curriculum for English majors and others enrolled in upper-level undergraduate courses; and research instruction and support for the English graduate students at both the MA and PhD program levels. In addition to teaching, the profile of liaison librarians at the U of S further encompasses the areas of collection development and management, the provision of reference and research facilitation, and conducting their own research as librarian practitioners.

In universities worldwide where English is the language of the university and therefore the language of instruction (notably universities in English-speaking Canada, the United Kingdom and other Commonwealth countries, and the United States), the English department in the undergraduate curriculum is central to courses and programs in the humanities. Whether or not an introductory English literature and composition course is a requirement toward the completion of an undergraduate degree, and in spite of a wide choice of possible entry-level courses, an English course is one of the most consistent choices for the fulfilment of these requirements (Bérubé, 2013; Jay, 2014). As both academics in English and academic librarians know, those who enroll in an English course need to be library literate for student success (Wong & Cmor, 2011).

17.2 PARTNERING TO ENABLE LIBRARY LITERACY FOR FIRST-YEAR ENGLISH LITERATURE STUDENTS

The fresh viewpoint of the newly assigned liaison librarian to the English department, coupled with having an incumbent installed in the University Library's recently created position of instructional designer were important project antecedents. A mutually positive approach of working toward the development of an online library literacy program for English students provided a sense of momentum to engage in the project's foundation.

In the next step, the librarian broached with key members of the English department the general idea and reasoning for streamlining and making a basic library instruction program for English undergraduates more widely accessible. This was accomplished as part of an initial meeting between the English liaison librarian and two senior faculty members from the Department of English—the department head and the department's official liaison to the University Library—during which a full agenda of University Library matters relevant to English were discussed. By education and inclination, faculty members in the English department are both book people and library advocates. The immediate reaction of the faculty representatives at that initial meeting was that some sort of online library literacy would be positive. A subsequent meeting between the English department's undergraduate chair and the English liaison librarian provided another indication of faculty enthusiasm and of specific outcomes desired by the faculty such as student success in finding traditional print books in the library stacks. In addition, the meeting began a philosophical conversation on whether the initial program would best be focused on the 100-level courses and the first-year students or on the English majors and the upper-level undergraduate courses.

A further meeting occurred soon after with the English liaison librarian and the entire undergraduate committee of the English department, after which the authors were ready to proceed toward concerted discussion with a representative advisory committee. The instructional designer and the librarian initiated the formation of the advisory committee, drafted terms of reference, and set up a timeline for the still nascent project. The advisory committee consisted of the following faculty members from the English department: the department head, the departmental liaison to the library, and the undergraduate chair, plus another member of the undergraduate committee. This began the journey of collaboration as well as the process of understanding the ownership of the program and the importance of recognizing the most workable middle ground between the library's and department's most important goals in order to achieve the best outcomes for students.

Both a terms of reference, which was eventually inclusively titled the "Guiding Document" and included a set of basic learning objectives that were later incorporated into the document as "learning outcomes," were essential to early progress. English faculty from the advisory committee initially drafted a set of learning objectives that the department considered essential teaching and learning points. That draft document was revised by the librarian and the instructional designer to incorporate further objectives considered essential within the context of current library literacy tools and as feasible within the technology platform to be utilized. The final amalgamated version is seen directly as follows.

17.2.1 Learning Outcomes for the English Undergraduate Online Library Instruction Program

On completing the instructional program, first-year students in English will have crossed the threshold of the University Library, online or physically (or both), to overcome unfamiliarity with the university's library system. With confidence, first-year students in English will be able to:

- access and utilize the University Library's website based on an appropriate approach in order to find resources required for the successful completion of first-year English courses;
- use the catalog to find a literary author, the call number for a book by that author, and the call number for a book of literary criticism about that author; and for on-campus students, actually find at least one of these books in the Murray Library stacks;
- search one of the University Library's subscribed databases for humanities library research such as *Academic Search Complete* or *MLA International Bibliography* to find an article in a scholarly journal;
- distinguish between primary and secondary sources;
- distinguish scholarly and peer-reviewed sources from popular and unreviewed sources of information, including understanding the appropriate uses of Wikipedia and other frequently used open access Internet tools such as Google and Google Scholar;
- locate the department of English *Requirements for Essays* through the University Library website; and
- write a paragraph describing a literary author, a work by that author, and a work of criticism about that author that uses MLA citation style as described in the *Requirements for Essays* (University of Saskatchewan English Department, 2014).

In working together with the English department, then, it was necessary to prioritize the project's requirements from both the library and department viewpoints. The collaboration and cooperation with the English department through the advisory committee was an important starting point. It was considered by the librarian and the instructional designer that the University Library should have ownership (or branding) of the program since the content of the program was to be about the concept of library as resource, a place and as facilitator and caretaker of the university's research resources. In that first half of the first year of planning, there was still an enormous amount of understanding and knowledge which needed to evolve. In this initial stage, it had already been decided that the program would be aimed at first-year courses, but it had not been decided in what form the material would be presented: whether as an English department coded course specific to library literacy for first-year English students, or through another presentation such as a compulsory component of particular courses.

Another aspect of working collaboratively with English and through the representative advisory committee was that initially the authors from the University Library felt the need to share their ideas democratically with the four faculty members from English. In retrospect, it is still felt that the early collaboration through the advisory committee with its membership of four English representatives and the two principals from the University Library provided both diversity of ideas and validation on a wider front. As the committee collectively evolved through the various preparatory stages, and although each member of the advisory committee was important, the librarian and instructional designer eventually came to rely almost exclusively on the chair of the English undergraduate committee for collaborative academic decision making. In effect, a cohesive subcommittee of three was formed; this group journeycd through the various steps in the development of the program from buy-in and other requirements to proceed to bringing the program to reality.

17.3 THE INITIAL DEVELOPMENT OF PROGRAM CONTENT, TECHNOLOGY, AND EVALUATIVE FORMULA

Initially, the content and teaching style of the program was based primarily on prior knowledge and teaching practice—i.e., in the style of in-person and classroom library literacy teaching, using live website demonstration, and including printed handouts for student reference, which were maintained up to that time by the English librarian. Style of teaching certainly needs to be informed by many factors, including (and in this case first and foremost) the overall format. But it is obvious that all of the live talk that an instructor might use to communicate teaching points in person to a classroom of students in a 50- or 80-minute classroom session cannot be directly transferred to an online program. What was needed was to translate the basic concepts and learning objectives into concise, attractive, and at least moderately entertaining modules of reasonable length. But to get to that point, our development team needed to determine what to teach and in which order, how many modules would be appropriate, and how to incorporate all of the learning objectives within the constraints of a Blackboard-based environment.

Which aspects of library literacy to teach first-year English students continue to be informed by a combination of traditional and still valid library instruction basics, such as how to find a book and how to find an article, combined with the teaching of the use of expanded technologies such as discovery services. For library instruction that is based in students' need to know certain aspects of library literacy because of specific course needs, the preparation for such library instruction has always been enhanced by knowledge of the course requirements that the instruction is to support (Blevins & Inman, 2014). That knowledge should minimally be through information from the particular course syllabus but is also enhanced through the engagement of the course instructor with the librarian instructor. Developing this online program was no different. In addition to the liaison librarian's background of previously providing in-person instruction for all of the 100-level English courses for multiple teaching faculty, it was also the knowledge gained by working with the advisory members from English, and especially working closely with the chair of the English undergraduate committee, that encouraged a collaborative approach (Melling & Weaver, 2013).

Important basics that influenced the outcome of the program included the faculty wish to actually get the students into the physical library building in order to find an appropriate traditional print book in the stacks and to promote the importance of basic academic writing skills. The authors considered how to position these important learning outcomes within the online teaching program. Using the Blackboard-based program, this translated into online teaching materials and multiple-choice quizzes as well as components beyond Blackboard. These were a book-finding and literary writing library exercise as a bonus assignment to be assigned by individual faculty teaching on a section-by-section basis. Ultimately titled "Find a Book in the Library," and including documentation with a selfie and written paragraph, this extra assigned part of the program allowed individual instructors, including those teaching at a distance, to tailor this part of the program only to individual course, section, location, and evaluation needs. In reflection of the entire program, this could be said to be the most innovative and "fun" part of the instruction; and its existence is due in large part to the collaborative aspects of this endeavor.

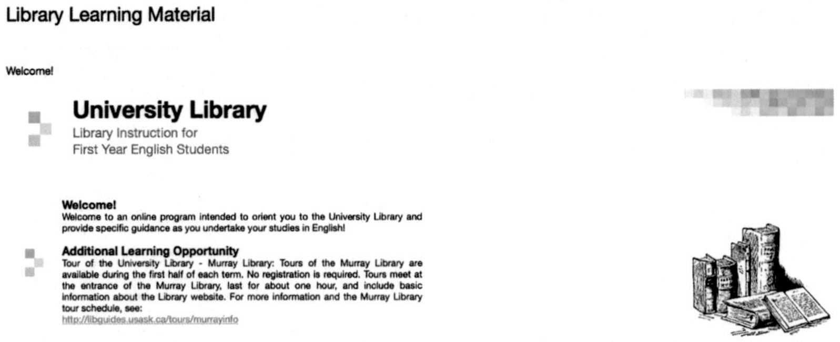

Library Learning Material

Welcome!

University Library
Library Instruction for
First Year English Students

Welcome!
Welcome to an online program intended to orient you to the University Library and
provide specific guidance as you undertake your studies in English!

Additional Learning Opportunity
Tour of the University Library - Murray Library: Tours of the Murray Library are
available during the first half of each term. No registration is required. Tours meet at
the entrance of the Murray Library, last for about one hour, and include basic
information about the Library website. For more information and the Murray Library
tour schedule, see:
http://libguides.usask.ca/tours/murrayinfo

Figure 17.1 Introductory page, online library instruction program for first-year students.

The instructional designer, familiar with various potential platforms on which to mount such an online program, advised that using a system already familiar to the campus community would be wise. That is why Blackboard was chosen. Both the library and the external campus community were already using this system. A commonly used and well-supported campus course management system afforded us numerous benefits, including the ability to import and export electronic content, back up and archive our work, and provide unified support for faculty members, many of whom were familiar with Blackboard.

The initial thoughts on evaluating students who would be using the program concerned simply finding a way to track student use of the program rather than to grade students for right or wrong answers. Thus, completing proposed end-of-module exercises was the key expectation, with students having the ability and encouragement to go back into the learning material and demonstrate their ability to report the correct answers in a formative way (Fig. 17.1).

17.4 THE PILOT OF THE PROGRAM AND THE EVOLUTION OF CONTENT, STYLE, AND DELIVERY

The timeline stated in the terms of reference was to get the program up and functional for a pilot run during the first half of fall term 2014. With the basic learning objectives in hand and positive encouragement from English and the Library, the librarian and instructional designer set out to draft the initial content of the program during the winter and spring of 2014. There were many basic considerations, principal among which were what to teach, how to teach it, how many teaching and learning sections the program would best be divided, what was to be accomplished through the assignments sections and what total number of exercises should be included, and where in the program certain elements would best be placed. A view of the introductory online program screen as viewed by faculty and students is shown as follows.

What to teach was based in the subject librarian's experience and knowledge of the English first-year courses and content in combination with the most suitable and available library resources for those course needs. How to teach in the online course environment was primarily influenced and shaped by the educational methods, technological experience, and knowledge of the instructional designer. The different strengths complemented each other, and the authors acknowledged that neither of them could have proceeded with the program individually.

An early decision was made to put the principal content into three distinct sections that were ultimately named *modules*. The rationale was that the basics had to be incorporated into three sections that would be workable within the students' spans of interest and achievable within a reasonable amount of learning time. This was estimated to be about 2 hours across the whole program for work to be done in 30-minute individual sessions (Knowles, Holton, & Swanson, 2014). Table 17.1 shows the substance of the program and its fundamentals of first-year library literacy.

The modules uniformly concluded with a summary of its teaching and learning points followed by corresponding required quizzes to be completed before advancing to the next module or section of the program. For the second module, which as noted previously was divided into two sections, it was decided to include teaching and learning summaries and quizzes at the end of each of the two sections. With each quiz containing five multiple-choice questions, this latter decision brought the number of distinct assignment sections to four, with a total of 20 questions.

Along with the three modules that were workably presented through Blackboard, a few other teaching and learning items were needed to complete the program and make it workable for use within all sections of every first-year English course, including those taught at a distance from the main campus in Saskatoon. First, it was necessary for the program to include all of the learning objectives agreed on jointly by English and the Library, including those learning objectives not feasible within the confines of the Blackboard-based environment as noted previously. Second, to ensure the same success for distance as well as on-campus students, a supplementary section for off-campus library access and use needed to be

Table 17.1 Module titles, estimated learning time, and learning objectives

Module number and title	Estimated learning time	Learning objectives
Module 1: Exploring the Library Website	30 min	The first module introduces the major aspects of the University Library's homepage as pertinent to the needs of first-year students: highlighting help sections for undergraduates; introducing the concepts of discovery service and also of the library catalog; further, as part of preliminary remarks on journal articles, pointing up the library's LibGuide-based subject research guides and ultimately focusing on that for English; and also explaining the Library of Congress classification system.
Module 2: Using the Library Website to Find Books and Journal Articles	60 min	The second module of the program is divided into two sections that each focus on multiple methodologies for finding books and articles in both traditional and electronic formats.
Module 3: Doing Research, Writing, and Citing	30 min	This module focuses on further library resources for English literature—based research, including more in-depth searching through the use of the library's discovery service; additional exploration of the English Research Guide; and including information on writing, citing, and citation management discoverable through that guide.

included. Third, a few other items were needed to round out the whole, including a welcoming statement and introduction to the program's objectives, a succinct student evaluation opportunity, and a final section in which to acknowledge those who participated in the development of the program.

The question of how to teach the required learning material was more complex because it dealt with the style, look, and feel of the program. In the early part of the development of the project, these concepts were still evolving. In 2014, the project was referred to as both a "program" and a "course" by the authors. And in terms of the presentation of

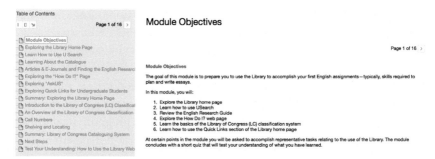

Figure 17.2 Sample content structure, online library instruction program for first-year students.

the learning materials, the authors aimed for a diversity of examples in various formats that, like the concept of the selfie in the library stacks, they hoped would provide an element of fun along with the learning. An example of how content was structured is offered in Fig. 17.2.

For the preparation and organization of the fall term 2014 pilot, the authors relied primarily on the chair of the English undergraduate committee to provide the volunteer instructors in whose courses and sections the program would run. Ultimately, seven English instructors collectively taught sections of courses that covered the full range of the five extant English 100-level courses as well as an off-campus course section. The chair of the English undergraduate committee also coordinated the student and instructor evaluations.

For that initial pilot during fall term 2014, the time period from introduction of the program in the classroom to student completion and evaluation was limited to the first 6 weeks (approximately the first half) of the university's term. At that point, the English department was considering that the best route for the program's presentation and compulsory requirement might be through a stand-alone, zero-credit English course, for which all students in all first-year English courses would be obliged to register. It was initially thought that such a course could be finalized through the various university committees and that the course would be able to be added the university course catalog as a compulsory part of the academic schedule for the following term—winter 2015. For that reason, the length of time that the students had to complete the program needed to be confined to only the first half of the term in order to meet a deadline for submission of proposed new courses for the following academic term. This was not ideal timing for the pilot, but there was no alternative.

Ultimately, presenting the program as a stand-alone English coded course proved unworkable, at least within the required timeframe for the program to proceed. It was then collectively decided that the program would proceed as a compulsory component of all first-year English courses, would be recognized in the syllabi of those courses, and would hold potential value within each course as a small percentage of the final grade. Further, it was recognized that making the program mainstream and compulsory for the following academic term was not possible. Due to the university's online registration system, which allows registration for courses for the entire academic year (fall and winter terms) in the summer prior to the year's start, introducing our program as compulsory midway through the academic year was likewise not workable.

All of these circumstances gave us the great gift of time for reflection, revision, and refinement. Although some student and faculty participants in the fall pilot complained that the period for completion of the program was too short, the authors were able to use the rest of the term reviewing the evaluations with an eye toward better content, look, and feel of the program. In addition, because the program was not going to be mainstreamed until fall 2015, the authors were able to go forward in winter term 2015 with an initially unplanned but much needed second pilot opportunity that allowed time to further refine the program by soliciting additional student observations and comments.

17.5 THE ROUTE TO SUCCESS: COLLABORATION, EVALUATION, AND THE IMPORTANCE OF PARTNERSHIP

From the debut of the first pilot onward to the start of academic year 2015—16 and the mainstreaming of the program within all English first-year courses, our created advisory committee continued to exist and represented an important support and advocacy group. As previously noted, the actual working subgroup that moved the project forward became consolidated as a group of three: the chair of the English under-graduate committee, the instructional designer, and the English liaison librarian, all of whom brought distinctive strengths to the task of refining the draft of the program in preparation for its mainstream debut.

Thanks to the student and instructor evaluations from both the fall 2014 and winter 2015 pilots, there were concrete aspects of program improvement on which to concentrate at each juncture. Evaluative information was made available to the authors entirely through the efforts of the English undergraduate chair, who fostered the commitment of the instructors who had volunteered their course sections for the pilot. From the seven course sections covered by the first pilot, there were approximately 100 student responses. For the second pilot, there were also seven course sections that utilized the program and from which about the same number of student and instructor comments were available. The ability of the English undergraduate chair to direct departmental clerical resources to organize the student responses for our analysis was invaluable. And because of individual course instructor commitment to the concept of the program, all participating English faculty provided valuable observations for improvement.

The 2015−16 terms marked the beginning of a new era of library instruction for English at the U of S. Much is still to come and we have much to learn. The idea for this project came about because only a tiny percentage of first-year English students were able to receive library instruction through the traditional faculty request and class-time and in-person approach. The number of students who need to be made library literate for student success in first-year English was and remains the major reason why the project was launched. So a need was seen, and with a newly minted instructional designer joining the library, it appeared to be the ideal time to move ahead.

It also seemed evident that English was the ideal subject area in which to embark on this journey, not simply because of the massive student numbers but also because of the percentage of students who choose an English course over courses in other humanities areas plus the natural affinity of the department and its faculty with libraries in general. And in this particular case, there was also the excellent working relationship that developed among the three principals who ultimately brought the program forward as a compulsory part of first-year English course student life.

Returning to the importance of collaboration and the question of program ownership, there remained a need to satisfy the teaching and learning goals of both parties—the University Library and the Department of English. It is not just about the University Library's objectives. Whether providing library literacy instruction in the classroom in response to faculty requests or virtually through this kind of online program, the librarian does not enter the classroom uninvited.

With an online library instruction program, however impressive, glitzy, entertaining, and technologically advanced any such program of instruction might be, it cannot be successful without endorsement by the subject department. An academic library and its resources get much more publicity from an online program incorporated into 80-plus capacity-enrollment courses on and off campus in a complete academic year than from traditional library awareness techniques due to the scope and scale of the approach. And an academic department gets much more influence over all aspects of this kind of program than it would over any in-person classroom session. Realistically, this is a winning situation for both parties.

The wording that follows is excerpted from the video and accompanying printed text of the English online library instruction program's welcome message for the 2015–16 academic year. This wording represents the collaboration and partnership that has successfully allowed this program, and in a larger sense the University Library, to play a more prominent role than ever before in the academic life of this university's first-year students:

> The English Department has introduced this instruction program, developed jointly with the University Library, as a required component within all sections of English 100-level courses... Enjoy getting to know the University Library and our resources for English!

17.6 CONSIDERING THE FUTURE WHILE REMAINING CONNECTED TO THE WIDER EDUCATIONAL COMMUNITY

Our experiences in widening the impact of undergraduate library instruction for English first-year students through the use of online instruction were based on the professional and academic experiences of the project team, as well as on best practices found in the literature (Davis, 2013; Gold, 2005; Leonard & McCaffrey, 2014; Mackey & Jacobson, 2011; Moniz, Henry & Eshleman, 2014). The impact of making instructional design available to those developing programs of library instruction is increasingly documented (Bell & Shank, 2007; Lo & McCraw Dale, 2009) as a means of ensuring currency in technology adoption and use and in supplementing the theoretical and design background already held by librarians. It is natural that increased specialization of skills would be introduced into university libraries as a means of allowing librarians to provide core academic research and teaching services without having to master all particular pedagogical or technological enablers along the way.

The establishment of partnerships by academic libraries with other campus entities is core to our academic activity as collaborators and leaders within the university setting (University of Saskatchewan Library Strategic Plan, n.d.). In this case, highly impactful pedagogical decisions were required by our wider group to ensure that our instructional program would not only be effective but also able to be enhanced over time based on multiple factors from changes to resources and technologies to faculty and student feedback. In working collaboratively to anticipate program planning and evaluation issues as well as solving problematic issues revealed during pilots of the program, it was learned that there is wisdom gained by the "crowd" (Surowiecki, 2004). Complexity can be best mitigated by having the right partnerships in place from project inception as opposed to ad hoc approaches when issues or problems arise.

In the experience detailed herein, it was paramount that all members of the project team shared a sense of the importance of program evaluation when initially designing and improving the instruction through the pilot phases. Thinking in terms of a "systems approach" that supported organizational learning (Bui & Baruch, 2010) was beneficial to our team, but not so systematic that we were bound by rigid documentation or plans.

As the authors consider the future, they are examining possibilities for broadening their online library literacy instructional approach to other areas within the humanities. Each time there is an endeavor to extend the reach of approach, a shared sense of risk is increased. With wider impact on students, faculty, and the library infrastructure, there is the need to be increasingly aware of the impact of the shortcomings of any program. It is also important to consider how to deal with various potential specific requests for customization based on real or perceived subject and academic program-based faculty requests. The authors are also aware that with wider deployment, what was simply a program becomes much more of an enterprise requiring more complex maintenance and infrastructure, including a means of tracking student completions and possible requirements of new partnerships with information technology professionals or those charged with centrally tracking student achievement from participants. All aspects may increase the complexity of such library instructional programs, as well as increase the potential for further library-centered campus-wide partnerships.

REFERENCES

Bell, S. J., & Shank, J. D. (2007). *Academic librarianship by design: A blended librarian's guide to the tools and techniques.* Chicago, IL: American Library Association.

Bérubé, M. (2013). The humanities, declining? Not according to the numbers. *Chronicle of Higher Education.*

Blevins, A. E., & Inman, M. (2014). *Curriculum-based library instruction: From cultivating faculty relationships to assessment.* Lanham, MD: Rowman & Littlefield.

Bui, K., & Baruch, Y. (2010). Creating learning organizations in higher education: Applying a systems perspective. *The Learning Organization, 17*(3), 208–227.

COPPUL Website (n.d.). <http://www.coppul.ca/licenses-products/fte-information>.

Davis, A. L. (2013). Using instructional design principles to develop effective information literacy instruction: The ADDIE model. *College and Research Libraries News, 74*(4), 205–207.

Gold, H. E. (2005). Engaging the adult learner: Creating effective library instruction. *Libraries and the Academy, 5*(4), 467–481.

Jay, P. (2014). *The humanities "crisis" and the future of literary studies.* New York, NY: Palgrave Macmillan.

Knowles, M., Holton, E., & Swanson, R. (2014). *The adult learner: The definitive classic in adult education and human resource development* (8th ed.). Oxford: Butterworth-Heinemann, Elsevier.

Leonard, E., & McCaffrey, E. (2014). *Virtually embedded: The librarian in an online environment.* Chicago, IL: Association of College and Research Libraries.

Lo, L. S., & McCraw Dale, J. (2009). Information literacy "learning" via online tutorials: A collaboration between subject specialist and instructional design librarian. *Journal of Librarian & Information Services in Distance Learning, 3*(3–4), 148–158.

Mackey, T. P., & Jacobson, T. (2011). *Teaching information literacy online.* New York, NY: Neal-Schuman Publishers.

Melling, M., & Weaver, M. (2013). *Collaboration in libraries and learning environments.* London: Facet Publishing.

Moniz, R. J., Henry, J., & Eshleman, J. (2014). *Fundamentals for the academic liaison.* Chicago, IL: American Library Association.

Surowiecki, J. (2004). *The wisdom of crowds: Why the many are smarter than the few and how collective wisdom shapes business, economies, societies, and nations.* New York, NY: Doubleday.

University of Saskatchewan English Department. (2014). *Requirements for essays.* <http://artsandscience.usask.ca/english/pdf/RequirementsForEssays.pdf>.

University of Saskatchewan Library Website. (n.d.). <http://library.usask.ca>.

University of Saskatchewan Library Strategic Plan. (n.d.). <http://library.usask.ca/info/strategicplan.pdf>.

Wong, S., & Cmor, D. (2011). Measuring association between library instruction and graduation GPA. *College & Research Libraries, 72*(5), 464–473.

CHAPTER 18

A Successful Reboot: Reimagining an Online Information Literacy Tutorial for a First-Year Experience Program

18.1 INTRODUCTION

Austin Peay State University (APSU), a four-year public, master's-level university in Clarksville, Tennessee, has been recognized nationally for efforts to improve student success, especially among its large number of first-generation, low-income, and post-traditional populations. In speaking before a congressional hearing in 2013 on the importance of innovation in higher education, former APSU President Tim Hall talked about the University's efforts to use tools and technology in ways that will increase student learning and engagement, and boost retention and graduation rates (Attaining a Quality Degree, 2013). This institutional initiative extends to APSU's Woodward Library, where faculty provide the means for APSU students to become information-smart citizens who can leverage technology to access, analyze, manage, and create knowledge in an increasingly complex information environment. This process begins in the first year of the university experience with the development of academic research and writing skills. To facilitate this process, librarians are active in the first-year experience (FYE) course, APSU 1000, Transition to the University. The semester-long one-credit hour interdisciplinary course attempts to provide students with the foundations for a successful university experience. As with most FYE courses, which have been a part of higher education for several decades, emphasis is "placed on the meaning, values, and goals of a liberal arts education as students engage in academic and career planning, use the tools for information literacy, develop good academic habits, and engage in campus life" (Austin Peay State University, 2013). The course is a requirement for all first-time freshmen, as well as students who enter the university with fewer than 12 hours earned in higher education. Although

Distributed Learning.
© 2017 T. Maddison and M. Kumaran.
Published by Elsevier Ltd. All rights reserved.

there are discipline-specific and interest-based sections of the course, some common core components common include university rigor, personal growth and development, a common reading experience, academic and career exploration, and library research.

The foundation of the library research component of the course is the six-module, self-paced, web-based Library Information Literacy Tutorial (LILT). Based on Searchpath, the tutorial created in 2001 by the board of trustees of Western Michigan University, incorporates material from the Texas Information Literacy Tutorial (TILT) developed by the Digital Information Literacy Office for the University of Texas System Digital Library. A team of librarians customized Searchpath and TILT for APSU in 2004. The six modules introduce students to basic information literacy concepts such as types of sources, effective searching (in both the online catalog and article databases), evaluating content, and citing sources.

It is a required coursework for APSU, students review the content in the six modules (which had been in place without any major modifications since they were first implemented almost a decade ago) and complete an accompanying quiz for each module. Students use this information as the foundation for putting into practice their own research in a "Library Assignment: Career Research Paper" in which they use the catalog, a library database, and other university resources to investigate their academic major and chosen career path. Students meet with a librarian assigned to their class on a scheduled library research day to help them find the required sources for the paper.

An opportunity to redesign the library component of the APSU 1000 course was presented by means of an internal, competitive Revitalize for Academic Success Initiative (RASI) grant. Administered by the university's Center for Teaching & Learning, the grant provides incentive money for "faculty-initiated and -developed projects that make substantial changes in pedagogy or delivery methods for the promotion of student retention and success" (Austin Peay State University, 2014). With this grant, a team of two librarians saw the potential to update both the content and the technological platform in order to better reach first-year students. In addition to the monetary incentive, the Center for Teaching & Learning provided resources such as supplemental workshops and ethics training. Both librarians responsible for the project completed the Collaborative Institutional Training Initiative required for research with human subjects, and their pilot study received the approval of the university's institutional review board.

18.2 SETTING THE SCENE: LITERATURE REVIEW

Course redesign involves taking a different approach to the curriculum in order to improve learning outcomes. A review of the literature found numerous articles on the prevalence of library instruction in the FYE and the growth of online tutorials to support this instruction. Best practices have been established and have influenced tutorial production from both the technological design and pedagogical levels. However, the focus of this project was broader than just the online tutorial, as opportunities arose that allowed for more student engagement within both the one-shot library instruction session, as well as Desire2Learn (D2L), the university's learning management system (LMS).

The FYE has been a part of higher education for decades. These courses help students in their transition to college life, providing an orientation to the university, introducing useful resources, and establishing or building on skills that allow students be successful academically. These skills also include those associated with information literacy such as academic research, the critical evaluation of sources, and the ethical use of information. As Boff and Johnson wrote in 2002, the "information literacy (IL) movement has ... grown and expanded parallel to the FYE movement" (p. 277). In their "first systematic effort to document the integration of library components in FYE course curricula," Boff and Johnson sought to "examine current practices within FYE courses" that could be used "to expand understanding of the role that libraries, and librarians play in this important transition period in a student's academic career" (p. 278). The results of their survey found that the majority of FYE programs had "some type of library component" with librarians "involved with curricular development" and "in the teaching of these library components" (p. 285). Librarians at APSU, who were among those who had made significant contributions in the FYE course at their institution, asked themselves how they could maximize the potential for instruction.

With course redesign, the process is often framed in terms of the possibilities of new technology. The prospect of using enhanced design tools and web applications appealed to the authors. Dewald wrote in 1999 of the web as "an exciting new tool for librarians to use to enhance the learning environment of students," (p. 31) emphasizing what to today's students is commonplace in our high-tech world—information when and where it is convenient for them. She also noted that "library instruction on the web can supplement and complement classroom instruction by expanding the librarian's teaching options and by expanding the student's options of time and place for instruction" (p. 27). Further, in applying this "new" technology to instruction, Dewald identified a number of best practices that influenced this project:

- instruction that is "course-related, and specifically assignment-related";
- emphasis on active learning to "reinforce instruction";
- information being offered in "more than one medium";
- the importance of "clear educational objectives";
- teaching "concepts, not merely mechanics";
- "the option of asking the librarian for help at any future time" (pp. 26–27).

Although the effect of tutorials on the information literacy competencies on different populations has been extensively explored in the literature, several researchers have naturally focused specifically on TILT. In a 2001 article intriguingly titled "Automating Instruction," Dupuis wrote about creating and implementing TILT at the University of Texas at Austin with her colleagues Cathy Fowler and Brent Simpson. She explained that "the goals were ambitious: to ensure that first-year students grasped basic research concepts, to best use librarians' expertise when in the classroom, and to provide basic information literacy skills accessible anytime and anywhere" (p. 21). TILT was intended as "a supplement to library instruction, not a replacement" and students completed the tutorial before library sessions, arriving "familiar with the basic library resources," allowing "librarians to build upon this basic understanding and advance students' skills in relation to projects and topics in their specific classes" (Dupuis, p. 22). TILT would go on to be the foundation of Western Michigan University's *Searchpath* and be implemented at a number of academic institutions, including APSU's Woodward Library as LILT.

Orme investigated the impact of TILT on the information-seeking ability of first-year students. He noted that "library instruction programs have traditionally worked most closely with freshmen populations, and it has long been an aspiration of many programs on large campuses to have the opportunity to reach the entire freshman cohort" (2004, p. 205). All students in APSU's first-year course are required to complete the library research component of the class, and the TILT and LILT content provided librarians with the opportunity to establish a foundation of information literacy concepts on which students could build with subsequent assignments. In Orme's interview with Dupuis and Fowler, "the persons most responsible for the development of TILT," they emphasized that TILT "was not intended as a replacement for classroom instruction but, rather, as a replacement for the more mundane aspects of classroom instruction so that librarians could transcend fundamental concepts and go beyond what they had been doing before" (p. 208). The librarians at APSU have taken this approach.

The students in Orme's study were in one of four different cohorts: those who received no library instruction, those who only used TILT, those who received classroom instruction without TILT, and those who completed TILT and received classroom instruction. The students were assigned a series of library-related tasks (such as locating a periodical title in an online catalog). Orme's research found that "a well-designed teaching tool such as TILT provides benefits that are not being realized from more traditional interactions between students and librarians" (p. 214), which supported the approach of those at APSU to the information literacy component of the first-year course: completing LILT, having a scheduled library research class period, and incorporating the online instructional modules into D2L.

Other efforts had been made to integrate TILT with Blackboard, an online course management system (CMS). In 2003, Roberts described an effort to combine the two elements in order to take advantage of Blackboard's registration and testing features for assessment purposes. Roberts outlined four reasons why "TILT and course management systems are a winning combination":

1. the content of TILT is leveraged against the registration and testing features of course management systems;
2. information literacy can be incorporated into course-specific work without reducing classroom time;
3. student information literacy can be measured and documented with CMS-based evaluations;
4. small institutions can leverage their investment in available course management systems without having to implement technology-intensive and expensive testing and registration systems (p. 12).

With this implementation, students accessed TILT and the accompanying quiz for specific courses. Anecdotally, the course instructors "saw an improvement in the quality of papers and bibliographies" and "public service librarians reported that students from the two TILT classes were more likely to utilize reference librarians for research questions" (Roberts, 2003, pp. 55–56). Librarians at APSU were also interested in taking advantage of the assessment opportunities provided by integrating the library tutorial into D2L.

More recent scholarship inspired the librarians who undertook the pilot to explore the use of technology for improving LILT. The librarians wanted to supplement, not replace, the library research day instruction and incorporate D2L, the university's LMS. The content in the web-based library tutorial that had been in place at APSU for almost a decade was still relevant and of high quality, but the platform and graphics were outdated and lacking. As Bowles-Terry, Hensley, and Hinchliffe noted, "online video tutorials offer several benefits" (2010, p. 19), which helped to shape the outcome of the pilot and reinforce for the librarians that videos would be the best way to move forward. Bowles-Terry et al. remarked that "video tutorials provide asynchronous library assistance, and students can view them on [their] own time at any hour of the day" and "the videos can be viewed as many times as necessary" (p. 19). Further, "closed captioning features can assist students who may be non-native English speakers or who have auditory disabilities" (p. 19). Finally, "videos engage visual and auditory learners, students who learn best through observation and listening, respectively" (p. 19).

18.3 THE PITCH: ESTABLISHING THE GOALS OF THE PROJECT

In the last several years, both LILT quiz results and anecdotal evidence from the instructors and librarians for the first-year course have shown that many students at APSU do not fully understand how to identify proper academic resources for college-level work, how to efficiently use technology to access information, and how to take information and correctly incorporate it into a writing assignment with proper source citations. Students cannot reach their full potential at the university if they do not master these information literacy concepts. The authors focused on four main objectives for the redesign project:

1. Enhancing the information literacy content delivery system to better engage students with different learning styles by replacing the web-based LILT with brief yet informative and entertaining videos.
2. Strengthening academic research and writing skills of students in the first-year course by emphasizing vital information literacy concepts that the students use in completing the "Library Assignment: Career Research Paper."

3. Making use of the LMS, which is required in the first-year course, to provide a better, integrated user experience for both the students and the course instructors.

4. Improving LILT quiz scores by use of this new content delivery system (as compared to those earned by students who completed the quizzes after reviewing the traditional, web-based LILT).

18.4 CASTING CALL: THE PILOT PARTICIPANTS

After reviewing the relevant literature and establishing an overall project timeline, the authors initiated the process of deciding which instructors would be asked to participate in the study. The goal was to include instructors who had taught the course previously so they would have experience with the traditional LILT, accessible via the library's home-page, and the library research day component. It was important to get approximately 20–25% of the students in the first-year course to partici-pate (which would be about 20–25 course sections), as well as to get a representative sample of students. The possible instructors were contacted by e-mail about the project and to obtain their level of interest in involve-ment. All of the instructors who were contacted tentatively agreed to participate in the project.

The final study would consist of 21 first-year course sections—including some discipline-specific sections that serve to introduce students to their majors—composed of 369 students taught by 18 different instruc-tors, as indicated in Table 18.1.

A total of 1605 students were enrolled in the first-year course (both on main campus and at a satellite campus), and the study included 23% of freshmen enrolled in the course. Although the sample size for the satellite campus was tiny, the authors still wanted to include those students in the sample. The 18 instructors for the pilot were contacted again via e-mail before the semester began to communicate the expectations of their participation:

- attendance at an instructor workshop and "screening" of the videos,
- class time to introduce the library component to students and have informed consent statements collected from those involved in the pilot, and
- evaluation of student performance and overall revitalized library component.

Table 18.1 Pilot course sections

Type of section	Number of sections	Number of students
Satellite campus	2	27
Biology	1	18
Business	1	23
Conditional	3	59
Education	2	31
English	1	14
General	3	65
Health and human performance	1	22
Honors	1	16
Mathematics	1	9
Nontraditional	1	12
Political science	1	16
Premedicine	1	17
Prenursing	1	23
Undeclared	1	17
Total	**21**	**369**

Each instructor committed either orally or via e-mail to be a part of the pilot study.

18.5 ACTION!: THE VIDEO PRODUCTION

The next step was to plan for the production of the videos and create or update the documents to support the project. In addition to the "Library Assignment: Career Research Paper" directions, librarians prepared supplemental handouts on citing sources. They also developed a checklist for use by course instructors to track student success in achieving the established learning outcomes in their completion of the "Library Assignment: Career Research Paper." In a 10-question survey, instructors were asked to comment on their experience of using the LILT tutorials in prior semesters and the new revitalized content.

The project timeline allowed for two to three weeks per module to write the script, design the storyboards, and produce the videos. Additional time was allotted to conduct focus groups and perform post-production editing. Once all of the content was completed, it would be loaded into D2L and checked (and double checked) before being made available to the course instructors and students.

Obviously, the most significant component of the project was creating and producing the videos. The traditional, web-based LILT had been in place at the university since 2004 with minor updating and maintenance throughout the years. This version of LILT contained text, images, and a series of links to navigate from page to page. The pilot project required the authors to transform the content of the six modules into a series of engaging videos of approximately 10 minutes in length. This process was guided by the writing of scripts and the creation of graphics, screen casts, and storyboards produced and edited using Adobe CS6 software. The videos would include oral narration, animation, and the option for closed captioning. These various means of communicating the information literacy concepts in the videos could reach students with a diversity of learning preferences as opposed to content delivery in the solely text-based web platform. To be able to have a solid comparison between the traditional LILT and its revitalized version, it was determined early in the project that a majority of the content for each module would be kept and the same questions in the quizzes would be used.

The material in each original module and the subsequent videos provides an introduction to basic information literacy concepts:

- Module 1—"Starting Smart" covers types of information; students learn to identify and describe a variety of information sources including those found on the web versus library resources.
- Module 2—"Choosing a Topic" introduces students to the APSU 1000 "Library Assignment: Career Research Paper" and covers topic selection, as well as constructing a search strategy using keywords, Boolean operators, and truncation.
- Module 3—"Using Austin" introduces how to search the library's online catalog. Students learn to identify the different types of materials contained in Austin, ways to search for materials in Austin, where to locate materials in Woodward Library, how to identify what a call number is and use it to find an item in the library, and how to search Austin for *The Occupational Outlook Handbook*, one of the major resources for the "Library Assignment: Career Research Paper."
- Module 4—"Finding Articles" covers how to use article databases to find periodical articles on a subject. Students learn to recognize the differences between popular and scholarly periodicals, what an article database is and what information it contains, how to choose an article database appropriate to their information need, and how to identify a scholarly article for the "Library Assignment: Career Research Paper."

- Module 5—"Using the Web" suggests how to best use the web for research. Students learn how the Internet is organized, how to use search engines to effectively search the web, and how to identify ways to interpret and evaluate information.
- Module 6—"Citing Sources" explains how to avoid plagiarism by introducing citation basics and style guides. Students learn how to identify the different parts of a citation, when to cite sources used in their work, and five tips for avoiding plagiarism.

One of Dewald's best practices was the importance of using "clear educational objectives" (1999, pp. 26-27) in online instruction, so the authors chose to begin each video by incorporating that content and letting students know what they would learn to accomplish by the end of the clip. The only major change from the web-based tutorial to the video was the introduction of the RefWorks bibliographic management system in the "Citing Sources" video. Librarians felt that this would be an appropriate time to present this tool to the students and then cover it in further detail in the library research day session.

An important benefit to this project was the ability to customize the video content specifically for APSU. The authors believed that this would help the first-year students get more familiar with the campus and feel more connected to it. APSU's mascot, "The Governor," is featured in each video, and library faculty and staff provided the narration for the videos. Three popular faculty members introduce the style guides most commonly used in their disciplines, as does the dean of students, who is mentioned in the video covering plagiarism and academic dishonesty. Screen casts captured navigation of the library's website in order to search the online catalog, Austin, as mentioned previously, and the database required for the career research assignment. At the end of each video, students are reminded to take the quiz, with an image of the quiz link in the D2L shell.

18.6 ON SET: THE FLIPPED (LIBRARY) CLASSROOM

The authors also took advantage of this project to reevaluate the classroom activities occurring on the library research day class meetings. Traditionally, the librarians introduce the career research assignment and walk the students through the process of accessing the required resources for the paper. With the implementation of the new videos, the authors recognized the opportunity to create a different experience for the library research day using the flipped-classroom approach. The authors believed that students would get more out of the library day meeting if the majority of the instructional content was offered outside the classroom via the videos, allowing students to collaborate with their peers on active learning exercises under the guidance of the librarian. With the videos, the instruction would be "course-related and specifically assignment-related," (1999, p. 26) which again is one of Dewald's best practices. The research assignment was introduced in the first video, "Starting Smart," with an overview of the assignment and an image of the handout; that image was branded for use throughout the other videos so that students would recognize the image and know to pay specific attention to instructions. When the library's online catalog in "Using Austin" and databases in "Finding Articles" were covered, searches of those resources were modeled. The students were instructed to locate the information they would need for their own career papers and bring those sources to class with them. If students brought the materials to class, librarians would have adequate time to incorporate additional engaging activities in the class.

In reviewing *Classroom Assessment Techniques* by Angelo and Cross (1993), two activities were decided on that would enhance the library research day—a peer-review project and a "One Minute Paper." For the peer review, students would be instructed to get into pairs or small groups to briefly analyze a classmate's scholarly journal article. The students would determine whether or not the article was, indeed, scholarly and support their answers with two identifiable criteria. Librarians would roam around the classroom to answer questions and provide guidance. At the end of the session, students would fill out a "One Minute Paper" form assessment of the class time, addressing the most useful thing learned and, if applicable, an important question that was not answered.

18.7 TEST SCREENING: FOCUS GROUPS

As part of the development process, the authors used student focus groups to view the videos and provide feedback. A useful resource for learning more about focus groups was *A Practical Guide to Information Literacy Assessment for Academic Librarians* (2007) by Radcliff, Jensen, Salem, Burhanna, and Gedeon. As Radcliff et al. note, "the apparent ease with which focus groups can be conducted can mislead us and cause us to overlook the planning process. Do not fall prey to the tendency" (2007, p. 73). Using the cited book for direction, the authors spent several weeks developing and preparing for the focus groups, including refining the questions that would be asked. Since the preparations for the pilot took place during the summer, librarians had a small pool of students from which to select participants. The focus groups were composed of approximately 10 university students, the majority of whom were student assistants at the library. These students had previously completed the traditional LILT modules for the purpose of the focus group or in their own FYE classes. The authors created a form for the students to complete after viewing the videos. Among the questions asked were "Which is the most effective format—web or video?" and "What would you recommend to improve the video?" Focus groups were facilitated by a library associate and a library assistant. In analyzing the focus group feedback in the process of creating the six new videos, the majority of the student comments were positive. A few minor modifications were made to selected videos such as slowing down the animation in the first one, "Starting Smart," and adding more descriptive images in the fifth video, "Using the Web," which diagrams a URL and explains domain names. Overwhelmingly, the focus group students preferred the new videos to the web-based content.

18.8 PROJECT DISTRIBUTION: MAKING THE MOST OF THE LMS

A number of benefits could be realized by embedding the new library content into an LMS. First, the entire library research component would be easily accessible in one place for all users. The D2L Content tab would contain the links to the six videos in YouTube, as well as a transcript for each video. Students had a choice of how to access the content, depending on individual learning preferences; they could watch the video, read the transcript (either on-screen or in print), or both. At each library research day, librarians distributed three print handouts to the students:

- the assignment instructions and basic scoring rubric,
- an introduction or refresher on style guides and the importance of citing sources in research, and
- sample citations for some of the resources required for the research paper in the most widely used style guides for the assignment—*The MLA Style Manual, APA Style*, and *The Chicago Manual of Style*.

By including PDFs of these handouts in D2L, the students would have that information accessible throughout the semester.

Students could also easily access the six quizzes via D2L. Although the LMS allows for restrictions to be placed on the quizzes—such as limiting the number of times the students can take the quizzes—the students were able to repeat the quizzes for this study multiple times. The six quizzes were composed of from 9 to 13 multiple-choice and true—false questions, with immediate feedback given. The LMS platform allowed the librarians access to the scores, which includes valuable information such as the "Question Stats" and "Question Details." In D2L, the "Question Stats" displays the class average, the percentage of questions that achieved that score, and the average score for each question. The "Question Details" displays each question from the quiz, the average grade, the correct answer, and the percentage of times each answer was selected. This data informed the instruction provided on the library research day session. If statistics indicated that students scored below 75% on a quiz question, the librarian addressed that information literacy concept in class. The highest quiz score for each student was automatically uploaded to the course instructor's gradebook in D2L.

An additional advantage of the LMS is the ability of the assigned librarian to be a presence in the course throughout the semester. In previous iterations of this course, librarians met with the class for the single, one-shot library research class and did not see the students after that session. With the LMS, librarians could post announcements and reminders for all students in the "News" section and also be available to answer questions via the D2L e-mail feature or via a course discussion board that could be dedicated to library research issues. In addition, each of the six videos ended with a reminder to students to bring any questions to the library's research assistance desk either in person or electronically. Again, this addresses one of Dewald's best practices, "the option of asking the librarian for help at any future time," (1999, p. 27) an option the authors felt strongly about incorporating into the instruction.

18.9 THE REVIEWS ARE IN: ASSESSING THE STUDENT QUIZ RESULTS AND INSTRUCTOR FEEDBACK

For the pilot program, after viewing each of the six new videos, the students completed an accompanying quiz in D2L with immediate feedback given. The students were able to evaluate how well they performed on the quiz in grasping the information literacy concepts covered in the video. More importantly, in assessing the overall project success, the authors compared the quiz results of students viewing the videos containing revitalized content against the quiz results of students utilizing

Table 18.2 LILT quiz scores comparison

LILT module	Traditional LILT (%)	Revitalized LILT (%)	Percent increase (%)
1 Starting Smart	89.8	91.7	2.1
2 Choosing a Topic	87.8	91.1	3.8
3 Using Austin	89.4	89.7	0.3
4 Finding Articles	88.8	91.2	2.6
5 Using the Web	94.2	95.9	1.8
6 Citing Sources	93.5	95.2	1.8

traditional content conveyed through the LILT tutorial. As stated previously, the quizzes in both the traditional and revitalized LILTs were identical in that the same questions and answers options were used. With the video-based LILT, scores increased on 83.5% of the questions, ranging from less than one-half of a point to more than eight points per question. In addition, the average student quiz scores for each of the six quizzes increased shown in Table 18.2. The data indicate an overall improvement in the LILT quiz results for each of the six modules.

The "Career Research Paper Checklist" was a means to evaluate how well students succeeded at putting what they learned from both LILT and the library research day into practice. Three major areas were investigated:

1. Sources—Did students find the required sources for the paper using the online catalog and a library database?
2. Content—Did students execute at least one correct paraphrase and direct quotation?
3. Citations—Did students cite the three required sources for the paper in the instructor-specified style guide with no more than three errors?

Librarians had access to the quiz scores, but not the final career research papers, so instructors were asked to complete the checklist and return it. Completed checklists were received from nine of the 18 instructors (three of the instructors taught two sections each). In these sections, there were 197 career research papers submitted by students. Based on the checklists from the instructors, more than three-quarters of the students were able to locate the required resources, with 85% of the students successfully using Austin to access *The Occupational Outlook Handbook*, while 76% found a scholarly journal article from the *General OneFile/InfoTrac* database. When it came to properly incorporating the information from the sources into the paper, 73% of students were able to execute a correct paraphrase with an in-text citation, but only 67% could do the same with a direct quotation. In looking at the citations for the "Library Assignment: Career Research Paper," five of the instructors required MLA style, while the other four required APA. Approximately 75% of the students could cite *The Occupational Outlook Handbook* with no more than three errors, while 68% cited their journal article correctly. However, only 61% of the students correctly cited the University's *Undergraduate Bulletin*.

To get constructive feedback from the participating instructors, the authors developed a 10-question survey that was loaded into SurveyGold and the link e-mailed to the instructors toward the end of the pilot semester. Eight of the 18 instructors completed the survey. All of the respondents indicated that the new video modules were their preferred format for providing the LILT content to their students. Further, the majority of instructors agreed that the LILT video modules had a positive impact on the success of students in their FYE courses. The instructor comments showed that the students found the videos "engaging," with the students communicating to the instructors how much they liked the videos and how "entertaining" they were. Finally, the instructors appreciated the use of the LMS, especially the grades being automatically imported into the course gradebook. As one instructor summed up the pilot: "I definitely feel it is an improvement . . . better grades, easy access, Gradebook. . . . more interesting!"

18.10 DIRECTOR COMMENTARY: DISCUSSION

There was one major goal of the study that was not met. As mentioned previously, one of the interests of the authors was in redesigning the instruction provided by librarians to students on the library research day class meetings. The students were instructed in the videos to find and print out two sources—the section of *The Occupational Outlook Handbook* that relates to their careers and a scholarly journal article from a library database focusing on a trend in their fields—and bring them both to class. If students brought the materials to class, the librarians would have adequate time to incorporate a peer review of the scholarly journal article and have students complete a "One Minute Paper" assessment. In only a few instances did students bring their materials to class—and those sessions were wonderful! The students had time to do the peer review of the journal article, so there was more interaction and engagement among the students (and with the librarian), and the "One Minute Paper" assessments were positive. Students reported they received useful instruction and, when there were unanswered questions, librarians could post these questions and answers in a D2L library issues discussion board for all students in the course. Unfortunately, the vast majority of students did not complete this homework assignment. When asked why they did not have the information, a few students admitted that they did not watch the videos, while others said that they were not confident in their ability to find a relevant article. For almost every library research day meeting, the authors had to revert to the traditional instructional method. Perhaps, if the librarians had asked the course instructors in the pilot to mandate *The Occupational Outlook Handbook* and scholarly journal article as required homework and assign points for them as part of the course grade, more students would have been prepared for the class.

18.11 THAT'S A WRAP! CONCLUSION

The results of this redesign indicate great success in improving the library component of the FYE course. First, both the students and instructors preferred the "engaging" and "entertaining" videos to traditional web-based LILT, and they liked the content being readily accessible via D2L. LILT quiz results improved on all six modules. Because of these successes, all APSU 1000 sections now access the library component of the course (LILT video links, transcripts, quizzes, librarian availability) via D2L. One major change is that the homework assignments from videos 3 and 4, "Using Austin" and "Finding Articles," have been removed, and librarians continue to conduct the library research day class meetings in the traditional way. In the future, the authors might revisit this component with course instructors and ask them to require the homework in order to take advantage of the opportunities afforded by the flipped-classroom approach.

A question for further exploration is how librarians might better help students put into practice those information literacy skills that are covered in the LILT videos. Average LILT quiz scores indicate mastery of these concepts, yet the results of the "Career Research Paper Checklist" show that approximately two out of three students are still struggling with correctly executing direct quotations and properly citing some of their sources. Possible solutions include spending more time on these concepts during the library research class meetings—as well as librarians building relationships with their students in these sessions—and promoting library services such as the research assistance desk and the writing center. Strengthening students' understanding of the connections among information literacy, reading, and writing is vital to their success at the university and beyond.

REFERENCES

Angelo, T. A., & Cross, K. P. (1993). *Classroom assessment techniques: A handbook for college teachers*. San Francisco: Jossey-Bass Publishers.

Attaining a Quality Degree. (2013). Innovations to improve student success In: *Hearing before the Committee on Health, Education, Labor, and Pensions, Senate, 113th Cong.* Testimony of Tim Hall.

Austin Peay State University. (2013). APSU 1000. In: *Austin Peay State University: Transitions*. Retrieved from <http://www.apsu.edu/transitions/apsu-1000>.

Austin Peay State University. (2014). What is RASI? In: *Austin Peay State University: Center for Teaching & Learning*. Retrieved from <http://www.apsu.edu/ctl/what-rasi>.

Boff, C., & Johnson, K. (2002). The library and first-year experience courses: A nationwide study. *Reference Services Review, 30*(4), 277–287. <http://dx.doi.org/10.1108/00907320210451268>.

Bowles-Terry, M., Hensley, M. K., & Hinchliffe, L. J. (2010). Best practices for online video tutorials in academic libraries. *Communications in Information Literacy, 4*(1), 17–28.

Dewald, N. H. (1999). Transporting good library instruction practices into the web environment: An analysis of online tutorials. *Journal of Academic Librarianship, 25*(1), 26–32.

Dupuis, E. A. (2001). Automating instruction. *Library Journal, 126*(7), 21–22.

Orme, W. A. (2004). A study of the residual impact of the Texas Information Literacy Tutorial on the information-seeking ability of first year college students. *College & Research Libraries, 65*(3), 205–215.

Radcliff, C. J., Jensen, M. L., Salem, J. A., Burhanna, K. J., & Gedeon, J. A. (2007). *A practical guide to information literacy assessment for academic librarians*. Westport, CT: Libraries Unlimited.

Roberts, G. (2003). The yin and yang of integrating TILT with Blackboard. *Computers in Libraries, 23*(6), 10–56.

CHAPTER 19

Rethinking Plagiarism in Information Literacy Instruction: A Case Study on Cross-Campus Collaboration in the Creation of an Online Academic Honesty Video Tutorial

19.1 INTRODUCTION

Librarians at Austin Peay State University (APSU) in Clarksville, Tennessee, understand that it is essential for students to ethically access, analyze, use, and manage information in an increasingly complex environment. Teaching and informing students about academic honesty can be a positive experience and does not have to focus on the punitive. Too often students feel accused of plagiarism before they even begin their first college research assignment. Librarians became aware of this when the university began a nationally recognized institutional initiative focusing on efforts to improve student success and retention. This initiative extended to the library, where librarians provide the means for students to become information-smart citizens and lifelong learners. When the librarians began reevaluating the types of tools and instructional materials used to teach academic honesty to students, they soon realized that their instruction frequently emphasized negative consequences over making the right choices from the start. Further, librarians did not take into consideration students' understanding of the research process. Armed with this new information, APSU librarians sought to tackle these issues in a positive manner by completely updating a web-based antiplagiarism tutorial. Ultimately, they achieved positive results.

Distributed Learning.
© 2017 T. Maddison and M. Kumaran.
Published by Elsevier Ltd. All rights reserved.

19.2 BACKGROUND: DISCOVERING THE NEED FOR A PLAGIARISM TUTORIAL

In 2011, the Tennessee Board of Regents (TBR) issued a systemwide rule on academic and classroom misconduct that states students enrolled in TBR institutions must be afforded due process in cases of academic and classroom misconduct. This ruling resulted from legal cases of cheating and plagiarism within the TBR system. The rule states that "students who received a lower grade or other discipline as a sanction for academic misconduct have a right to choose due process among other options" (Tennessee Board of Regents, 2011). It also states that "due process" must include as a minimum:

- notice to the student in writing of the conduct violation(s);
- notice to the student of the time, date, and place of the hearing allowing reasonable time for preparation;
- the right of a student to present her or his case;
- the right of a student to be accompanied by an adviser;
- the right of a student to call witnesses and to confront witnesses against the student; and
- the right to be advised of the method and time limitations for appeal (Tennessee Board of Regents, 2011).

APSU had several issues to address. First, the university did not have a policy on student academic misconduct per se. Academic misconduct was covered in the *Code of Student Conduct*, not in a separate policy. The consequences for academic misconduct were a grade of F for the assignment, F for the course, a reprimand, probation, suspension, or expulsion. The code did not include a procedure for due process in cases of academic misconduct. Moreover, at that time, faculty could assign a grade of F for the course and tell the student not to return to class, which could be viewed as a punitive measure. It was the permanent removal of the student from the classroom without due process that was a point of major concern for APSU. Therefore, APSU incorporated due process into its guidelines and policies.

The faculty handbook, policy committee, and dean of students worked together to write APSU's policy on student academic misconduct (Austin Peay State University, 2015). The policy was distributed and discussed across campus beginning with the faculty senate and continuing to student affairs, the deans' council, academic council, and finally, the university policy committee before it was submitted to TBR. There was an effort across campus to disseminate information about the specifics of this new policy to all constituencies, especially to faculty and students. The consequences of academic dishonesty remained the same for the most part. However, a significant change affected a student whose grade was lowered because of academic dishonesty. In that event, the policy required that the student, department chair, and dean of students be notified. The student had to remain enrolled in the course until the university settled the matter; the faculty member could not remove the student from the classroom, and the student could not drop the class.

The second paragraph of the policy begins, "All members of a University community have responsibility for ensuring academic integrity" (Austin Peay State University, 2011). With the campuswide effort and the high priority TBR placed on the issue of academic misconduct, the new student policy was quickly put in place and effective within 1 month. In contrast, approval of new student policies is a lengthy process that often takes up to 18 months.

Some APSU faculty members had concerns about the implementation of this policy because it required faculty to document and report cases of academic dishonesty, adding additional work. Once the new student academic misconduct policy went into effect, the number of cases for the academic year increased significantly. Twenty-six cases of plagiarism were reported during the 2011−12 academic year compared to 113 cases reported in 2012−13, a more than 300% increase in the number of plagiarism cases reported to the dean of students (Table 19.1).

A variety of initiatives were introduced across the APSU campus informing and educating students about academic honesty with a focus on plagiarism. The Office of Student Affairs instituted Academic Integrity Week during the second week of the fall semester. Students also receive an

Table 19.1 Violations: five-year comparison by academic year

Academic year	Plagiarism	Other academic misconduct	Totals
2009−10	26	1	27
2010−11	27	0	27
2011−12	26	5	31
2012−13	**113**	**48**	**161**
2013−14	99	17	116

Greg Singleton, Dean of Students at APSU, 2014 (Singleton, 2014).

informative guide during the first week of school, which deals solely with academic honesty and integrity. Student affairs administrative staff looked to the librarians for help planning the programming for the week-long event. Librarians provided workshops on citing sources and using RefWorks in the library, and they cohosted an open house with the writing center, which is housed in the library. Since plagiarism and academic integrity are addressed throughout our first-year experience (FYE) course, APSU 1000—Transition to the University, these campuswide changes prompted the librarians to evaluate their web-based plagiarism tutorial. Librarians saw the need to create a video tutorial that was engaging, fun, and informed students about proper use of information and avoiding plagiarism.

Going back about a decade, the APSU information literacy instruction program had been using the "Plagiarism: The Crime of Intellectual Kidnaping Online Tutorial" by Edith Crowe and Pamela Jackson of San Jose State University. The tutorial included web-based content with an accompanying quiz. The content was excellent and included detailed definitions, information about style guides, and how to cite a direct quotation and paraphrase, but the library had been using it for many years without any real revisions. Although it had some outstanding content, it also had outdated graphics that utilized old software like Flash Player and focused heavily on the punitive. For example, the tutorial begins with a student being kicked out of the university for plagiarism. He is literally "kicked out," with the foot on his bottom and the door slamming shut behind him. As if that was not enough, the specter of death appears in the shadows to signify the death of his academic career (San Jose State University Library, 2004) (Fig. 19.1).

Figure 19.1 Specter of death.

19.3 LITERATURE REVIEW: WHO'S CREATING LIBRARY VIDEOS?

One current trend in academic libraries is to engage students with instructional videos that supplement or enhance face-to-face classroom instruction. This trend is largely the result of limited librarian time in the classroom and the desire to connect with students face-to-face, as well as online. A review of the literature shows that a variety of methods can be used in library video production. In addition, best practices for video production is not a one-size-fits-all approach but appears to differ, depending on the needs of the individual library and available resources.

Thornton and Kaya (2013) created a series of instructional and informative videos to assist students in learning about information literacy skills while simultaneously marketing their library to a wider university community. They share their foundation for their project, problems they encountered, and assessment. After studying 50 library videos available on Google Videos, YouTube, and library- and university-owned sites, Thornton and Kaya found that every video was unique in having a distinctive style, topic, and use of characters. They also provided direction by offering suggestions from their study on what makes a successful video. They reported that students preferred shorter videos with a narrator. Similarly, "the format and style of the videos partly depends on the subject matter: presentation of individual databases requires a factual use of screen-casting, whereas a more 'human' problem in the library might lend itself more to a dramatic movie representation" (p. 75). In many of the videos, they found that librarians were either narrating the script or voicing the role of a character. Moreover, they found that "librarians could serve to reveal the face—in some cases, the friendly and fun face—of the library to the students" (p. 75).

Scales, Nicol, and Johnson (2014) reported on redesigning two comprehensive library tutorials at Washington State University. Although their tutorials had been revised many times by rewriting text and updating screenshots or graphics, they recognized that the online tour set of tutorials "had become an 'information dump' of important, but uninspiring, content" (p. 243). They decided to take their online tour, which was originally produced by using screenshots, graphics, and HTML frames, and create eight videos. They made improvements by moving from a "text-based interface to an appealing, pedagogically sound multimedia resource for students" (p. 243). Their qualitative analysis of the new video tutorials showed notable results. They discussed incorporating visual cues such as the use of arrows, highlights, and other means to engage students as part of their tutorial redesign project. They also go on to include music clips, sound effects, and voice narration as another means to get the students' attention. In one example shared, they used the school's fight song and added a "robotic voice declaring the value of librarians in the first few seconds of the tutorials" to direct "the learner's attention to the auditory channel" and prepare "them to listen to the upcoming presentation" (p. 245). Like Thornton's and Kaya's video approach, they place an emphasis on oral narration. In their analysis of the visual components used in the tutorials, they discovered that screen casting was the most used visual, followed by animation and then other elements such as photographs and key pieces of text.

Clapp and Ewing (2013) reported on their video project, which was entirely student-centered from beginning to end. Their videos not only were developed for students but also created, narrated, and performed by students. They referred to this as "revenue positive labor" (p. 335), a concept adapted from Roy Lansdown at California State University at Long Beach. Librarians worked with student assistants to develop 12 topics that were based on common questions frequently asked at the library service desks. The librarians noted that they "were mostly unaware of what the students were doing until they shared the final video" (p. 335). Since they had limited resources, student assistants used personally owned equipment as simple as flip video cameras and personally owned software such as Apple's Final Cut Pro to create their videos. The librarians uploaded the videos to their institutional repository for archival purposes, and they also used YouTube to make the videos readily accessible. These videos are publicly available and can be embedded on websites and watched on many devices. A partnership between the librarians and the dean of students' office was formed to evaluate three of the videos in the series that were being used in the FYE course. A six question pre- and post-viewing survey provided to students during their FYE course showed an increased awareness and confidence about use of the library. Using the YouTube analytics data from each video, coupled with the survey results, they found that the library videos were highly useful.

Bowles-Terry, Hensley, and Hinchliffe (2010) provide best practices with a report on how the University of Illinois's Library created brief instructional videos using a combination of the Camtasia and Audacity technology. They were able to use this low-cost and easy-to-learn software to produce a video in 1 day. They advised that "the technology learning curve is not too steep for most librarians, and after becoming familiar with the process, a 2-minute video can be created (from script to final product) in 4 to 6 hours" (p. 18). They also developed a set of emerging best practices for creating library videos that evolved from their research study on investigating library video tutorials and their usability, findability, and instructional effectiveness. These best practices on creating library videos based on their usability study include suggestions on pace, length, content, look, and feel; video versus text; findability; and interest in using video tutorials.

When Amanda Hess (2013) reported how Oakland University Library made its online tutorials more focused on users' needs, she discussed how they approached design restructuring. She stated that "the central purpose of this initiative was to refocus the University Library's online learning objects on users while making librarians' creation and maintenance of these objects as simple as possible" (p. 333). While gathering feedback before starting the project, it was discovered that many colleagues were frustrated that the web tutorials were not user-centered. Most of the time students had to navigate to another webpage to find the tutorials. Hess's colleagues also wanted tutorials that could be edited and updated without having to re-create a new video. Most of the free and lower-end technology tools allow users to create a video but with few if any editing features. Regarding their previous learning objects, she observes that "librarians had created the online learning resources with a variety of technology tools, from the very labor-intensive and interactive (i.e., Adobe Captivate) to the free and quick screencapture (i.e., TechSmith's Jing); these tutorials had one thing in common almost regardless of the format: they were difficult (or, in some cases, impossible) to edit after creation" (p. 334). Having accessible tutorials through a variety of means and devices is necessary. More important, with a changing library environment, having the ability to edit the videos is vital when updates are made on a regular basis.

There are a variety of approaches to creating videos with just as many purposes. Some libraries are using videos to supplement instruction, assist students when the library is closed, provide a review, or introduce a new service. The common theme in reviewing this literature is that no matter what approach libraries take in creating videos, they all want to engage their students with videos that inform and are visually appealing. It is obvious that libraries need time to plan, identify their resources and skill sets, and become informed about best practices before moving forward with a video project. Technology has allowed libraries to grow and be creative in the area of tutorial development. Creating library video tutorials is an exciting experience for those involved, but it also has its challenges.

19.4 METHOD: PRODUCING A USEFUL AND INFORMATIVE VIDEO

The process for updating the tutorial on how to avoid plagiarism was based on a successful project in which librarians reimagined the online information literacy tutorial. Two librarians applied for and received an internal grant at APSU, the Revitalize Academic Success Initiative (RASI), to redesign the library component of the FYE course, APSU 1000. The library used the six-module Library Information Literacy Tutorial (LILT) based on the Texas Information Literacy Tutorial and Searchpath. LILT introduced students to information literacy concepts before a one-shot, face-to-face library research day in which librarians met with students and showed them how to find the resources to complete a career research assignment required of all first-year students.

Librarians took the outdated web-based tutorial they used for years that covered types of sources, search strategies, using the catalog and databases, searching the web, and citing sources and created a series of YouTube videos customized specifically for APSU. They also prepared transcripts of the videos. After production, the librarians embedded the video links, transcripts, and the accompanying quizzes in Desire2Learn (D2L), the university's learning management system (LMS). The result was a widely successful summer project culminating in a pilot program in 20 courses. Based on feedback and surveys in APSU 1000, students loved the tutorial and found the video versions much more entertaining and engaging. More important, the student scores on the accompanying quizzes went up across the board, by as many as eight points a question.

Once librarians committed to redesigning the library's antiplagiarism instruction in the same vein as LILT, they needed to determine the vital material to be covered. Librarians knew they wanted to include basic content about plagiarism such as definitions, prevention strategies, and when and what to cite, but they also needed to introduce the research process to students to help put their assignments in context. It was also important that this new project be customized specifically for APSU students by focusing on the institution's policies and procedures, services and resources, and faculty and staff. The librarians incorporated information from a customized presentation that one librarian taught for academic honesty sessions for APSU 1000 classes as an instructor covering information literacy concepts such as the ethical use of information, as well as what motivates students to cheat. Librarians wanted to emphasize the idea of personal responsibility and accountability in this new effort to let students see the significance of making the right choices from the start.

As indicated in Fig. 19.2, the video project flowchart tracks the project from start to finish. Librarians worked in three stages: preproduction, production, and postproduction. The most time-consuming part of the preproduction stage was getting the team together to create a plan. This team included an instruction librarian, digital services librarian, and the digital services assistant. Together they began meeting on a regular basis to develop learning objectives and outcomes based on the content from the old tutorial and the new material they wanted to include. A proposed calendar was created with deadlines to keep the team organized and accountable for the work.

The team elected to use the Adobe Creative Suite 6 (CS6) software because it included all of the tools needed for the project from Photoshop for graphics to After Effects for production and editing. Of the 15 CS6 applications, half of them were used in the creation of the videos as seen in Fig. 19.3. The team members discovered quickly that there was a steep learning curve in using the software, but they consulted Adobe's online tutorials and worked through issues by trial and error to get the desired effects. At that time, it cost $600 for one license that could be installed on two computers. Adobe now provides the Creative Cloud version for educators and charges a monthly fee for access to the software, which the team highly recommends (Adobe, 2015).

The instruction librarian led the team in the creation of the scripts, writing and documenting every detail to include the synopsis, direction, and at times animation. For example, the scene for the beginning of the video clip is set up before any narration occurs. There is a description from the student's point of view near the top of the page where the script takes a troubled student back in time. The viewer sees the student sitting in the waiting area of the dean of students holding his failed paper waiting to be called into the office. This setup leads to the description of the first

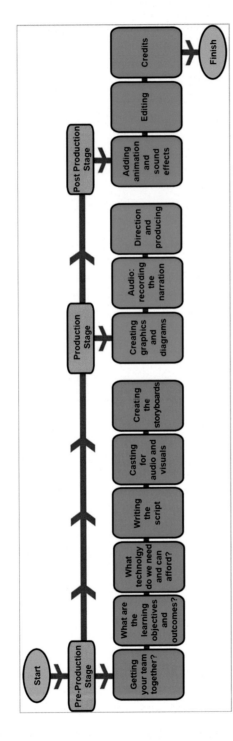

Figure 19.2 Video project flowchart.

Adobe Creative Suite 6

Ps   Photoshop

Au   Audition

Ae   After Effects

Pr   Premiere Pro

Br   Bridge

Figure 19.3 Creative Suite 6 applications used.

PLAGIARISM TUTORIAL OUTLINE

<u>SYNOPSIS:</u>
POV is a student who is in trouble for plagiarizing and imagines going back in time to learn more and make the right choice.

APSU campus; student is waiting in Dean Singleton's Office, holding a paper with a big red "F" on it. "Plagiarism?! I am so busted. What did I do wrong? I wish I had paid more attention. I wish I had known! I wish I could GO BACK!"

REWIND – VIDEO OF THE STUDENT WALKING OUT OF UC AND INTO LIBRARY

"Friendly Voice" (Sean the Librarian)

Research is a process. It takes time. It involves choosing a topic, devising a question or thesis, collecting research, identifying resources, synthesizing your research, and expressing your findings. Whenever you refer to the work of others in your assignments, you must indicate that you consulted those sources and make it easy for your readers to find them for three important reasons:
    So they can verify your interpretation of the sources
    So they can use the sources in their own work
    So you can avoid academic dishonesty by giving credit to the original author.

Figure 19.4 Plagiarism video tutorial script outline.

voice. Notice the positive use of the "friendly voice." A sample section of the script is shown in Fig. 19.4.

Once the scripts were written, the digital services librarian led the team in creating storyboards, a visual representation of the video, using Microsoft Word. The storyboards were extremely important when it came to visualizing the video and creating a graphical representation based on the written script. This was another time-consuming effort but ultimately worth the investment. The storyboards set up the plan for production by pinpointing the timing of audio and transitions, arranging images and text, and inserting animation. To illustrate, there were 18 horizontal $8\frac{1}{2}$- × 11-inch storyboards with images on the top half and notes on the bottom half for the first 3 minutes of video. At first the storyboards were used digitally during the planning stage but were printed when used again during the postproduction editing stage (Fig. 19.5).

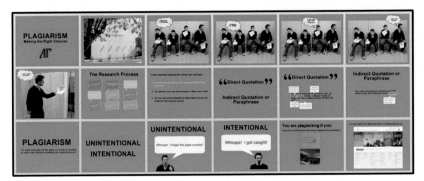

Figure 19.5 Plagiarism video introduction storyboards.

Recording the audio became the next challenge. The team learned a lesson during the production of the LILT videos that proved helpful during this project. In recording the narration for the LILT videos, team members used a microphone that was attached to a headset to capture the audio, but it picked up extra noises when the narrator inhaled and exhaled or said any word with the letter "p" in it (the beginning of "p" words sound like miniexplosions!). The team tried to cover the microphone by snipping the end of the finger section of an archival glove and placing it as a filter over the device. Although there was a slight improvement, the team decided it was not professional enough for the video production. After investigation into this noise, they purchased a new microphone and a pop filter that fixed the extra noises. A low-end studio microphone, a Blue Yeti USB microphone, was purchased for professional sound in subsequent video projects.

The digital services assistant led the production and editing efforts using Adobe After Effects. This was the most complicated application to use because the content is created in a series of layers, key frames, and numerous panels that resemble instruments in the cockpit of an airplane, as you can see in Fig. 19.6.

At this point, the audio is synchronized to the images and the animation is created. There is no such thing as an easy edit in this stage. However, the software will allow a brief preview of the video before moving on to the next frame or section. The digital services librarian served as a director during this stage and worked closely with the team to ensure the video remained true to the original script. It took approximately 3 weeks to complete the video from start to finish, with additional time being provided for a screening of the video to a focus group for feedback.

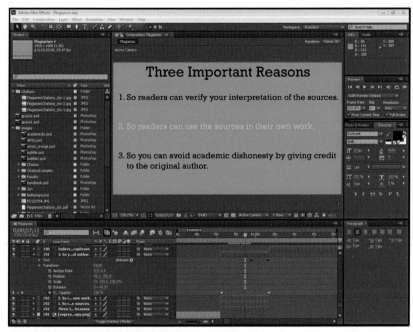

Figure 19.6 Adobe After Effects editing console.

When the video segment was completed and ready for viewing, the team rewrote the quiz questions from the original "Plagiarism: The Crime of Intellectual Kidnapping Online Tutorial." Librarians added new questions, addressing APSU policies and procedures and the research process, as well as basic academic honesty concepts that students should know as college freshmen. Using the D2L LMS, the video link, transcript, and quiz were uploaded in the same module. When students access the module, they can choose to access the content by watching the video embedded from YouTube or by reading the transcript. These options work for students with different learning preferences. After students access the content in their desired format, they are prompted to take the quiz via D2L.

19.5 LESSONS LEARNED: LOOKING BACK AND MOVING FORWARD

Organization is a key component throughout the entire project. It became clear at the beginning of the production phase the team needed a central staging area to access and maintain files. A folder was created on the library's server that was accessible from the two computers being used by the team. One computer was used for production, while the other was used to create images in Adobe Photoshop, search for Creative Commons audio files and sound effects, and make available the script and storyboards. In addition to aiding in collaboration, this provided a permanent location to store the files, which the team realized would be important when future edits were necessary. Rather than producing a new video when updates are required, the team could simply update the current video, which only required access to the original After Effects composition, as well as all of its constituent photo, sound, and video files.

The team encountered problems naming files and moving them to the central location on the server. If an edit was made to an image such as fixing a typo or changing the size, it was important that the new image be identified so that both users were working from the correct files. If the team changed a file, they removed the old file rather than updating it. This kind of discipline with the central location helped during the original editing, as well as when performing updates to the videos.

The team learned quickly about editing video once the project was underway. The first lesson was realizing that users can edit video using two Adobe CS6 applications—Premiere Pro and After Effects. Premiere Pro creates full production videos for film, TV, and the web. This video-editing program puts video clips together and can add text and effects, create transitions, and even edit audio. After Effects is used to create digital motion with graphics, visual effects, and compositing software. It creates special effects for titles, shapes, and two-dimensional animations. After Effects is the better tool for this job, as Video School Online notes, "although you can create motion in text, images, and any other layer in Adobe Premiere Pro—it is a lot easier to do in After Effects" (Video School Online, 2016). The team found that if it wanted to edit straight video, it would use Adobe Premiere Pro; to add special effects, however, it used After Effects. Adobe Dynamic Link allows the creator to use both programs at the same time, enabling the import of After Effects compositions into Premiere Pro. This is a powerful combination of tools! Using just one application limits the videos' potential.

On a large video project it is important to pay attention to timestamps. The plagiarism video has a separate audio file, a recording of the written script, that is synchronized with the video. If an edit is made to the video, it will affect the audio, so it is important to make sure the edits are clearly noted with timestamps. This way adjustments can be made appropriately to keep files in synch so the video is seamless. Remember that every edit made affects the video from that point on. Both the Adobe After Effects and Premiere Pro software visually show the layers of files on the timeline and make it possible to edit precisely. The need to "change time" became important especially when making edits.

19.6 ASSESSMENT: MEASURING IMPROVEMENTS

Once the video and the quiz questions were completed, the team piloted the video project in the APSU 1000 course. Members created a staging area named Sandbox in D2L by incorporating the content, which included the video link, transcript, and quiz. They loaded it to Sandbox so that it could be copied into specific sections of APSU 1000. The librarians accessed the D2L tools to view statistics and create reports for the quiz results. They were able to view and analyze user stats, question stats, and question details. Librarians used these statistics in a variety of ways, one of which was to tailor instruction to a specific course. The question stats were used to assist them in customizing academic honesty instruction for each APSU 1000 section because the librarians learned that every class had different needs. For example, one class might need additional instruction on defining plagiarism, while another might need help creating citations. If question statistics fell below 75%, then the librarian knew they should plan to explain that topic in class to reinforce the academic honesty concept to students. This point is illustrated in Fig. 19.7.

The question details option in D2L assisted the video project team in analyzing the content in the tutorial and the composition of the quiz questions. If students answered specific questions incorrectly, the team reviewed the video to see if the way the information was delivered needed to be addressed. Was the content clear? Did the question need to be reworded? Moreover, the team evaluated the structure of the quiz questions and had an opportunity to edit, delete, or create new quiz questions based on these statistics. The APSU 1000 course instructors liked the fact that the quiz results would automatically be exported to the gradebook in D2L (Fig. 19.8).

The pilot of the new "Plagiarism: Making the Right Choices" video tutorial was so successful that word quickly spread around campus. The librarians were getting requests from faculty members who did not teach APSU 1000 but wanted to be able to use the video tutorial in their courses. The digital services assistant developed a web application for the

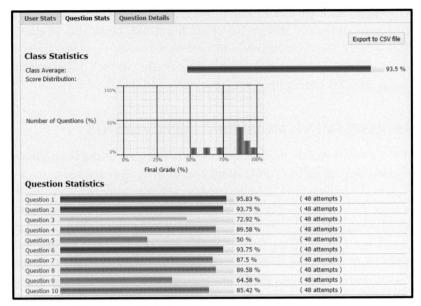

Figure 19.7 Question stats, D2L view.

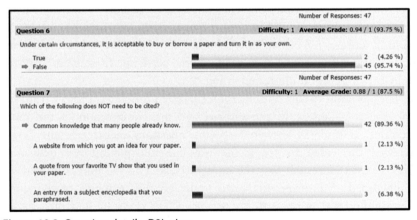

Figure 19.8 Question details, D2L view.

tutorial that is hosted on the library's website. Instructors can enroll their classes by completing an online form providing their e-mail address, name, course name, and section number. Once the course was available, students accessed the tutorial but were prompted to register before they accessed the video and quiz. An access point for instructors was also

Plagiarism Quiz Results

HIST 2010-09e -

Sort by: First Name | Last Name | Score

Name	Email	Date/Time Taken	Score(s)
		August 25, 2015 8:47pm	100
		August 28, 2015 2:28pm	100
		August 25, 2015 7:20pm	100
		August 27, 2015 6:58pm	100
		September 11, 2015 1:19pm	90
		August 30, 2015 3:47pm	100
		August 28, 2015 8:25am	100
		August 28, 2015 11:03am	100
		August 27, 2015 9:50am	100
		September 06, 2015 9:56pm	100
		September 01, 2015 7:16pm	100
		August 26, 2015 12:15pm	100
		August 30, 2015 11:53pm	80
		August 25, 2015 4:49pm	100
		August 24, 2015 6:54pm	100
		August 28, 2015 9:36am	100
		August 24, 2015 4:04pm	100

Return to Class Selection

Figure 19.9 Plagiarism quiz results, library website view.

created, where they could retrieve quiz results for their classes by semester and class, and then have an option to sort by names or scores as illustrated in Fig. 19.9.

In addition to the more than 1600 APSU 1000 students taking the tutorial via the course D2L shells, the library fielded requests from faculty members through the library's website to enroll more than 50 other courses with almost 1000 students. Classes enrolled ranged from ENGL 1010— English Composition I, which is the freshman composition course, to graduate courses in education, in which the instructor thought the students could use a good refresher on what plagiarism is and how to avoid it.

Assessment became important for students, instructors, and librarians. First, students receive immediate feedback with every submission, allowing them to evaluate their answers. For example, if students incorrectly answer a question, they see the specific question, their answer, and the correct answer with an explanation. The default setting for the quiz is to allow students unlimited attempts to take the quiz; however, this can be changed by individual instructors. The quiz is automatically graded, and the grade can be accessed by students and instructors. These options are similar in the web version. The only difference is that the instructor can sort the results by best score or all scores.

Librarians wanted feedback about the new antiplagiarism content and how it was presented in the academic honesty video, and they needed to be able to measure what students learned from the video based on the quiz results. The project team planned for a focus group composed of faculty, staff, and students to assess the video clip to include the content, graphics, audio, credits, and overall presentation. Members of the focus group were required to complete the old "Plagiarism: The Crime of Intellectual Kidnapping Online Tutorial," which included taking the accompanying quiz, before watching the new video, "Plagiarism: Making the Right Choices." The team was proud of the video and wanted to treat the focus group members as if they were attending a movie premier, so they provided free popcorn, snacks, and drinks at the event. The focus group viewed the video, had a quick break, and then returned for a 90-minute discussion about the video.

The focus group was led by the instruction and digital services librarians who facilitated the open-ended discussions that provided feedback and recommendations on improving the video. Further, librarians observed the group's reactions, watching to see if the group laughed at the right times, looked engaged, or indicated obvious issues with the video that the production team did not notice. Based on the focus group's feedback, the team did make some edits to the video clip. The feedback was helpful because it provided suggestions to improve the video and recommendations for future enhancements. For example, some of the feedback stated that the animation was too fast and needed to be slowed down. The team working on the project did not notice the animation was too fast because they were too close to the video. Once they slowed the animation down, they saw how it helped to improve the overall quality of the video.

19.7 CONCLUSION: ALL GOOD THINGS...

In 2004, librarians at APSU realized they needed help informing students about plagiarism leading them to adapt the older "Plagiarism: The Crime of Intellectual Kidnapping" tutorial. During that time, the FYE APSU 1000 course, with its required library component, became mandatory for every freshman with 12 or fewer credit hours. Before this tutorial, librarians needed to introduce information about academic honesty, as well as information literacy concepts, in a 55-minute library session. What started out as an adapted textual and graphic tutorial that concentrated on the punitive aspects of plagiarism has turned into an informative, fun, and engaging video that focuses on students making the right choices with their time and efforts during the research process. Using updated content and newer technologies, the "Plagiarism: Making the Right Choices" video was created and has proven to be successful based on quiz scores and feedback from students and instructors alike. Having the module accessible in both D2L and on the website allows for greater student participation from a variety of courses. Likewise, students can access the video on multiple devices and have a choice as to whether to watch the video, read the transcript, or both.

The librarians recognize that the video is essentially a work in progress and will require ongoing maintenance because technologies, policies, and information evolve on a regular basis in higher education. The production team meets a few times a year to discuss the tutorial, and edits are made during the summer to minimize time constraints and because there is less traffic on the website. Due to their previous experience in successfully redesigning the LILT tutorial into informative and engaging video clips and embedding it in the university's LMS, librarians were able to take advantage of the opportunity to undertake a similar project for our academic honesty module. Just as researching, reading, evaluating, and writing are important to being an information-smart individual so, too, is being ethical and "making the right choices."

REFERENCES

Adobe Systems Incorporated. (2015). *Adobe creative cloud.* Retrieved from < http://www.adobe.com/creativecloud.html>.

Austin Peay State University. (2011). Student academic misconduct (Policy 3:035). *Austin Peay State University Policy and Procedures.* < http://www.apsu.edu/sites/apsu.edu/files/policy/3035.pdf>.

Austin Peay State University. (2015). Academic and classroom misconduct. *Student disciplinary system.* < https://www.apsu.edu/sites/apsu.edu/files/student-affairs/Handbook_ 16_Academic_and_Classroom_Misconduct.pdf>.

Bowles-Terry, M., Hensley, M. K., & Hinchliffe, L. J. (2010). Best practices for online video tutorials in academic libraries. *Communications in Information Literacy, 4*(1), 17–28.

Clapp, M. J., & Ewing, S. R. (2013). Using students to make instructional videos: Report of an initial experiment. *The Reference Librarian, 54*(4), 332–340. Available from: <http://dx.doi.org/10.1080/02763877.2013.806192>.

Hess, A. N. (2013). The MAGIC of web tutorials: How one library (re)focused its delivery of online learning objects on users. *Journal of Library & Information Services in Distance Learning, 7*(4), 331–348. Available from: <http://dx.doi.org/10.1080/1533290X.2013.839978>.

Plagiarism. The crime of intellectual kidnapping online tutorial. Retrieved from <http://tutorials.sjlibrary.org/tutorial/plagiarism>.

Scales, B. J., Nicol, E., & Johnson, C. M. (2014). Redesigning comprehensive library tutorials: Theoretical considerations for multimedia enhancements and student learning. *Reference & User Services Quarterly, 53*(3), 242–252.

Singleton, G. (2014). *Violations: Five year comparison by academic year.* Clarksville, TN: Austin Peay State University Division of Student Affairs.

Tennessee Board of Regents (2011). *Academic misconduct provisions in institutional student disciplinary policies.* Nashville, TN: Tennessee Board of Regents.

Thornton, D. E., & Kaya, E. (2013). All the world wide web's a stage: Improving students' information skills with dramatic video tutorials. *Aslib Proceedings: New Information Perspectives, 65*(1), 73–87. Available from: <http://dx.doi.org/10.1108/00012531311297195>.

Video School Online. (2016). *What is the difference between Adobe Premiere Pro and Adobe After Effects? And why I use both! Video School Online.* <http://www.videoschoolonline.com/what-is-the-difference-between-adobe-premiere-pro-and-adobe-after-effects-and-why-i-use-both> Accessed 13.01.16.

CHAPTER 20

Adapting to the Evolving Information Landscape: A Case Study

20.1 THE EVOLUTION OF AN INFORMATION LITERACY COURSE: A CASE STUDY

Fontbonne is a small, 4-year, liberal arts Catholic university of approximately 1500 students, founded in 1923 by the Sisters of St. Joseph of Carondolet. It is located in the heart of St. Louis, Missouri. Primarily a commuter campus, the university community includes traditional, nontraditional, and international students, with a student-to-faculty ratio in 2015 of around 11:1 (Fontbonne, 2015). In 2002, all librarians began teaching an information literacy course as part of the general education requirements of the university.

20.2 INFORMATION LITERACY COURSE HISTORY

In the year 2000, prior to the arrival of the current librarians, many changes occurred at the university that led to the creation of the information literacy course:

- Fontbonne automated its library holdings for inclusion in the online catalog of the Missouri Bibliographic Information Users System (MOBIUS), the recently developed statewide consortium of academic libraries.
- The Coordinating Board for Higher Education of Missouri adopted "Guidelines for Student Transfer and Articulation among Missouri Colleges and Universities," and Fontbonne became a signatory institution. This is also referred to as the Missouri State Articulation Agreement.
- The addition of new technology on campus, including student computers and digital resources, spurred a realization among the

Distributed Learning. © 2017 T. Maddison and M. Kumaran. Published by Elsevier Ltd. All rights reserved.

English and Communication faculty that additional coursework was needed to help students develop new skills to succeed in the changing information landscape.

According to the chair of the English department at the time, "rhetoric instructors have traditionally taught library research skills. However, as electronic resources have become more pervasive, teachers of rhetoric find themselves increasingly lacking in expertise, even as teachers and students alike devote more and more time to keeping up with the rapidly changing virtual landscape" (B. Moore, personal communication, December 12, 2006). In response to changing needs, the library director at the time proposed the development of an information literacy course that would be taught completely online using Blackboard. It was approved by the Faculty General Assembly in 2002 and officially adopted as a general education requirement. Since that time, the course has solidified the importance of information literacy instruction and the assessment of student performance in managing information. Consequently, the one-credit information literacy course remained an integral part of general education requirements through the university-wide general education revision in 2012.

Fontbonne's information literacy course as it currently exists is an 8-week, one-credit online class—Information Literacy in Higher Education (LIB199). Blackboard is the learning management system (LMS) used to deliver content that librarians develop, teach, and regularly update. Students complete asynchronous modules that incorporate a required textbook (Badke, 2014), videos, discussions, podcasts, narrated PowerPoints, articles, and weekly assessments. LIB199 provides students with a basic introduction to skills and concepts for using the library and other resources to locate, use, and evaluate information. Originally, the Association of College and Research Libraries (ACRL) Information Literacy Competency Standards for Higher Education (2000) and the Missouri state general education goal for managing information shaped the content of the course. The role of the ACRL's Framework for Information Literacy in Higher Education will be addressed later in this chapter. Changes to course titles and descriptions over the past 12 years reflect the numerous revisions, some more extensive than others, and provide evidence of the ever-evolving nature of the curriculum and its subject matter (Appendix A).

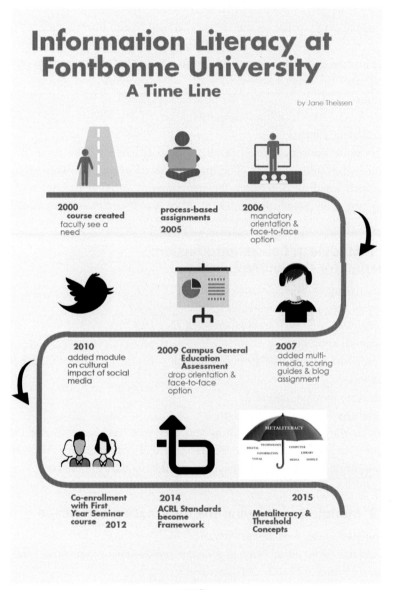

Information Literacy at
Fontbonne University
A Time Line

by Jane Theissen

2000
course created
faculty see a
need

process-based
assignments
2005

2006
mandatory
orientation &
face-to-face
option

2010
added module
on cultural
impact of social
media

2009 Campus General
Education
Assessment
drop orientation &
face-to-face
option

2007
added multi-
media, scoring
guides & blog
assignment

Co-enrollment
with First
Year Seminar
course   2012

2014
ACRL Standards
become
Framework

2015
Metaliteracy &
Threshold
Concepts

powered by

20.3 A QUESTION OF FORMAT

In its initial stage, prior to the authors' arrival to the library, the information literacy course was developed and delivered in an online format. As the course evolved, alternative optional modes of delivery were offered to students:
- online,
- face-to-face with online components, or
- equivalency exam

The same content is used in all modes of delivery. The focus of this chapter is the online format. The modules listed in the following sections, along with learning outcomes and assignments, provide examples of the course structure (Note: the course number was INT199 at this point):

20.3.1 Module 1: Course Introduction: Preparing for the Journey

After completing this module, students will:
- *explain their personal information seeking preferences*
- *acknowledge that they are ready to begin INT199*

Assignments:
1. Watch the video *INT199: Introduction*. (Transcript is provided below the video.)
2. Read textbook Chapter 1.
3. Read the Parlez Café handout.
4. Go to the **Discussions** button and add your responses to the Parlez Café Discussion Board. The Parlez Café will count as your Attendance Assignment and let me know you are ready to start the course.

20.3.2 Module 2: Surveying the Information Landscape

After completing this module, students will:
- *articulate specific ideas pertaining to the ever-evolving information landscape and the future of libraries in a discussion*
- *retrieve, analyze, evaluate, and use information*
- *evaluate website content for purpose, authority, content, accuracy and currency*

Assignments:
1. Read textbook Chapter 2.
2. Read *Surveying the Information Landscape* in this folder.
3. Watch YouTube video *Evaluating Sources on the World Wide Web*.

4. Watch Films on Demand *TED Talks: Eli Pariser—Beware Online "Filter Bubbles."*
5. Complete Blog assignment.
6. Listen to NPR Podcast *Information Overload.*
7. Complete Skills Check over Module 2/Chapter 2.

20.3.3 Module 3: Navigating the Wayward Web

After completing this module, students will:
- *articulate ideas and knowledge gained about the pros and cons of Facebook and Twitter*
- *support opinions on issues related to social networking*
- *utilize web sources as appropriate*

Assignments:
1. Read textbook Chapter 3.
2. View the handout: *Databases vs. Websites: The Difference.*
3. Read Facebook articles.
4. Complete Discussion Board on "Facebook & Privacy."
5. Read Twitter articles.
6. Complete Discussion Board on "The Cultural Impact of Twitter."
7. Read *Surfing the Web* in this folder.
8. Watch YouTube video on Google Scholar.
9. Complete Skills Check over Module 3/Chapter 3.

20.3.4 Module 4: Searching the Information Landscape

After completing this module, students will:
- *identify the differences between library databases and the "web"*
- *recognize the advantages and disadvantage of keyword searching*
- *demonstrate an understanding of Boolean operators AND, OR, NOT to conduct searches*
- *demonstrate an understanding of the concept of truncation and use it appropriately and effectively*

Assignments:
1. Read textbook Chapter 4.
2. View YouTube video *Database Search Tips.*
3. View YouTube video *Boolean Logic.*
4. Complete Skills Check over Module 4/Chapter 4.

20.3.5 Module 5: Using the Library Catalog

After completing this module, students will:

- *locate a book in the Library*
- *use the MOBIUS catalog to access books*
- *know when to use Fontbonne's online catalog and the MOBIUS catalog*
- *increase awareness of services and facilities at the Jack C. Taylor Library at Fontbonne*
- *explain how e-readers are changing how we read*

Assignments:

1. Read textbook Chapter 5.
2. View tours of the Jack. C. Taylor Library:
   - Tour the Library Building at Fontbonne University
   - The Digital Library via the Web
3. View the web page *Understanding LC Call Numbers.*
4. Read the New York Times article *"Do we still need libraries?"*
5. Click on the **Discussions** button and complete the Discussion Board on the "Do we still need libraries?"
6. Click on the **Discussions** button and share your responses to the "Print vs. Online Reading" articles.
7. Complete Skills Check over Module 5/Chapter 5.

20.3.6 Module 6: Getting your Hands on Articles

After completing this module, student will:

- *select appropriate multi-disciplinary and/or subject specific databases for topic searches*
- *recognize academic, scholarly, peer-reviewed sources*
- *identify and differentiate between resource types and formats*
- *locate, access and navigate periodical indexes to retrieve results relevant to a specified topic*
- *utilize **Journals A-to-Z** to determine the availability of journal titles and retrieve full-text articles*
- *utilize interlibrary loan to request articles unavailable from the Fontbonne Library*

Assignments:

1. Read textbook Chapter 6.
2. View YouTube videos:
   - Scholarly vs. Popular Articles.
   - EBSCOhost Advanced Search Tutorial.
   - PsycINFO on EBSCOhost: Subject Field Search.

3. Complete the *Points of View Reference Center* assignment.
4. Review handouts and tutorials:
   * Peer-reviewed scholarly journals vs. popular magazines.
   * Journals A-to-Z handout.
   * Requesting Materials via Interlibrary Loan.
5. Complete Skills Check Module 6/Chapter 6.

20.3.7 Module 7: Finding Answers-Digital & Print

After completing this module, students will:
* *locate and retrieve a book from the shelf of the Library using the call number*
* *use the index of a book to find the answer to a question*
* *select the appropriate resource to fulfill the information need*
* *evaluate resources and apply them appropriately to complete an assignment*
Assignments:
1. Read textbook Chapter 7.
2. Complete *Exploring Global Issues in Context* assignment.
3. Complete Skills Check over Module 7/Chapter 7.

20.3.8 Module 8: Staying Honest

After completing this module, student will:
* *identify types of sources based on a citation*
* *grasp how to avoid plagiarism*
* *apply copyright guidelines to information use*
Assignments:
1. Read textbook Chapter 8.
2. Watch the copyright video *Copyright Basics*.
3. Read the excerpt from *the Cult of the Amateur* and the article *Battlefield SOPA* under **Intellectual Property-Use and Misuse.**
4. Read the handout *Intellectual Kidnaping*.
5. Complete the Comprehensive Skills Check. (This Skills Check covers the entire course and will take approximately 2 hours to complete. Plan accordingly.)

20.4 PERCEPTION BECOMES REALITY

Soon after the authors arrived on campus in 2004, it became obvious that students often did not react positively to this course. Misunderstanding the course, many students felt that they knew all about computers and had no need for a course about information literacy. Since they did not understand the definition of "information literacy," many students procrastinated in taking LIB199 until just prior to graduation.

The 2012 general education revision helped address this procrastination problem. Students are now enrolled in LIB199 during their first semester on campus. Faculty supported this move because they had come to understand the importance of information literacy and to appreciate the value and relevance of LIB199 to student success in other courses. Students now gain valuable exposure to information literacy skills early in their college careers.

20.5 A QUESTION OF CONTENT

The fact that LIB199 is one-credit leads students to perceive it as an easy course. Although they often had positive comments such as "this course introduced valuable information that will be helpful to me currently as a student and in the future when I begin my career" once they completed the course, they are often shocked initially when they realize there are readings, podcasts, writing assignments, and so on that require higher-order thinking skills and an unexpected amount of time to complete. The question of content is perennial. As instructors, faculty librarians feel constrained by the one-credit limit when contemplating the addition of new content. Student comments on course evaluations routinely indicate that the course requires too much work for a one-credit course. The persistent question is, "What determines the amount and nature of content appropriate for a basic course versus an advanced course?" Recent efforts to raise the credit designation have failed to garner faculty approval. Consequently, course revisions must balance the addition of new material with the credit-hour limitation.

20.6 THE EVOLVING LANDSCAPE

As the information landscape continued to evolve, it became apparent that the cultural effects of social media should be included in the information literacy course. In 2010, readings and discussions that focused on the impact of social media on the wider culture were added, along with this introduction to "Module Three: Navigating the Wayward Web":

Facebook and Twitter represent a tremendous shift in culture. Is the social networking phenomenon redefining privacy? Is the willingness of users to share and overshare information about themselves without realizing how much their personal data is worth to someone's bottom line a good thing or a bad thing? Is Twitter a communications revolution or waste of time? Is Facebook too much in your face grabbing and sharing your preferences with marketers? Learn the pros and cons of social networking and privacy and then decide. Express your opinions in the discussion board forums happening in this module (Ridlen, 2010).

Articles from *Time, Information Today,* and *American Libraries,* along with a podcast from National Public Radio provided content and formed the basis for class discussions. This module became especially appealing to students. Firsthand experience with social media connected students to its impact, not only on society but also on their personal access to information. Course evaluations revealed this to be the favorite part of the course for many students.

Initially, there was little support among faculty for addressing social media within the course in any meaningful way. In fact, one faculty member noted that parents would be appalled to know that students were learning about Facebook in lieu of traditional research skills. The participatory nature of social media, however, continues to have a major impact on how students find and evaluate credible resources. Eventually, faculty came to understand that the collaborative nature of social media and the advancement of technology require updated skills to locate and assess information and are now highly supportive of this addition to LIB199.

20.7 NEW WAYS OF THINKING

Threshold Concepts

Ideas that are passageways or portals to enlarged understanding or ways of thinking and practice in a discipline.

(Framework, June 2014 Draft, Line 25-27)

It was difficult to align content identifying the cultural impact of Facebook and Twitter with ACRL's 2000 standards. Not until 2012 did ACRL call for a task force to revise the standards. The revision process took a couple of years, and the faculty librarians read with interest the various drafts disseminated during that time. The 2015 Framework for Information Literacy for Higher Education addresses the shifts in this ever-evolving information landscape. As many instruction librarians understand, it was a struggle to revise content formerly focused on competency standards into a course aligned with the new framework. It was clear that threshold concepts as presented in the framework are gateways to new modes of thinking. A review of the literature indicated that a simple transfer of the standards to the framework was not advisable or even possible, since the framework requires the instructor to be invested in students' learning rather than to simply measure the transfer of information.

The 2015 framework presented a welcome new approach to course revision that year. Transitioning from the standards to the framework required incorporating elements such as "format as a process," metaliteracies, and "authority as constructed and contextual" as meaningful components of the course.

After much discussion (and reading), the librarians refocused current content to address the six frameworks and student surveys on levels of information literacy skills, pre- and post-course, were added. New content included conversations about metaliteracy, source types, copyright, and net neutrality as explained in the table below.

New framework	Content added
Authority is constructed and contextual	New videos on how to evaluate information and choose your news sources
Information creation as a process	New Prezi on information formats and a PowerPoint on how & why digital and physical formats differed
Information has value	Updated copyright information
Research as inquiry	Revised PowerPoints on using the digital and physical library
Scholarship as conversation	New video on peer review and content on scholarly resources
Searching as strategic exploration	New and updated videos on search strategies

The following modules were developed based on the frameworks and the information in table above.

20.7.1 Module 1: Information in the Digital Age

Student Objectives:

- *Define and explain the advantages of being information literate within the context of a digital environment.*
- *Identify and discuss other literacies associated with information literacy.*
- *Analyze and interpret statistics about students' ability to apply literacy skills.*

Assignments:

1. Read Chapter 1 p. 4–9 of the textbook.
2. Read Chapter 1 p. 22–26 of the textbook.
3. Read the Applying Information Literacy Skills infographic displaying statistics from a study conducted about students and the application of information literacy skills. You will be asked to react to these statistics and reflect on your progress in becoming information literate again in the Final Reflection Assessment at the end of the course.
4. Complete the "Assessing Your Literacies" poll.
5. Introduce yourself to your classmates on the Parlez Café Discussion.
6. Compete and submit Module 1 Skills Check.

20.7.2 Module 2: Determining the Information You Need

Student Objectives:

- *Recognize various ways to access and locate information relevant to a specific discipline, topic, or research question.*
- *Identify and describe digital and physical information source formats.*
- *Identify and define various information retrieval systems.*
- *Distinguish between primary and secondary information sources.*
- *Identify scholarly, popular, and trade publications.*

Assignments:

1. Read all of Chapter 2 of the textbook.
2. Watch Prezi presentation "Information Formats."
3. Watch the "Tour the Fontbonne Digital Library PowerPoint."
4. Complete "Discovering Points of View" assignment.
5. Review "Types of Periodicals" handout.
6. Watch the PowerPoint presentation "Periodicals–Physical & Digital."
7. Complete "Discovering Global Issues" assignment.
8. Complete the Module 2 Skills Check.

20.7.3 Module 3: Accessing the Information You Need

Student Objectives:
- *Utilize Library's Search Discovery and Classic Catalog to locate and access information.*
- *Utilize digital resources to locate and access information.*
- *Utilize effective search techniques to retrieve relevant information on a specific discipline, topic, or research question.*

Assignments:
1. Read all of Chapter 3 of the textbook.
2. View the PowerPoint "Tour Fontbonne Library's Building."
3. Review the handout "Using Interlibrary Loan."
4. Read over the information on the webpage "Understanding Library of Congress Call Numbers."
5. View Infographic "Top 10 reasons for choosing a paper book over an e-book."
6. Watch YouTube video "Database Searching Tips."
7. Watch YouTube video "Boolean Operators."
8. Complete the Discussion "Visiting the Library of Congress."
9. Review the handout "Using Journals A-to-Z."
10. Complete the Module 3 Skills Check.

20.7.4 Module 4: Evaluating the Information You Found

Student Objectives:
- *Articulate specific ideas pertaining to the ever-evolving nature of information.*
- *Utilize specific criteria to evaluate information and information sources.*
- *Retrieve, analyze, evaluate, and use information.*
- *Evaluate website content for purpose, authority, content, accuracy, and currency.*

Assignments:
1. Read all of Chapter 4 of the textbook.
2. Watch YouTube video "Evaluating Resources."
3. Watch YouTube video "Peer Review in 3 Minutes."
4. Watch YouTube video "How to Choose your News."
5. Participate in the Discussion "Getting the (Right) News."
6. Watch YouTube video "Beware Online Filter Bubbles."
7. Complete the "How are Things Going?" survey.
8. Complete the Module 4 Skills Check.

20.7.5 Module 5: Social Networks, Internet Privacy and Net Neutrality — Who Controls the Internet?

Student Objectives:
- *Define data mining and identify data mining techniques.*
- *Examine and critique laws related to data mining of social networks and privacy protection.*
- *Articulate ideas and knowledge gained regarding Facebook and Twitter.*
- *Formulate and support ideas related to net neutrality.*

Assignments:

There is no reading from the textbook for this module.

1. Read the article "Analyzing Data in Social Networks–An Ethical Dilemma?"
2. Read Time article "Advertisement Starring. . .You."
3. Watch the video "Happy Birthday Twitter — You're Blocked."
4. Complete Discussion on "Social Networking and Internet Privacy — Facebook & Twitter."
5. Read article "Welcome to the New Internet."
6. Peruse the website "Network Neutrality."
7. Watch YouTube video "Internet Citizen: Defend Net Neutrality"
8. Complete Discussion on "Net Neutrality: Should there be a "tiered" Internet?"
9. Complete Module 5 Skills Check.

20.7.6 Module 6: Intellectual Property — Use and Abuse

Student Objectives:
- *Apply copyright guidelines to information use.*
- *Analyze the impact of the Internet on intellectual property rights.*
- *Identify types of sources based on a citation.*
- *Analyze and interpret infographic about critical thinking.*

Assignments:

1. Read Chapter 6 pages 189–205 of the textbook.
2. Review website "Copyright Basics: A Beginner's Guide."
3. Watch YouTube "Anti-Piracy Video."
4. Watch YouTube video "Copyright Basics in the Workplace."
5. Read and complete activities on the website "Plagiarism: What Every Student Needs to Know."
6. Watch Films on Demand video "Plagiarism 2.0: Information Ethics in a Digital Age."

7. Watch video "Know Your Citations."

8. Complete Module 6 Skills Check.

20.7.7 Module 7: Final Reflection/Assessment

Assignment:

1. Complete the Final Reflection/Assessment.

The authors agreed that the requirement of a textbook would substantially validate the course from a student's perspective and serve as an important authoritative source for providing content. The decision to search for an updated text turned out to be more difficult than anticipated. Finding a current textbook that emphasized the essential skills of locating, evaluating, and using information ethically was a challenge. Serendipitously, the librarians discovered the fifth edition of *Research Strategies: Finding Your Way through the Information Fog* by William Badke. The authors realized they had heard Mr. Badke speak at a EBSCO-sponsored presentation at the ACRL Conference in Mar. 2015. This endorsed his credentials and validated the selection of his book. Assigned readings and references from the book were added to the course.

20.8 INFORMATION LITERACY AND VISUAL THINKING STRATEGIES

In an effort to broaden students' exposure to credible sources of information beyond Fontbonne's Library, an assignment that asked students to browse the Library of Congress website, find something of interest, and share it with their classmates in a discussion was added to LIB199 in 2014. The assignment, "Visiting the Library of Congress: A Discussion" in module four, illustrates the connection between information literacy and a visual thinking strategy (VTS). According to Nelson (2015), "Visual Thinking Strategies is a teaching method that encourages interaction with a work of art through a facilitator led discussion using three simple questions" (p. 26). After attending a conference session in Oct. 2015 titled "Visual Thinking Strategies: From Museum to Library," the authors realized that visual thinking strategies had already been incorporated into the online information literacy course within the Library of Congress assignment.

VTSs are easily transferrable from the field of art to the field of information literacy. The same types of questions asked by students using

VTSs can also be applied in this assignment. As students explore the Library of Congress website, they ask themselves:
- What visual access points and links do I see?
- Will these links lead to more useful information?
- What more can I find out about this?
- Where else do I look?

VTS invites students to be more aware of their surroundings, including websites and other visual sources of information. It can enhance the skills students need to relate to website navigation and evaluation because the process requires them to observe, analyze, and think about what they see. As Nelson (2015) notes, "The use of VTS in information literacy sessions creates an interactive and engaged learning environment that simulates in-depth investigations while encouraging learners to make personal connections" (p. 26). One student shared her "aha" moment:

> I found the complete works of Abraham Lincoln's papers at the Library of Congress. The documents they have are all his speeches, notes, and anything he wrote down. I found this interesting because I love history and really like Abraham Lincoln.... One of my favorite things was that he kept his speeches short and ... to the point where people listening to him would still be intrigued. Some of the papers that they have include a draft of the Emancipation Proclamation, draft of his second inaugural Address, and his memorandum saying what he expects of being defeated in the ... upcoming presidential election.... [T]o see those papers he wrote in real life would be amazing. It is like breathing in history when you are that close.
>
> http://memory.loc.gov/ammem/alhtml/malhome.html

Apart from enabling VTSs, the Library of Congress assignment provides an opportunity to experience primary sources. The Library of Congress website is a treasure house of primary sources and historical documents. Instructors in various disciplines often require students to use primary source materials in assignments but provide few concrete examples, and students have often noted that they are confused about how to identify these materials. To "connect the dots" for students, librarians point out that what students have discovered (and find so interesting!) includes primary source materials in the form of speeches, videos, photographs, reports, and audio files. As instructors, it is gratifying to witness a transformation of attitude as students complete this assignment. They articulate an unexpected enthusiasm on discovering historical documents and artifacts to be living vibrant resources. This assignment clearly enables students to demonstrate an understanding and appreciation of primary sources. Many students often express a desire to visit the Library of Congress in person after completing this assignment.

20.9 ASSESSMENT

General education assessment has been an annual fixture on Fontbonne University's campus for many years. Universities nationwide have noted the challenge assessment poses (Patthey & Thomas-Spiegel, 2013; Beyer & Gilmore, 2007). It takes an enormous amount of effort and time to gather, compile, and analyze assessment data that can be used to enhance course content to impact student achievement. In every course that is taught as part of the general education curriculum at Fontbonne (including LIB199), the instructor selects an artifact or assignment that addresses the learning outcomes of the course. The artifact is assessed to ensure that students are meeting course learning outcomes, and this assessment data is forwarded to the University's Office for Institutional Research.

As noted previously, LIB199 is rooted in the ACRL standards (2000) and the Missouri State Articulation Agreement (in the competency for managing information). From these two sources, the following course learning outcomes were developed by library instructors and were approved by the administration. These outcomes continue to guide course content and assessment:

Students who complete this course will be able to:

- articulate or discuss the ethical and legal use of information,
- demonstrate the ability to access information to address an issue,
- discriminate among types of information for their intended purpose, and
- synthesize information and formulate a response to address an issue.

Alignment of threshold concepts of the 2015 framework with these learning outcomes is presented in Appendix B.

The assessment artifact used in LIB199 is the comprehensive skills check (in Module 7), a final test that requires students to demonstrate they can successfully apply the skills they have acquired during the course and reflect on what they have learned. As a cumulative assessment, questions require students to access information in a database, provide metadata on catalog records, differentiate between primary and secondary sources, explain the use of Boolean operators and truncation, identify effective search strategies, select the appropriate database for a given topic, and demonstrate an understanding of how to use information ethically to avoid plagiarism and abide by copyright law. This assessment constitutes at least one-third of the total grade, so students cannot pass the course without completing the comprehensive skills check.

In LIB199, as in all assessments on campus, grading and assessment are separate processes. As already noted, the comprehensive skills check is the artifact and assignment used for assessment in LIB199. After all coursework has been graded, the comprehensive skills check for each student is assessed according to a performance-based scoring guide that was developed by our instructional librarian (see Appendix B). Learning outcomes are aligned with weekly modules in the course. Comprehensive skills check questions address the learning outcomes in each of these modules, with the total number of points on the comprehensive skills check evenly divided among the four learning outcomes (see Appendix B).

20.10 ASSESSMENT ANALYSIS

In 2012, several part-time librarians were hired to staff evening and weekend shifts and work on numerous reference projects. In addition to providing reference service, these librarians have become valuable partners in the assessment process. Processing assessment data for 400 students each year would be impossible without these part-time librarians, who dedicate time to compile, review, and disseminate this information. Collecting assessment data is pointless unless it is used to improve the course. While the data is also forwarded to the university's Office of Institutional Research, analysis of this data indicates where students are excelling and where they are struggling. Prior to each semester's revision, assessment data is internalized and specific changes focused on improving student achievement are made, as shown in this example from the report on the 2012–13 assessment data compiled by Peggy Ridlen (2013) reference and instruction librarian:

20.10.1 Summary

We have data from the comprehensive final skills check for 332 students in 18 sections (nine fall, seven spring, and two summer).

Students are doing consistently well in most areas, roughly two-thirds (66%) to more than three-quarters (86%) of them scoring a four (4) in four of the six areas (total score 84%, comparison 66%, evaluation 86%, and final reflection 77%). Outcome 5 (synthesis) barely misses the two-thirds goal, but it is the first outcome—ethics and plagiarism—that stands out as our biggest concern. Although more than three-quarters (86%) still score in the top half of the range, a considerable majority are not performing as well as we would like in this important area.

20.10.2 Recommendations

The students' struggles with the legal and ethical materials (Learning Outcome #1) are not entirely new. Part of the problem may lie in its placement at the end of course and the fact that it is not addressed by a specific assignment or another Skills Check. It's an important area, however, and we would like to see improvement over the next year. We recommend a review of the question pool and the content from Module Eight to address the weakness.

In addition, at the request of administration, a long-term assessment survey was recently added to upper-level bibliographic instruction sessions. The survey asks students to reflect on their level of information literacy after having completed LIB199. Data from this survey which is yet to be received and analyzed, will provide insight into the retention of information literacy skills learned in LIB199.

20.11 THE EVOLUTION CONTINUES

Over the last decade, as librarians have taught and refined Fontbonne's undergraduate information literacy course, recurring requests from graduate students in all disciplines of the university for information literacy instruction have been received. With the merging of computer, visual, information, and numerous other literacies into the broader concept of metaliteracy, the time seemed right for Fontbonne librarians to move forward with the idea of an advanced research course on campus. The website of the National Forum for Information Literacy (2015) reinforced the need for this offering:

In today's workplace, be it a classroom, retail store, government agency, ware-house, and/or corporate headquarters, information literacy is a key driver, if not the key driver, in achieving success in whatever job you perform now and, most likely, in the future... Information literacy practice is a college and career readiness success strategy that develops critical and creative thinkers, learners and workers who know how to problem solve in conjunction with utilizing a variety of information resources that produce quality results (Workplace Information Literacy).

While still a work in progress, the goal of The Skillful Researcher, Fontbonne's advanced information literacy course, is to move students beyond basic information literacy. The purpose is to:

- address metaliteracy and threshold concepts as presented in the framework,
- develop information literacy skills focused on specific disciplines or academic majors, and
- teach students to apply information literacy skills to the workplace environment.

The course description can be found in Appendix A.

20.12 CONCLUSION

The current Fontbonne University information literacy course looks far different than it did when the authors arrived on campus in 2004, but then so does the information landscape. The campus community has accepted LIB199 as a vital part of undergraduate education at Fontbonne. Realization by students and faculty that the information landscape is ever changing supports the need for an up-to-date approach to information literacy. Revisions to this course will be a continual endeavor.

APPENDIX A

Fontbonne's information literacy course (LIB199) is continually revised to reflect the ongoing evolution of the information environment, as evidenced in the following changes to the course title and description over the years:

2002—Information Literacy/2007—Libraries and Information Research

This course is an introduction to basic skills and concepts for using diverse information sources, systems, and search strategies to locate, evaluate, and use information. Topics include classification and organization of information, online search techniques, information search tools and strategies, evaluation and analysis of information, and responsible and ethical use of information.

2009—Information Navigation and Evaluation

This course is designed to improve the skills and knowledge necessary to conduct library research in an academic setting. Areas of focus include utilizing various types of information sources and formats, developing effective search strategies, critically evaluating information, differentiating scholarly from popular sources, and using information ethically and responsibly.

2015—Information Literacy in Higher Education

As global citizens students must understand the parameters of the knowledge they consume, as well as the knowledge they create. In this course students will develop basic skills to engage ethically, socially, personally and intellectually with digital resources with attention to collaborative production, sharing and evaluation of information in participatory digital environments. It will prepare students to function and thrive not only in the higher education environment in which they currently participate, but also in the broader world in which they live and work.

2016—The Skillful Researcher (New graduate level course)

This course is designed to address the information needs of graduate and advanced undergraduate students. It will move quickly beyond the basics of LIB199 *Information Literacy in Higher Education* into development of subject-specific research skills and an exploration of what ethical and productive collaborative involves in today's information society. Students will engage in topics such as searching online and print resources in their specific discipline, the process involved in creation of scholarly information, and the ethical use and privacy of information in a collaborative environment.

APPENDIX B

Fontbonne University Embedded Assessment of General Education Learning Outcomes. Managing Information.

SCORING GUIDE

General Education Learning Outcome (GE LO) Rating Scale

GE LO rating	1	2	3	4		
Student score	0–12 points	13–25 points	26–37 points	38–50 points		
ACRL frames[a]	Fontbonne University general education learning outcomes	Corresponds to content in LIB199 (Module)	Question number on LIB199 final assessment artifact	Total points for learning outcome	Student score	GE learning outcome rating
Information has value	1. Articulate or discuss the ethical and legal use of information	Ethical use of information (7); citation components (5)	1–9	50		
Research as inquiry Searching as exploration	2. Demonstrate the ability to access information to address an issue	Information fog and primary and secondary sources (1); database searching and Boolean operators (3); controlled vocabulary (4)	10–20	50		
Authority is constructed and contextual Format as a process	3. Discriminate among types of information for their intended purpose	Net neutrality (2); types of sources (5); evaluating information and global issues assignment (6)	21–31	50		
Scholarship is a conversation	4. Synthesize information and formulate a response to address an issue	Research is a conversation (2); points of view assignment (3); comprehensive skills check (7)	32–37	50		

[a]From Frameworks for Information Literacy in Higher Education.

REFERENCES

Association of College and Research Libraries. (2000). Information literacy competency standards for higher education. Retrieved from < http://www.ala.org/acrl/standards/informationliteracycompetency>.

Association for College and Research Libraries. (2015). Framework for information literacy in higher education. Retrieved from < http://acrl.ala.org/framework/>.

Badke, W. (2014). *Research strategies: Finding your way through the information fog.* Bloomington, IN: iUniverse.

Beyer, C. H., & Gillmore, G. M. (2007). Longitudinal assessment of student learning: Simplistic measures aren't enough. *Change, 39*(3), 43–47.

Coordinating Board for Higher Education for the State of Missouri. (2000). Credit transfer guidelines for student transfer and articulation among Missouri colleges and universities. Retrieved from <http://dhe.mo.gov/files/policies/credittransfer.pdf>.

Fontbonne University. (2015). Quick facts. Retrieved from < http://www.fontbonne.edu/infocenter/quickfacts>.

National Forum on Information Literacy. (2015). Workplace information literacy. Retrieved from < http://infolit.org/workplace-information-literacy/>.

Nelson, A. (2015). Visual thinking strategies: From the museum to the library: What's going on in this picture? In *Kansas Library Association and Missouri Library Association Joint Conference Program*, Kansas City, MO.

Patthey, G. G., & Thomas-Spiegel, J. (2013). Action research for instructional improvement: The Bad, the Ugly, and the Good. *Educational Action Research, 21*(4), 468–484.

Ridlen, P. (2010). Module three: Navigating the wayward web—Introduction. *INT199 Information Navigation and Evaluation.*

Ridlen, P. (2013). INT199: Information navigation and evaluation assessment data, 2012-2013. Unpublished manuscript. Jack C. Taylor Library, Fontbonne University, St. Louis, MO.

SECTION V

Innovations

CHAPTER 21

Gaming Library Instruction: Using Interactive Play to Promote Research as a Process

21.1 INTRODUCTION

For years, the Gardner-Harvey Library (GHL) at Miami University's regional campus in Middletown, Ohio, focused on providing library instruction within five categories: (1) face-to-face (f2f) individual course instruction sessions, (2) embedded librarianship through the learning management system (Canvas), (3) video tutorials posted on the library website, (4) a video tutorial on research consultations with students, and (5) an elective 200-level two-credit research skills and strategies sprint course that has two offerings each semester for students. None of the instructional materials or services are mandatory for faculty or students to use but are optional additions to help with their courses or research. The most popular resources for library instruction are embedded pages in Canvas and f2f sessions, which are completed on request of the faculty or course instructor. Video tutorials are mostly screen casts of library tools that are updated as needed and posted on the library website. Links to resources and video tutorials are also often embedded into Canvas.

Following state and national trends, Miami University pushed for an increase in the number of online courses, which could impact the amount of instructional interaction librarians have with students. Currently, the library reaches approximately 1300—1500 students per semester through various forms of instruction. These numbers include some duplication for students who have multiple embedded courses or participate in f2f sessions in different classes. These numbers indicate that the library is potentially reaching half of our total student population, as full-time equivalency (FTE) regularly hovers around 3000. However, each semester there are still more than 1000 students who are not receiving any type of

Distributed Learning.
© 2017 T. Maddison and M. Kumaran.
Published by Elsevier Ltd. All rights reserved.

library instruction or research assistance. Ideally, librarians would like to meet with every student and provide personalized help, but with a staff of only three librarians it is not feasible financially or physically. Stiwinter (2013) notes that for many libraries, reductions in resources, including worker power and time, is one reason why libraries create online tutorials.

Faced with a possible decline in the f2f interactions with students and their lack of engagement with the library, the library made the decision to expand the impact of online tutorials. While there is often concern about the effectiveness of online tutorials, studies by Anderson and May (2010), Ganster and Walsh (2008), Silver and Nickel (2007), and Nichols, Shaffer, and Shockey (2003), show that online instruction is just as effective as classroom instruction and that a significant number of students prefer the former.

A quick email request from librarians to faculty at Miami University Middletown on library tutorial topics revealed that the number one concern was students' ability to navigate through the library resources and the research process. Middletown faculty elaborated that a large number of their students seemed unsure of how to develop research topics and search for relevant information using the library. On a regional campus with a student population that is composed of a large number of nontraditional students and a growing number of international English as a second language students, this came as no surprise to librarians. Every semester faculty members found themselves spending more and more class time reviewing steps to complete the research for each assignment. They wanted self-paced online tutorials that would not take up class time and would be easy for students to use. Building these new tutorials also benefit online faculty members, as they could share links or embed the videos within their online course as needed.

With a topic chosen and a decision made to create new online tutorials, the question was the format of these tutorials. Creating screen-cast tutorials seemed like the best option as they could be embedded into a course or posted online and watched repeatedly. However, librarians wanted an engaging video on the research process so students could interact with the multiple steps of the research process, seeing how their decisions could influence the outcome of their work. Librarians had many questions: How do we create an engaging interactive tutorial? How do we build and maintain it? How much learning is involved for librarians?

How do we maintain the necessary accessibility standards for online tools? When faced with all of these questions, this team of three librarians remembered a childhood book series called *Choose Your Own Adventure*. Each book in the series gave a reader control of the outcome, allowing her or him to build the story. This story model was used to build a game in the style of *Choose Your Own Adventure* in which each choice influences how a student progresses and whether she or he will ultimately succeed.

21.2 GAMIFICATION AND INTERACTIVITY

Using games as part of library instruction is not a new concept. Margino (2013) notes that "rather than learning through listening, games provide students with opportunities to engage through competition, participate in hands-on activities, and learn from mistakes" (p. 334). Furthermore, "games promote group work, direct interaction with content either in-person or virtually, and permit users to learn-by-doing" (p. 334). Examples of popular in-person library instruction games include Penn State University and Georgia State University's *Jeopardy* adaptions, Utah Valley University's *Get a Clue*, and University of Florida's *Humans vs. Zombies* (Margino, 2013; Smale, 2011; Smith & Baker, 2011; Leach & Sugarman, 2005). These games have a specific purpose and defined learning outcomes that support their use in class. They often introduce students to the library and help build skills that can be used in later research. In-person games allow for quicker implementation and adaption. Digital or online games may take longer to build but are accessible to a greater number of students at any time because they are not restricted to a one-time meeting.

Game-based learning is often referenced in connection with using video games to support teaching and learning. Libraries often focus on digital games rather than video games. Broussard (2012) highlights the variety of games that have been used in library instruction throughout the years, including trivia, role playing, and casual games, games that mix physical and virtual elements, alternate reality games, and social games. While these types of instructional tools are puzzle games, Broussard points out that they still add game elements into instruction that builds a level of active learning that can make tutorials more fun. Phetteplace and Felker (2014) note how gamification has two sides as it can be "both a strategy for engagement and a framework for immersive learning and play" (p. 20). Kim (2015b) further notes that gamification in libraries can help with both improving library instruction, as well as providing a greater awareness of available library services. With their application to library services and instruction, gamification seems ideal for reaching more students.

Kim (2015a) provides a summary of five different definitions of gamification, but the most common definition is that gamification applies game elements to real world, non-game situations. Whether in-person or digital, all games share four key elements: a goal, rules, a feedback system, and voluntary participation (McGonigal, 2011, p. 21). When looking at gamification, these game traits are all repeatedly referenced, often with the addition of fun. In fact, out of all of these elements, the focus on fun seems to be the real key to success. Gamification is a useful tool for modifying something that may seem boring into something interesting.

Looking more specifically at the link between gamification and learning, Faiella and Ricciardi (2015) note that studies focus on three factors—student motivation, engagement, and learning outcomes—to show the influence of gamification. Kim (2015b) and Walsh (2014) highlight how libraries in particular are using gamification to help engage and motivate students to use the library and its various services. Kim (2015b) covers a variety of examples of how gamification can be used in libraries, including gamifying the library experience, gamifying library instruction, gamifying library orientation, and building a library gamification mobile app. The games within each category vary in price of production, with some of them built in house and others bought from vendors; the majority, however, share some of the most common competitive game elements within their gamified platforms, including the use of leaderboards, badges, and prizes or rewards.

Walsh (2014) focuses on one game in particular called *Lemontree*, offered by the University of Huddersfield in the United Kingdom, that gamifies the library experience to increase the use of library resources. Created by an outside vendor (Rith Ltd, http://rith.co.uk/), *Lemontree* is another example of gamification that rewards users with badges and points for successfully completing specific goals—in this case, common library activities, including visiting the library, borrowing and returning books, and using online resources. However, unlike some of the other games, *Lemontree* focuses on creating a light-hearted and fun interface with no visible library or university branding to reduce the stress often associated with using the library. User feedback shows that students enjoy the gamified platform and found it to be engaging, entertaining, fun, and rewarding.

Gamifying different features from orientation to the research process can help libraries reach students in new and exciting ways that are both interactive and entertaining. The implementation of different gamification platforms such as *Lemontree*, may be evidence that users are open to the use of games and fun within different library services.

21.3 CREATION OF THE TUTORIAL AT GHL

The original plan for the research adventure at GHL was to build a role-playing game (RPG). In eight-bit style, librarians wanted to build or create a game in which students have to pick a character and navigate through the basement of a castle, only managing to escape if they chose the correct doors as part of their navigation through the research process. As the game-development process began, it was clear that that the time and work required to build the game was huge. Another concern was whether the game would meet accessibility requirements. According to university policy, all online materials are required to meet accessibility standards as set in the Web Content Accessibility Guidelines 2.0 (https://www.w3.org/TR/WCAG20/). These guidelines include adding alternative text to images, providing a longer text alternative for infographics, closed captioning all videos, using descriptive link labels, and using headings within the content. Noting both the accessibility and timing requirements, the focus shifted from creating game-based instruction to gamification, which means instead of using a true video game in instruction, game elements are used to build the tutorial. A decision was made to use existing librarians' skills and knowledge with different online tools and start with a text-based story model and build a website.

21.4 TEXT-BASED

The text-based model began with the development of a storyboard. The research process needed to be covered—but how? According to Broussard (2012), the best practices when building a game include keeping it simple, knowing how to market your game, having a plan for using it, and, as noted previously, making it fun. With these practices in mind, GHL librarians started by creating a structural outline of the story. Taking the seven steps of the research process as a simple beginning, the team came up with the following chapters for our story:
- Chapter 1: Choosing a Topic

- Chapter 2: Finding Background Information
- Chapter 3: Finding Books
- Chapter 4: Finding Articles
- Chapter 5: Finding Digital Media
- Chapter 6: Evaluating Web Resources
- Chapter 7: Citing Sources

Choose Your Own Adventure books began in the 1970s as a way to involve the reader in the story. By playing an RPG in book form, the reader has the option to choose which path the character takes as he or she progresses through the tale, ultimately controlling how the story ends. The seven chapters of the story outline provided the basis for the branches of the story, with each chapter having decision points where the player can choose from different paths to progress through the story, with each progressive step then having its own set of paths to follow. A total of 28 different steps were created as part of the overall story, including an introduction, correct answers, misleading answers, and a conclusion. Each step was set to be concise yet informative. There is always a risk of losing a player's attention if there is too much to read on the page. The topic choice for the story was also of great importance as the team wanted to provide a current and evolving topic for the research adventure and sustain students' interest. The "legalization of marijuana" was chosen as the first topic because it is current and relevant and most people have an opinion about the subject.

While the game is text based, the team did not want to use just words on the screen. Aside from keeping the text short and choosing an interesting topic, other attempts to maintain student interest in the story included adding images and video to different paths within the chapters. These additions meant that the website builder hosting the story would have to allow for embedded images and videos.

21.4.1 Google Sites

The team chose to build the story using Google Sites, as it is already attached to the university email and is a tool regularly used for assignments within the library online credit course. With its widespread use across campus, students should have no problem accessing the website. The team's knowledge of how to build sites within Google also meant that screenshots and videos could be included fairly easily within the site to help students through the game.

As the first model of our game, the "Choose Your Own Research Adventure" Google site had the benefits of being free to use and easy to build and update. It offered a clear layout with a menu for students who wanted to start at various chapters, and it allowed for the addition of screenshots and videos as desired to provide more detail within different paths and chapters. The homepage of the website contained the introduction, explaining what the story is about and invited students to begin (see Fig. 21.1). The adventure had a subtitle of "Surviving the Research Process" because the team hoped to develop future stories that focused on completing research for specific courses or assignments if the first adventure was useful. Players can choose which path to follow on each page by selecting the appropriate links (or icons) at the bottom (see Fig. 21.2).

While the Google site offered a place to include images and videos, the team did not like the restrictive nature of the website formatting, particularly with moving the images to different locations on the page. Other factors (Table 21.1) also influenced the decision on whether to continue using Google Sites.

While the team was comfortable with the ease of navigation and layout of the first website, a decision was made to have a more focused

Figure 21.1 "Choose Your Own Research Adventure" Google site homepage (https://sites.google.com/a/miamioh.edu/choose-your-own-research-adventure/).

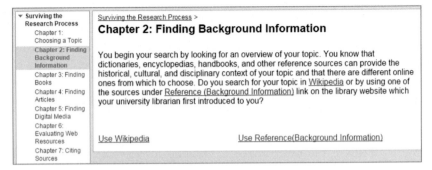

Figure 21.2 Chapter 2 with example links.

Table 21.1 Google Sites
Google Sites

Pros	Cons
• Free hosting • Easy to build if for those who know Google • Easy to update • Good menu options if starting players at a specific stage of the process	• Restricted layout • External website leading away from the library website • Menu or links the only way to progress; cannot move from one chapter or video without them

Choose Your Own Adventure display to the game and less of a classic website look.

21.4.2 Twine

Twine is a free, "open-source tool for telling nonlinear stories" (http://twinery.org/). This browser-based resource allows a user to build stories or games in the *Choose Your Own Adventure* style and then download them as an HTML file. These files can be uploaded to a personal website or a suggested hosting site for stories. The discovery of Twine was a happy accident just as the team was starting to look at other website builders for the second text-based story model. As it is built for the story style librarians had in mind, using Twine allowed for keeping the online story based format but without the website framework. The team felt that the storyline and topic were still sound, so the chapter text and paths previously written in Google Sites were reposted into Twine storyboard. Copying and reposting each pathway and chapter into Twine showed

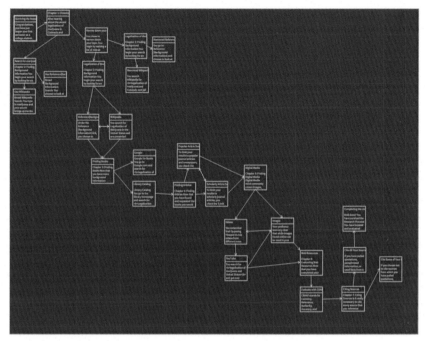

Figure 21.3 Twine blueprint showing the paths of our story.

how one chapter led to the next and revealed a full layout of the story (see Figs. 21.3 and 21.4).

To add images and videos, Twine requires HTML tags, so every page with an image or video on the Google site had a corresponding HTML tag included in the paths of the Twine game. For those who are not comfortable with HTML, both an informative wiki (http://twinery.org/wiki/) and a highly knowledgeable online community (http://twinery.org/forum/) can help answer questions and troubleshoot issues that may arise for those using Twine. Although Google took the project away from the library website, because of its HTML possibilities Twine allowed for the building of a finished game as a page on the library website (see Fig. 21.5).

Although Twine's layout, ease, and features are preferred over the Google site, it still has limitations (see Table 21.2).

Overall, our biggest concern was the layout of the story. Although Twine allows for moving back a step by providing back arrows on screen to return to a previous page, if students use the back button in the browser versus the back arrow in the game, they will return to the beginning of the game, not just go back a step. This was frustrating to users.

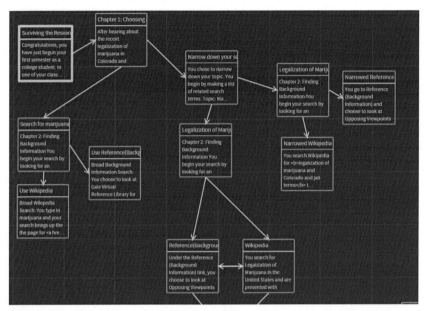

Figure 21.4 Close-up section of the Twine blueprint showing the beginning paths of our story.

Figure 21.5 "Choose Your Own Research Adventure" Twine homepage (http://www.mid.miamioh.edu/library/chooseyourown/).

Table 21.2 Twine
Twine

Pros	Cons
• Open source	• Must save HTML file or it can be
• Still online based, but not a website	lost when browser history is cleared
• Built for choose your own format	• No menu option for starting at a
• Can easily access online because it	later chapter
is browser based or downloadable	• Had to search for information to add
• Useful community to help with	images and videos; not intuitive;
questions about color change or	HTML tags were needed
video and image insertion	• Can post on your own website
• Suggested story hosting that counts	
visits	
• HTML output if you want to post	
on your own website	

21.5 VIDEO BASED

Despite its issues, Twine was chosen over Google Sites for the text-based game due to its more classic *Choose Your Own Adventure* layout. And while plans to create an RPG are currently shelved due to the amount of staff time required to learn and build this format, the team would still like to provide the game beyond the text-based mode. Ideally the team would like to offer more video tutorials, particularly for our students who prefer to learn by watching versus reading. The screen casts created often highlight different online resources, including GHL catalog and databases, and incorporating the storyline of *Choose Your Own Research Adventure* game into this format could help build a more visual story. One concern with a video-based story, however, is the loss of interactivity available in Twine that allows students to choose how they progress through the research process.

21.5.1 Adventr

Adventr is an online platform that allows the use of video clips to create interactive videos (http://www.adventr.tv/). As a web-based product, Adventr does not require materials to be downloaded or any additional knowledge of HTML. Its cost varies based on a user's needs—from a free Creative setup that can be embedded anywhere but may include advertisements and lacks analytics to a Plus ($10), Pro ($60), and Enterprise (variable price) accounts, each with an increasing selection of features.

Figure 21.6 Adventr editor view.

GHL maintains a free account but may advance to a Pro account if the view count and performance analytics are needed. Users must first sign up for the account that best suits their needs and then proceed to upload any videos they wish to link together. For users without their own videos, Adventr offers its own small collection. Once videos are uploaded, a user can begin to build a path from one to another, just as GHL librarians did with the links in the text-based Twine story (see Fig. 21.6).

When a creator wants to connect videos, he or she creates a button link at the end of the current video to allow the user to choose which video will be watched next. Creating a button link at the end of a video allows the user to choose the next video. While Adventr allowed the team to take the written story a step forward, it still has both advantages and drawbacks that have to be considered (see Table 21.3).

At GHL, the development of the Adventr videos is still ongoing. As librarians began creating screen casts, they realized that simply reading text from the original story does not add much excitement to the game. The videos almost blend from one to the other and lose that key element of fun that is so important for the enjoyment and effectiveness of our tutorial. Thanks to the talented work of a student employee, Jordan Martin, the team is now looking into using animated gifs (graphic

Table 21.3 Adventr
Adventr

Pros	Cons
• Visual representation of the story • Video and screen-cast focus • Built for choose your own style • Builds links to pages at end of each video • Easy to upload and use • Provides analytics (paid versions)	• Links are limited with the amount of words due to space • Lots of videos to record and update (not an Adventr con just a video con) • No back button or menu • May have a waiting period for account to be available • Some features only come with paid versions

Figure 21.7 Character development of animated gifs.

interchange format files) to show changes between stages (i.e., moving onto the next task) using Adobe Photoshop (see Fig. 21.7). GHL librarians are also reviewing the use of web-based products to create videos using cartoons such as those at Muvizu (https://www.muvizu.com/) or live action characters versus simple screen casts. Cost and ease of use continue to be factors in choosing products. Whatever the final decision, librarians plan to use both the text- and video-based games as options for librarians and faculty to share with students not only to help cover the preferences of different types of learners but also to meet accessibility requirements.

21.6 FEEDBACK AND CONTINUED GAME DEVELOPMENT

The use of interactive and unconventional methods of instruction can create a lasting and we hope positive impact on students. The inclusion of a text-based game in the embedded classes and as a highlighted research tool on the library website has raised two important questions: Are students actually using these tutorials? Do these tutorials help students improve their research skills? As one game type is complete and the team continues to develop other versions, librarians want to ensure that these gaming tutorials will be used. An early online survey of students has provided useful feedback on how the tutorial is presented along with suggestions for improvement.

Some of the survey questions included the following:

- How long did it take you to complete the tutorial?
- Approximately how many missteps did you have?
- Is the language of the tutorial understandable?
- Did the tutorial hold your attention?
- Would a video tutorial be better or worse?
- Do you have any suggestions on how to improve the tutorial?
- Do you think completing this tutorial will help you successfully complete research in the future?

Of the 34 students surveyed, including both domestic and international freshman and sophomores, more than half show a desire for a video-based game. Although more than 75% of students found the language understandable and felt the tutorial held their attention, they said they supported videos because the videos helped clarify some of the language confusion as students could hear the words and see the images connected to the topic. One suggested improvement was more game elements, specifically the inclusion of badges or rewards when the game is successfully completed. As other effective gamification models have integrated badges into their games, this suggestion seems like a natural step with the GHL game. Overall, early feedback shows that approximately 80% of students found the tutorial helpful for completing future research. Although not a substantive number of students provided feedback, it is positive. Another survey of the same students should be conducted in the future to learn if students retained what they learned.

In addition to surveying students, we also reviewed analytics from the GHL website to see how often the tutorial is viewed, which can show usage but not effectiveness. A lower number of views may be influenced by the voluntary nature of the tutorial and our minimal promotion of this resource. The number of views of the tutorial spike when it is highlighted or referenced specifically within a class or added as an embedded feature within the learning management system, with an average of 5—10 more views per reference. Future assessment plans include holding small group sessions to test out the tutorial and follow-up conversations with students as the team reviews changes to the tutorial. The ultimate goal for gamification is to create an interactive guide that students will find both educational and enlightening. GHL librarians hope to retain an element of fun within the tutorial to help with this.

21.7 CONCLUSION

A recent survey from the Pew Research Center shows that 49% of American adults play video games. These include games played on a computer, TV, game console, or portable device such as a cell phone. Perhaps it is not surprising that game players are more likely to agree with the positive aspects of games versus the negative. Sixty-five percent of those who play games believe that most (18%) or some (47%) video games help develop good problem-solving and strategic thinking skills (Duggan, 2015). Based on these numbers it is no surprise that students may support the use of online game elements within education in general and libraries in particular.

The gamification of the research process through the creation of our *Choose Your Own Research Adventure* has three main goals: (1) to reach, engage, and motivate more students; (2) to support different types of learners; and (3) to fill faculty requests for a more involved tutorial. A growth in online courses and the potential decline in f2f instruction sessions have necessitated the review and expansion of GHL online tutorials, but the interactivity and variation in those tutorials was a focused choice made by the library. The popularity of online games and the positive feedback for the library game received so far support its continued use and further development.

In addition to video and animated improvements, increasing the visibility and promotion of the research game along with other types of tutorials would help support the diversity of learners we have on campus. Creating a badge or reward system for all tutorials may help further motivate students to learn about different library resources on their own.

The team hopes to complete the majority of the work over the summer term when members have more time for projects. Student workers will also be trained on the relevant technology to help supplement staff time. Regular reviews and assessment will be added to the schedule for the fall and spring semesters to help evaluate any changes made in the summer. As a GHL project, the cost of any staff time and resources will be covered by the library budget.

Overall, the lesson learned from the creation of this game is adaptability. Gamification of library services is a field that continues to grow. Maintaining and improving games are challenges for both in-person and digital games. Technology and resources are constantly evolving, and librarians must be willing to change or even scrap things that do not work. When building games, creators will spend more time than planned as they learn new software or technology, regularly assess the game, and use suggestions to make improvements. Despite these challenges, the benefits from expanding on how student needs are met and library instruction is provided can only increase as libraries continue to promote fun and play as additional elements to support learning in higher education.

REFERENCES

Adventr. (n.d.). Retrieved from < http://www.adventr.tv/ > .

Anderson, K., & May, F. A. (2010). Does the method of instruction matter? An experimental examination of information literacy instruction in the online, blended, and face-to-face classrooms. *The Journal of Academic Librarianship, 36*(6), 495−500.

Broussard, M. S. (2012). Digital games in academic libraries: A review of games and suggested best practices. *Reference Services Review, 40*(1), 75−89.

Duggan, M. (2015, December 15). *Gaming and gamers.* Retrieved from Pew Research Center, Pew Internet & American Life Project: < http://www.pewinternet.org/2015/12/15/gaming-and-gamers/ > .

Faiella, F., & Ricciardi, M. (2015). Gamification and learning: A review of issues and research. *Journal of E-Learning & Knowledge Society, 11*(3), 13−21.

Ganster, L. A., & Walsh, T. R. (2008). Enhancing library instruction to undergraduates: Incorporating online tutorials into the curriculum. *College & Undergraduate Libraries, 15*(3), 314−333.

Kim, B. (2015a). Gamification. *Library Technology Reports, 51*(2), 10−16.

Kim, B. (2015b). Gamification in education and libraries. *Library Technology Reports, 51*(2), 20−28.

Leach, G. J., & Sugarman, T. S. (2005). Play to win! Using games in library instruction to enhance student learning. *Research Strategies, 20191−20203.*

Margino, M. (2013). Revitalizing traditional information literacy instruction: Exploring games in academic libraries. *Public Services Quarterly, 9*(4), 333−341.

McGonigal, J. (2011). *Reality is broken: Why games make us better and how they can change the world* (p. 2011). New York, NY: Penguin Group.

Muvizu. (2016). Retrieved from < https://www.muvizu.com/ > .

Nichols, J., Shaffer, B., & Shockey, K. (2003). Changing the face of instruction: Is online or in-class more effective? *College & Research Libraries, 64*(5), 378–388.

Phetteplace, E., & Felker, K. (2014). Gamification in libraries. *Reference & User Services Quarterly, 54*(2), 19–23.

Running in the halls: Apps, games and interactive experiences. (n.d.). Retrieved from < http://rith.co.uk/ >.

Silver, S. L., & Nickel, L. T. (2007). Are online tutorials effective? A comparison of online and classroom library instruction methods. *Research Strategies, 20,* 389–396.

Smale, M. (2011). Learning through quests and contests: Games in Information Literacy Instruction. *Journal of Library Innovation, 2*(2), 36–55.

Smith, A. L., & Baker, L. (2011). Getting a clue: Creating student detectives and dragon slayers in your library. *Reference Services Review, 39*(4), 628–642.

Stiwinter, K. (2013). Using an interactive online tutorial to expand library instruction. *Internet Reference Services Quarterly, 18*(1), 15–41.

Twine. (n.d.). Retrieved from < http://twinery.org/ >.

Walsh, A. (2014). The potential for using gamification in academic libraries in order to increase student engagement and achievement. *Nordic Journal of Information Literacy in Higher Education, 6*(1), 39–51.

Web Content Accessibility Guidelines (WCAG) 2.0. (2008, December 11). Retrieved from < https://www.w3.org/TR/WCAG20/ >.

CHAPTER 22

Implementing Flipped Classroom Model Utilizing Online Learning Guides in an Academic Hospital Library Setting

22.1 INTRODUCTION

Credited with lowering failure rates, introducing flexibility into education, improving learner attitudes, and even raising test scores, the flipped classroom is a hot topic among educators in secondary and postsecondary academic settings (Bergmann & Sams, 2012, 2014). This blended learning instructional model reverses the traditional construct of a classroom in which an instructor presents content and then leads activities or exercises to reinforce acquired skills or knowledge. The flipped-classroom model employs technology to deliver content to learners before class "so that classroom time is spent on discussion, analysis, and problem-solving activities" (Youngkin, 2014, p. 368), including interactive group learning. Studies suggest that learner preparation is a key factor for flipped-classroom success (Khanova, Roth, Rodgers, & McLaughlin, 2015). This chapter outlines how the flipped classroom works in the specific context of a teaching hospital library that educates a diverse group of staff and learners from a variety of medical and health disciplines. The authors will highlight some of the challenges they have encountered in adopting this model and document how it helped them accommodate the needs of learners and busy professionals and create a learner-centered experience.

22.2 HISTORY OF THE FLIPPED CLASSROOM

Bergmann and Sams (2014) introduced the flipped-classroom model to their high school science classes in 2006 to avoid repeating material for

Distributed Learning.
© 2017 T. Maddison and M. Kumaran.
Published by Elsevier Ltd. All rights reserved.

learners who missed class. The flipped classroom became a routine part of their lesson planning that freed up class time to assist learners apply course material to real-world problems. Bergmann and Sams (2014) note that this style is not completely original; they traced the idea from an economics professor at Miami University in the 1990s who emailed PowerPoint presentations to learners before his lectures.

While the liberal arts have long demanded preparation before class, the flipped-classroom model differs from traditional learning in that it employs new technologies and new kinds of learning objects (Bishop & Verleger, 2013). Flipping the classroom emphasizes that the teacher's time is best used helping learners rather than simply delivering material that learners can access in their own time and at their own pace. The learning objects do not replace teaching; they serve as resources for learners to review material, as well as prepare for class (Bishop & Verleger, 2013).

Bishop and Verleger (2013) traced the roots of the flipped classroom to Bloom's Constructivist Taxonomy of Learning (Bloom, 1956). Bloom's taxonomy asserts that knowledge does not come packaged in professors' or learners' heads for transmission to one another; instead, instructors and learners possess information, not knowledge. Knowledge must be constructed or reconstructed by individuals making sense of new information in terms of what they already may know (Gilboy, Heinerichs, & Pazzaglia, 2015). In the flipped-classroom model, the teacher does not present facts for learners to absorb but works with learners to construct knowledge through experiential activities.

In theory, the flipped classroom saves precious class hours while allowing both instructors and learners to customize learning by focusing on specific learner questions and interests. According to Pannabecker, Barroso, and Lehmann (2014), "this method allows learners to have prior demonstration and explanation of content or skills and an opportunity to apply their new skills and knowledge in a risk-free, experimentation-friendly environment where instructors can provide further support when questions arise" (p. 140). It reduces the amount of time spent in "mind-numbing lectures" and frees up time for more useful discussion (Vogel, 2012).

There is at least one recent scoping review on the topic of flipped classrooms (O'Flaherty & Phillips, 2015) and the literature on the use of flipped classrooms for information literacy often reports the results of learner and instructor feedback surveys. Learner feedback is generally positive (Morgan et al., 2015), and instructors find that it reduces learner passivity (Lochner, Wieser, Waldboth, & Mischo-Kelling, 2016). The flipped classroom is becoming increasingly common despite a dearth of evidence to support its value (Bendriss, Saliba, & Birch, 2015; Chi & Verghese, 2014) and is particularly popular in medical education and other health professions where learners have significant time constraints due to clinical responsibilities (Youngkin, 2014). It can save both teachers' and learners' time and add flexibility to their schedules (Arnold-Garza, 2014; Milman, 2014; Morgan et al., 2015; Rivera, 2015), and it is well suited to procedural learning—i.e., knowledge about how to do something (Milman, 2014). Learning to search databases is largely procedural and therefore suited to the flipped-classroom instructional model.

Most of the literature on the use of flipped classrooms in libraries focuses on teaching learners how to use research databases (Youngkin, 2014; Pannabecker et al., 2014). Learners follow a prerecorded lesson on devising search strategies and finding vocabulary before class, leaving time for questions and hands-on experience when meeting with the instructor in class (Fawley, 2014). The majority of the literature on librarians using flipped classrooms in the health sciences continues to grow from three widely cited articles: Kurup and Hersey (2013), Pannabecker et al. (2014), and Youngkin (2014). In the spring of 2016, the authors identified five articles on health sciences librarians employing flipped classrooms in a library setting (Bendris, Saliba & Birch, 2015; Conte et al., 2015; Fawley, 2014; Pannabecker et al., 2014; Youngkin, 2014), but none addressed teaching in a hospital library. The authors explore the implementation of the flipped-classroom model to meet the unique needs of St. Michael's Hospital's health sciences library.

22.3 BACKGROUND

St. Michael's Hospital in Toronto, Canada, is a relatively large, multicultural inner-city teaching hospital affiliated with the University of Toronto. As of 2015, St. Michael's had slightly more than 6000 staff members, 780 physicians, and more than 3500 student placements a year (St. Michael's Hospital, 2016). The hospital is one of two regional trauma centers and has a large research unit that includes programs in both basic science and clinical research. The health sciences library is staffed by 4.5 full-time equivalency librarians and 3.5 library technicians. The library organizes workshops on a variety of topics approximately eight times a month for staff, physicians, and students. Participants at library workshops are a combination of working health-care practitioners, scientists, administrative and research personnel, and health care trainees and can range from high school interns to late-career professionals. Attendance is entirely voluntary and based on an individual's learning needs. Workshop instructors do not have prior knowledge of the participants' proficiency with computers or other skills, much less their learning needs.

In 2013, one of the library's instructors experimented with the flipped-classroom model to teach Medline (Ovid) to a specific group of learners from one department. These learners already had significant experience with Medline but felt they had gaps in their knowledge. The instructor distributed class handout materials and links to vendor-produced instructional videos by email before class. She then met with the learners as a group to answer questions that arose from the hand-outs and tutorials. By distributing preclass materials, the instructor was able to focus the in-class workshop on specific questions rather than imparting general information. Learners who attended this workshop said they appreciated the format. This experience inspired the library instructional committee (comprising all librarians at the hospital) to explore the possibility of formally implementing the flipped-classroom model for some of their standard workshops. The instructional com-mittee meets regularly to review and revise workshop curricula, review workshop evaluations, plan for special educational events, and schedule teaching assignments. One issue the instructional committee identified was the ideal length of their workshops to match clients' schedules. Many of the workshops were up to 2 hours long, but most learners expressed a preference for shorter workshops with learners often leav-ing workshops before they ended. Building on the success of the exper-imental Medline workshop previously mentioned, the instructional committee decided to redesign some of the workshops to increase learner participation and improve individualized learning by using the Coursera flipped-classroom field guide to help facilitate this process (Coursera, 2014).

22.4 METHODOLOGY

Early in 2014, the instructional committee identified three of its existing workshops that would suit the flipped-classroom approach: Maximizing Your Search Skills (an introduction to the library's resources and search skills), PubMed, and Medline (Ovid) . All three workshops contained materials or content on effective database searching such as identifying controlled vocabulary (medical subject headings), choosing effective keywords, applying limits, exporting search results into bibliographic software, and accessing articles. Toward the end of each workshop, the instructors introduced other features that suited the needs of individual learners such as using PubMed's citation matcher to track down articles from incomplete citations and applying search filters to find the most relevant results to help answer treatment questions.

22.4.1 Preworkshop Learning Objects

Using Springshare LibGuides as a platform to deliver the preworkshop content, the instructors created online resource guides (learning objects) for each of the three planned flipped workshops, enabling them to free workshop time for review and skill-building exercises. Updates to the guides are fairly simple to do, and tools that come with the platform such as the one that automatically checks for dead links, ensure quality control. Librarians reviewed the content from the previous traditional workshops and decided what should be carried over to the new platform for resource guides. These guides cover much of the same content that was previously delivered in class and are designed to be used in conjunction with the in-class workshop or for independent learning. The authors used the software to create visually appealing and user-friendly guides (see Fig. 22.1). Once the resource guides were complete, librarians presented them to the instruction committee for feedback.

Learners register for the workshop using the hospital's learning management system (LMS). At registration, participants are sent a link to the applicable resource guide for the workshop. Participants might register as far in advance as a month or up to the day before a workshop. Those who register far in advance are sent as many as two reminders prior to the workshop about the guides, which contain foundational workshop information that should be reviewed by learners before attending the in-class session. (See Fig. 22.2 for the Medline guide, with instructions for workshop preparation.)

All preworkshop guides have a link to the workshop handout, which contain a miniquiz designed by the instructors. Participants are encouraged to complete this quiz and print the handout before class. The resource guides provide opportunities for self-paced learning and review with a balance of text, screen captures, and videos to address a range of learning needs and styles.

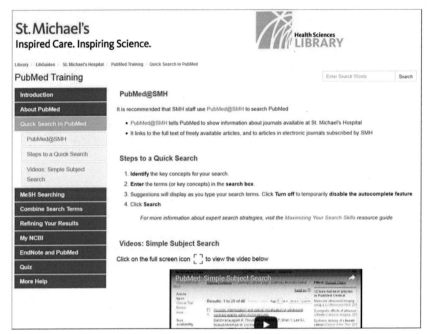

Figure 22.1 Integrating videos such as this one from the National Library of Medicine into the resource guides accommodates different learning styles.

Figure 22.2 The online resource guide containing the prework for the Medline workshop.

22.4.2 In-Class Workshop

The in-class workshops are between 60 and 90 minutes and taught in the library instruction lab, which accommodates as many as 15 learners at personal computers and is equipped with a presentation console. Each workshop is facilitated by one of the librarians who shares responsibility for delivery of the workshops on a rotating basis. In classes larger than six learners, a second librarian is available to assist and provide one-on-one assistance as necessary.

At the beginning of each workshop, the instructor engages participants in a conversation about the workshop structure, which sets the tone for the workshop and provides an opportunity for the instructor to gauge whether participants have completed the preworkshop exercises. Instructors typically present a minireview or summary of the information contained in the resource guide and a review of the preworkshop quiz. Participants are encouraged to ask questions about the content from the resource guide to ensure that everyone has sufficient foundational knowledge before moving on to the active learning exercises. If the instructor is concerned that learners are not able to function at the same level, then she or he may decide to do in-class demonstrations. When it is time to move on to active learning exercises, participants can engage in practice exercises (see Fig. 22.3) or explore their own research questions. Some learners take this opportunity to work on their own research question, which can be a challenge for both the instructor and the learner as they must work together spontaneously without the opportunity to have prepared in advance. While this may be difficult, it helps ensure that the material is learner centered and focused on their actual research questions, not simply questions that work well for demonstrations.

22.4.3 Communicating the Flipped Workshop to Learners

Communicating the workshop model to learners is vital to its success. Despite ensuring that all promotional materials (flyers, bookmarks, hospital-wide email announcements, and so on) for the new workshops contained information on the expectations of participation, initially the authors found that learners were frequently unclear of this model's expectations and arrived unprepared. The authors reviewed this problem and added information about the prework expectations to their online workshop page and in the LMS, where participants

MEDLINE with the Health Sciences Library

Practice Exercises

Health Sciences
LIBRARY

1. To what Medical Subject Heading (MeSH) does shingles map to?

2. If you explode the term "phytotherapy", what additional terms are you also searching?

3. What ages are included in the term, "Child?"

4. What is the preferred MeSH term for "chewing?"

5. Look for citations dealing with liver transplants and living donors. Are there recent clinical trials available on this topic?

6. Find references that discuss the relationships between circadian rhythms and either cortisol or melatonin in humans.

7. Find references about surgical procedures. Combine the results with the previous search to find references about surgery, circadian rhythms and cortisol or melatonin in humans.

Figure 22.3 Practice exercises for the Medline workshop. Questions were adapted from PubMed Online Training, US National Library of Medicine: https://www.nlm.nih.gov/bsd/disted/pubmed.html.

register for the workshops. They also programmed the LMS to send learners information about the preworkshop learning expectations and a link to the online resource guide. The authors also developed an automatic second reminder email sent 48 hours before the workshop. Subsequent to implementing these automatic emails about the preworkshop expectations, they noticed an increase in the number of learners who arrived at the workshop more prepared because they had reviewed the resource guide and completed the miniquiz.

22.5 EVALUATION

To evaluate this project, the authors explored the experiences and opinions of both learners and instructors. Four primary questions regarding the learners' experience were investigated:
1. Did learners use the resource guides?
2. Did they find the resource guides useful?
3. Did learners attend the in-class workshops?
4. What was their experience of the in-class workshops?

Instructors were asked to individually reflect on their experience by completing a structured reflection tool (Appendix A) in which they provided their feedback on what went well, what needed improvement, whether they felt the flipped model worked better for any of the subjects, their preference for facilitating the flipped workshop, and their perceptions of learner engagement. Instructors also commented on any successful strategies they employed to address challenges and any changes they would like to see made to the flipped workshops.

22.5.1 Resource Guide Usage

LibGuides usage statistics showed that between May 2014 and June 2015 the resource guides that accompanied these flipped workshops were viewed 3736 times. These statistics suggest that the resource guides are quite popular (being viewed three times per day on average). In addition to using them for the prework portion of the learning experience, learners are also referring to them after the in-class workshop and may be using the guides for independent learning.

Page views of resource guides May 2014–June 2015

Maximizing your search skills LibGuide	1649
Medline LibGuide	844
PubMed LibGuide	1243

22.5.2 Workshop Attendance

Between May 2014 and June of 2015, 29 flipped workshops were offered. A total of 189 learners attended one or more workshops, with 72 learners attending the maximizing search skills workshop, 58 attending the Medline workshop, and 59 attending the Pubmed workshop.

Workshop attendance May 2014–June 2015

Maximizing your search skills workshop	72
Medline workshop	58
PubMed workshop	59
Total learners attending workshops	189

22.5.3 Learner Feedback

For the purposes of instructional improvement, participants were asked to complete an anonymous evaluation form that included Likert-type scales and open-ended questions. The evaluation form was modified midway through the project to include an additional question relating to the resource guide. Out of a total of 189 participants, 167 completed an evaluation; 58 participants completed the original form, and 109 participants completed the modified form (Appendix B). Not all learners completed all questions on the form.

Learners were asked to rate the workshop overall on a scale of 1–10, with 1 being unacceptable and 10 being exceptional. The average rating was 8.6/10. They were also asked to rate the content and the instructor's knowledge of the subject matter and ability to communicate on a five-point scale of poor, fair, good, very good, and excellent. When asked how they rated the subject matter of the workshop, 76 learners rated the workshop as excellent and 69 rated it as very good. The majority of learners rated the instructor as excellent ($N = 88$). When asked if they found the resource guides useful, only one respondent indicated the guides were not useful.

Learner evaluations

Resource guides	Useful	Somewhat useful	Not useful	
If you reviewed the resource guide before class, did you find it useful?	58	22	1	

In-class workshops	Excellent	Very good	Good	Fair
How do you rate the subject? (interest, benefit to your work, etc.)	76	69	7	0
How do you rate the instructor? (knowledge of subject matter, ability to communicate, etc.)	88	55	10	2

22.5.4 Qualitative Data From Learners

Open-ended questions on the evaluation form provided further opportunity for participants to elaborate on their experience. They found the workshops engaging, informative, practical, useful, and easy to follow. They often mentioned liking the highly interactive hands-on nature of the workshops, the supplementary materials (worksheets, handouts, tip sheets, examples, practice exercises), as well as having an online guide to pull the material together and reference later. They commented on the workshops being highly user-friendly, clear with step-by-step instructions, and "highly relevant and useful," and they liked the amount of individualized attention they received.

Several people specifically commented on liking the flipped structure of the workshop. One learner said, "I really appreciated the opportunity to review this before the course." Other comments included, "[I] liked … practice exercises and 'hands-on' working with computer. I liked that it's simple and clear, not too much crammed in…." Those who did not like the flipped format complained that the in-class workshop is difficult for those who did not have time to prepare ahead of the class. One learner found the prework "a lot to expect considering [our] busy schedules." However, another participant stated that the "biggest advantage was the ability to stop, pause and replay the video as many times [as needed]."

Conflicting comments were also received about the in-class workshop. Some learners said it could be more detailed or more challenging. Others found that "doing questions as a class made it easier to understand"… and that "the interactive components were very helpful." Some learners found it too short and others too long or too rushed and they desired more time to work on the examples. Some learners commented that they would have preferred more time to do their own searches rather than the practice exercises. Learners also commented that the workshop would improve if all learners did the prework.

Learners also had differing opinions on the quality of instruction. Some learners found the instruction too fast or that the instructor seemed disorganized. Others noted that the instructor was "very knowledgeable and responded to questions" and "explained concepts and talked in a clear and detailed manner." Overall, a majority of learners were pleased with the instruction, commenting on the clarity, enthusiasm, and flexibility of all of the instructors. One learner noted that it was great that the instructor could "modify the class for individuals who were not prepared."

Evaluative comments can vary for a variety of reasons. Learner opinions may reflect differences in how individual instructors approached the class or on changes that instructors made as they learned and modified how they facilitated workshops. The variety of experiences expressed may also be reflective of different learning needs and styles. However, qualitative comments received were progressively more positive throughout the year. Therefore, the authors concluded that there was enough evidence to suggest that retaining the flipped-classroom model was preferable for the majority of their learners.

22.5.5 Instructor Reflection

Instructors of the flipped workshops also shared their experience by completing a structured reflection tool. Instructor perceptions of the success of the workshops varied, and some instructors believed it worked better for one or another of the workshops mentioned (PubMed, Medline, or Maximizing your Search Skills). In terms of what they felt went well, instructors noted that workshop participants seemed to enjoy the practice exercises and having the support of the assistant teachers. They believed that learners were more engaged and attentive when applying their learning to actual questions. Learners appeared willing and able to ask questions, which was noted as more helpful to their learning. Instructors commented that this less scripted way of teaching seemed to enhance participants' learning by making the class more relevant because the instructors could adapt the workshop content to the learners' interests and abilities.

The most common frustration noted by instructors was that learners did not complete the prework before the workshop. Even among those who had done the prework, the level of understanding of the material varied. Learners often told the instructors they were unaware of the prework, and some learners seemed to resent that they should do any prework. The instructors also noted that it could be even more complicated when some learners come prepared by completing the prework and those who did not, required more basic instruction. Adapting to the unscripted nature of these workshops, especially when learners did not prepare as instructed, was an additional identified challenge. Learning to facilitate a class that did not rely on the delivery of a set curriculum but was focused more on responding to learners' needs was a new experience for the instructors.

Over the course of the year, instructors experimented with many ways to facilitate these workshops. Some instructors delivered minilectures or summaries at the beginning of class with mixed results; it allowed those learners who did not do the prework to acquire some basic knowledge, but it frustrated other learners who had taken the time to learn the material. Many of the instructors looked for ways to assess learners' understanding of the concepts covered in the prework and then adapted their approach in the hope that this will satisfy learners who came prepared. Some began the session by asking for questions from the group and reviewing the mini-quizzes from the prework before moving on to practice exercises or helping learners work through their own research questions. One instructor reported applying "more visual exercises and involving participants in action tasks that allowed me to gauge their understanding of the subject matter."

Instructors were asked to reflect on their perception of the learners' level of engagement in these workshops (1 = very engaged, 2 = moderately engaged, 3 = somewhat engaged, and 4 = not very engaged). The majority (four out of six) of the instructors felt that learners were moderately engaged, and two of the instructors found them somewhat engaged. Almost everyone noted that learner engagement varied greatly from learner to learner and group to group, but that learners who were unprepared were generally less engaged.

While only two of the six instructors reported a preference for facilitating flipped workshops, all instructors found value in the increased amount of interaction and hands-on exercises these workshops provided for learners. The instructors continue to reflect on how to improve these workshops and are continually proposing and implementing modifications to both workshop facilitation methods and the prework learning objects. For example, they worked together to develop shorter standardized summary presentations for each workshop that instructors can use if they feel it is necessary. A recent review and enhancement of the resource guides added more engaging content, including increased audiovisual content, to the prework element. Due to a recent upgrade in the LibGuide software, the organization, usability, and appearance of the online content also improved.

The hospital also is planning on implementing a new LMS in the next year, and this will provide further opportunity to improve the prework portion by better integrating the miniquiz elements and other possible interactive components that might increase participation. A new LMS will also allow the library to explore the advantages and disadvantages of moving the entire online component on to an e-learning platform.

Moving to an such a platform will provide more statistical data, including the possibility of collecting detailed data not only on whether learners completed the preparatory work but also if they seemed to have difficulty with any of the specific concepts introduced in the prework. This will allow instructors the option to restrict access to the workshop to only those learners who completed the prework or to better gauge their understanding of specific concepts before the class and make any necessary modifications to how they facilitate the workshop.

BOX 22.1 Recommendations for Practice

For librarians considering implementing the flipped-workshop model for their own information literacy instruction, the authors offer the following recommendations for practice:

- Take the time to prepare visually appealing and usable resource guides and learning objects. If possible, conduct usability testing on your learning objects.
- Introduce the concept of the flipped classroom at the time of registration and send learners reminders to complete the prework.
- State the learning objectives for the workshop in the class description and prework materials.
- Emphasize that learners are responsible for their own learning and that in-class time will focus on questions and hands-on practice.
- Create a welcoming and positive learning atmosphere regardless of whether or not the prework was done. Make it clear that learners are there to try things out and discover how the databases work in a relaxed, nonthreatening environment.
- Move away from being the "sage on the stage." Be a facilitator rather than a lecturer. Learners' questions provide the opportunity for everyone to learn and practice. Encourage learners to answer each other's questions.
- Avoid repeating content that is in the prework and resource guides, but be prepared to answer questions about it.
- Have plenty of exercises, questions, and memorable examples on hand to fall back on in case learners are less participatory.
- Structure the exercises in such a way that they build on each other and can be practiced by learners at different levels. An interesting clinical question or scenario makes a good example to work through. Learners may also be willing to share their own research questions.
- Consider adding small group work.

(Continued)

> **BOX 22.1 Recommendations for Practice—cont'd**
> - Leave time at the end of the class for individual consultations. This is especially useful for more reserved learners who are less likely to speak up in front of their peers.
> - Evaluate, modify, implement, evaluate.

22.6 CONCLUSION

Like the just-in-time-teaching model (Novak, 2011), the flipped workshop requires a certain amount of flexibility. The authors' experience is that the model provides even more opportunity for creating teachable moments than a standard workshop model. For example, when using participants' research questions rather than prepared scenarios, the chance of unexpected results is much higher. This demystifies the literature search process, which is frequently iterative and includes trial and error. The workshop facilitator is more of a learning partner than "the expert." Learners are encouraged to bring and share what they already know. This approach is more consistent with adult learning theory, which implores instructors to respect what adult learners already know and to start from where they are at (Merriam & Bierema, 2013). With no set in-class lesson plan, each class is different because it is a spontaneous interaction between learners and teachers.

The authors found that learners are challenged by this model in terms of understanding the requirement to do the prework, and learners reported an overall high level of satisfaction with the workshops. The resource guides were frequently accessed, and learners reported that they liked having access to the guides and found them extremely useful.

Implementing the flipped-classroom model was a significant change for the instructors at St. Michael's Hospital Library, but most feel that the learning benefits of a flipped-classroom model are worth the effort. Many reported that the sessions are more dynamic and responsive to learners' needs and individual questions, that they enjoy teaching more, and that their teaching has improved. One instructor noted, "I certainly have changed as a facilitator in general because of the experience. I feel more connected with the participants and more confident. I welcome questions because working through it together is part of the process."

Implementing the flipped-classroom model at St. Michael's Hospital library was an iterative process. Some of the insights gleaned during the course of this project included learning to move into a facilitator mode versus a lecturer mode. Moving away from "the sage on the stage" is not accomplished overnight; it requires the development of facilitation skills and greater confidence in the classroom. Likewise, it is important to find effective methods for ensuring that learners are aware that they are responsible to do the prework and that in-class time will be focused on questions and hands-on practice. (See Box 22.1 for further recommendations for practice.)

Further research on the value of the flipped classroom, facilitation skills in this context, and the creation of effective learning objects for information literacy instruction will be beneficial. Data collected from the evaluation of the flipped-classroom model for this pilot project supports the continued use of the flipped model at the St. Michael's Hospital Library and demonstrates growing potential for its adoption in hospital settings. As other libraries experiment with the model in their own settings and share their experiences, we will be able to further gather evidence as to its efficacy as a teaching model for academic hospital libraries and how it may be extrapolated to other library settings.

REFERENCES

Arnold-Garza, S. (2014). The flipped classroom teaching model and its use for information literacy instruction. *Communications in Information Literacy, 8*(1), 7−22.

Bendriss, R., Saliba, R., & Birch, S. (2015). Faculty and librarians' partnership: Designing a new framework to develop information fluent future doctors. *Journal of Academic Librarianship, 41*(6), 821−838. Available from: <http://dx.doi.org/10.1016/j.acalib.2015.09.003>.

Bergmann, J. S., & Sams, A. (2012). *Flip your classroom: Reach every student in every class every day.* Eugene, OR: International Society for Technology in Education.

Bergmann, J. S., & Sams, A. (2014). *Flipped learning: Gateway to student engagement.* Eugene, OR: International Society for Technology in Education.

Bishop, J.L., & Verleger, M.A. (2013, June). The flipped classroom: A survey of the research. In ASEE National Conference Proceedings, Atlanta, GA (Vol. 30, No. 9).

Bloom, B. S. (1956). *Taxonomy of educational objectives: The classification of educational goals.* New York, NY: Longmans, Green.

Chi, J., & Verghese, A. (2014). Clinical education and the electronic health record: The flipped patient. *JAMA, 312*(22), 2331−2332.

Conte, M. L., MacEachern, M. P., Mani, N. S., Townsend, W. A., Smith, J. E., Masters, C., et al. (2015). Flipping the classroom to teach systematic reviews: The development of a continuing education course for librarians. *Journal of the Medical Library*

Association, *103*(2), 69−73. Available from: <http://dx.doi.org/10.3163/1536-5050.103.2.00210.3163/1536-5050.103.2.002>.

Coursera. (2014). *Flipped classroom field guide.* < https://docs.google.com/document/d/1arP1QAkSyVcxKYXgTJWCrJf02NdephTVGQltsw-S1fQ/edit?pli = 1 >.

Fawley, N. (2014). Flipped classrooms: Turning the tables on traditional library instruction. *American Libraries*, *45*(9/10), 19.

Gilboy, M. B., Heinerichs, S., & Pazzaglia, G. (2015). Enhancing student engagement using the flipped classroom. *Journal of Nutrition Education and Behavior*, *47*(1), 109−114. Available from: <http://dx.doi.org/10.1016/j.jneb.2014.08.008>.

Khanova, J., Roth, M. T., Rodgers, J. E., & McLaughlin, J. E. (2015). Student experiences across multiple flipped courses in a single curriculum. *Medical Education*, *49*(10), 1038−1048. Available from: <http://dx.doi.org/10.1111/medu.12807>.

Kurup, V., & Hersey, D. (2013). The changing landscape of anesthesia education: Is flipped classroom the answer? *Current Opinion Anaesthesiology*, *26*(6), 726−731. Available from: <http://dx.doi.org/10.1097/aco.00000000000000410.1097/ACO.0000000000000004>.

Lochner, L., Wieser, H., Waldboth, S., & Mischo-Kelling, M. (2016). Combining traditional anatomy lectures with e-learning activities: How do students perceive their learning experience?. *International Journal of Medical Education*, 7, 69−74. Available from: <http://dx.doi.org/10.5116/ijme.56b5.036910.5116/ijme.56b5.0369>.

Merriam, S. B., & Bierema, L. L. (2013). *Adult learning: Linking theory and practice.* John Wiley & Sons.

Milman, N. (2014). The flipped classroom strategy: What is it and how can it best be used? *Distance Learning*, *11*(4), 9−11.

Morgan, H., McLean, K., Chapman, C., Fitzgerald, J., Yousuf, A., & Hammoud, M. (2015). The flipped classroom for medical students. *The Clinical Teacher*, *12*(3), 155−160. Available from: <http://dx.doi.org/10.1111/tct.12328>.

Novak, G. (2011). Just-in-time teaching. *New Directions for Teaching and Learning*, *2011* (128), 63−73. Available from: <http://dx.doi.org/10.1002/tl.469>.

O'Flaherty, J., & Phillips, C. (2015). The use of flipped classrooms in higher education: A scoping review. *The Internet and Higher Education*, *25*, 85−95. Available from: <http://dx.doi.org/10.1016/j.iheduc.2015.02.002>.

Pannabecker, V., Barroso, C. S., & Lehmann, J. (2014). The flipped classroom: Student-driven library research sessions for nutrition education. *Internet Reference Services Quarterly*, *19*(3), 139−162. Available from: <http://dx.doi.org/10.1080/10875301.2014.975307>.

Rivera, E. (2015). Using the flipped classroom model in your library instruction course. *The Reference Librarian*, *56*(1), 34−41. Available from: <http://dx.doi.org/10.1080/02763877.2015.977671>.

St. Michael's Hospital. (2016). *Facts about St. Michael's.* Retrieved from < http://www.stmichaelshospital.com/about/snapshot.php >.

Vogel, L. (2012). Educators propose "flipping" medical training. *Canadian Medical Association Journal*, *184*(12), E625−E626. Available from: <http://dx.doi.org/10.1503/cmaj.109-4212>.

Youngkin, C. A. (2014). The flipped classroom: Practices and opportunities for health sciences librarians. *Medical Reference Services Quarterly*, *33*(4), 367−374. Available from: <http://dx.doi.org/10.1080/02763869.2014.95707310.1080/02763869.2014.957073>.

APPENDIX A: INSTRUCTOR REFLECTIONS

1. What went well with teaching these workshops? What were any successful strategies, techniques, etc. that you employed to help these workshops succeed?
2. What did not work well with these workshops? Were you able to employ strategies to address these problems? If so, what were they?
3. Please comment on your perception of the learner's experience of the workshops. On a scale of 1–4 how would you rate their participation overall in the flipped classroom workshops you facilitated.
   1 = very engaged
   2 = moderately engaged
   3 = somewhat engaged
   4 = not at all engaged
Comments:
4. If you facilitated more than one of the 3 workshops offered (Medline, Pubmed and Maximizing your Search Skills) did you find that learners were more or less or differently engaged in one or more of the workshops? For example, did it seem to work better for the medline workshop than the others, etc. please comment.
5. Did you change your approach to managing the facilitation of the flipped classroom model over the course of the year? If so, how and why? Please refer to specific workshops if applicable.
6. How would you rate your preference to facilitating regular workshops versus flipped workshops?
   I greatly prefer facilitating regular workshops
   I prefer facilitating regular workshops
   I don't have a preference
   I prefer facilitating flipped workshops
   I greatly prefer facilitating flipped workshops
Comments on the above:
7. In the coming months, what changes would you like to see made to how we deliver the flipped workshops?
8. Please add any other comments you would like to make about your experience or Perception of these workshops.

APPENDIX B: LEARNER EVALUATION FORM

Health Sciences Library: Training Evaluation Form

Please give us your feedback, reactions and comments. The statements below concern specific aspects of this course and your input will help us to evaluate this session and improve future workshops.

Instructor: _____   **Course:** _____

1. How do you rate the **subject**? (Interest, benefit to your work, etc.)

Poor	Fair	Good	Very good	Excellent
O	O	O	O	O

2. How do you rate the **instructor**? (Knowledge of subject matter, ability to communicate, etc)

Poor	Fair	Good	Very good	Excellent
O	O	O	O	O

3. If you reviewed the resource guide before class, did you find it useful?

Not applicable	Not useful	Somewhat useful	Useful
O	O	O	O

4. How did you hear about this course? (check all that apply)

- o Daily Dose email
- o Intranet Event Calendar
- o Learning Centre course catalogue
- o Colleague
- o Other? _____

5. What did you like most about today's session?

6. How could today's session be improved to better facilitate your learning?

7. Overall Course Rating

1 = Unacceptable
10 = Exceptional

1	2	3	4	5	6	7	8	9	10
O	O	O	O	O	O	O	O	O	O

8. What other topics or resources are you interested in learning about from the Health Sciences Library?

Thank you for taking the time to share your comments on your learning experience

INDEX

Note: Page numbers followed by "*f*", "*t*", and "*b*" refer to figures, tables, and boxes, respectively.

A

Academic integrity, 343
Academic libraries, 13–17, 157, 222, 256
  collaboration, 36
  distributed learning, 34
Academic partnerships, 320–321
Academic Search, 251
Access technology, 21–24
ACRL. *See* Association of College and Research Libraries (ACRL)
Active learning, 409. *See also* Online learning
  in library instruction, 49
  limitations in, 50
  scholarship, 48–50
ADDIE model. *See* Analysis, design, development, implementation, and evaluation model (ADDIE model)
Adelphi University, 93–94
Adobe After Effects software, 352, 353*f*, 354
Adobe Captivate, 278
Adobe Connect tools, 145–146, 276–277
Adobe Creative Suite 6 software (CS6 software), 349
Adobe CS6 software, 331
Adobe Dynamic Link, 354
Adobe Flash Player, 276–277
Adobe Photoshop, 396–397, 397*f*
Adobe Premiere Pro software, 354
Adventr, 392–397, 396*f*, 397*t*
AECT. *See* Association for Educational Communications and Technology (AECT)
Alt text. *See* Alternative text (Alt text)
Alternative research assignments, 246–247
Alternative text (Alt text), 301–302

Analysis, design, development, implementation, and evaluation model (ADDIE model), 3, 75–76, 75*f*, 178–179. *See also* Kuhlthau's guided inquiry model
  design phase, 78–80
  development phase, 81–83
  evaluation phase, 84–86
  implementation phase, 84
  implementation using, 76–78
  module template in Canvas CMS, 81*f*
  task analysis for single rubric criteria, 80*t*
Animated gifs, 396–397, 397*f*
Animoto tool, 62
Apple's Final Cut Pro software, 346
APSU. *See* Austin Peay State University (APSU)
ARL. *See* Association of Research Libraries (ARL)
Assessment(s), 32–34, 33*t*, 118–120, 355–358, 376–377
  analysis, 377–378
  book reviews
    and book talk, 211–214
    using Goodreads, 212
  book talk using YouTube, 212–214
  book-publishing course, 208–209
  course evaluations, 119–120
  D2L view, 356*f*
  design, 118–119, 240–247
  editing Wikipedia assignment, 210–211
  management, 119
    research methods course, 209–210
  mapping, 119, 120*f*
  measures, 104–105
  plagiarism quiz results, 357*f*
  social media plan group project, 214–215
  visiting expert, 207–210

Printed in the United States
By Bookmasters